PHYSICIAN INVESTIGATOR HANDBOOK

GCP Tools and Techniques

Second Edition

Practical Clinical Trials Series

Deborah Rosenbaum, CCRC, CCRA
Fred Smith, MD

CRC Press
Taylor & Francis Group
Boca Raton London New York

CRC Press is an imprint of the
Taylor & Francis Group, an **informa** business

CRC Press
Taylor & Francis Group
6000 Broken Sound Parkway NW, Suite 300
Boca Raton, FL 33487-2742

First issued in paperback 2019

© 2009 by Taylor & Francis Group, LLC
CRC Press is an imprint of Taylor & Francis Group, an Informa business

No claim to original U.S. Government works

ISBN-13: 978-1-57491-124-4 (hbk)
ISBN-13: 978-0-367-39642-8 (pbk)

Visit the Taylor & Francis Web site at
http://www.taylorandfrancis.com

and the CRC Press Web site at
http://www.crcpress.com

CONTENTS

INTRODUCTION

The purpose of this book is quite simple—to bring together, in an easily accessible format, the essential information that you should know when contemplating clinical research participation as an investigator. This practical guide to clinical research will focus on the knowledge necessary to perform well as a clinical investigator and will provide checklists and forms that will help your research site function efficiently.

Reasons to Participate in Clinical Research

- Research is fun. There is great fulfillment in learning new skills and "being on the cutting edge" of medicine.
- Research is satisfying. Do not underestimate the gratification that comes from making a contribution to our profession.
- Research stimulates learning. Both industry-sponsored and independent clinical research enable the clinician to become an expert in the field of study.
- Research increases the skill of critical evaluation. Once you have engaged in clinical research yourself, you better understand the many pitfalls and are better able to critically evaluate the merits of published research.
- Research may become a satisfying new career. The pharmaceutical industry is continually searching for talented clinical researchers as well as for clinicians who wish to pursue a career within the pharmaceutical industry.
- Research improves office morale. Your office staff is proud of your contributions to science and is equally proud of your status as the local expert on your research topic.
- Research improves clinical skills. The clinical investigator must master the art of observation and the discipline of meticulous documentation of techniques and results and must gain a broad knowledge of the requirements of federal regulation.

- Research can be financially rewarding. The pharmaceutical industry pays clinical researchers well. Research participation usually increases public awareness of the physician's expertise, which may lead to increased patient demand and referrals.
- Research stimulates self-analysis. The clinician must "know thyself" and critically evaluate all phases of the practice before beginning clinical research. This heightened awareness of the strengths and weaknesses of the practice, by itself, will repay any physician who enters clinical research.

Goals

This book prepares you to function as a clinical research investigator by

- informing you of the pertinent federal regulations that govern physician activities in clinical research,
- enabling you to adhere to the federally mandated supervisory activities required of pharmaceutical companies or funding agencies (sponsors),
- guiding you in the preparation of the site to perform clinical research, and
- advising you of some considerations in clinical research design and protocol development.

To reach its goals, this book contains

- an overview of the research process;
- the pertinent Food and Drug Administration (FDA) regulations that govern the investigator's behavior during clinical research activities;
- an interpretation of those FDA regulations as developed through the interaction between the FDA and the pharmaceutical industry;
- guides and checklists for every phase of the clinical research process, from the initial thoughts and evaluation process, through the critical start-up activities, the actual clinical trial performance, and final product outcome at the project's conclusion; and
- in-text references to appropriate federal regulations throughout the process.

Although the focus of this book is on conducting investigational new drug research, the principles apply to all clinical research. Join us as we take you step-by-step through each of the special areas that the proficient clinical investigator must know.

Fred Smith
Deborah Rosenbaum
September 2001

1

OVERVIEW AND ELEMENTS
OF CLINICAL RESEARCH

Clinical investigation means any experiment that involves a test article and one or more human subjects and that either is subject to requirements for prior submission to the Food and Drug Administration under section 505(i), 507(d), or 520(g) of the act, or is not subject to requirements for prior submission to the Food and Drug Administration under these sections of the act, but the results of which are intended to be submitted later to, or held for inspection by, the Food and Drug Administration as part of an application for a research or marketing permit. The term does not include experiments that are subject to the provisions of part 58 of this chapter, regarding nonclinical laboratory studies.

—21 CFR 50.3(c)

Clinical research is a vital component of health care. Consider where we would be without the medical advances that we all take for granted—vaccinations for our children, insulin for the control of diabetes, and pacemakers for people with cardiac arrhythmias, among others. These developments required years of research by dedicated scientists and clinicians. The fight against cancer is another example; in the 1960s, children with acute lymphocytic leukemia had little chance of survival. Now, with sophisticated chemotherapy, many of these children can live to become active, productive adults. Significant strides have been made, but it is clear that there will always be much clinical research work ahead of us. The twenty-first century has dawned with a new arena of research—genetic research. With the successful mapping of the human genome, phenomenal implications in the treatment and prevention of diseases are on the horizon and will be our newest research challenge. The methods employed by physicians who participate in the clinical research process form the basis for this book.

GENERAL GOALS

The general goals of clinical research are to identify the mechanism of the disease process and to determine the effectiveness of intervention in that disease process, generally with drugs, surgery, devices, nutrition, or behavioral changes. In drug research, the specific goals of clinical research include

- determining the efficacy (effectiveness) and safety of a new drug,
- defining dose administration routes and frequencies,
- testing drug formulations, and
- exploring combination and adjuvant therapies.

These are accomplished through careful planning and implementation of clinical research trials. Other facets of patient care, such as quality of life and pharmacoeconomic issues, have also become topics of research.

Clinical trials are conducted under a very precise set of rules called the *protocol*. Many other documents peculiar to clinical trials supplement the protocol, such as the Investigator's Brochure, Case Report Forms (CRFs), and the records involving the investigational agent itself. Because clinical trials involving investigational drugs, biologics, and devices are strictly regulated by the Food and Drug Administration (FDA), specific study documentation must be maintained. The physician investigator and the Clinical Research Coordinator (CRC) play crucial roles in creating and storing this documentation.

WHO CONDUCTS CLINICAL RESEARCH?

Clinical research is conducted in a variety of venues. The National Institutes of Health (NIH) sponsor a large number of federally funded trials through the many different institutes. Even within the NIH, rules vary. NCI is very active in consolidating clinical trial information (http://cancertrials.nci.nih.gov). Most prominent is the research that is required by the FDA to provide evidence of safety and efficacy of an investigational product prior to approval for marketing. That is the focus of this manual. However, general research principles apply across the board.

THE DRUG DEVELOPMENT PROCESS

New drugs are developed through a series of laborious steps, as summarized in Table 1.1. New Molecular Entities (NMEs) or New Chemical Entities (NCEs) are novel compounds created in the laboratory by means that vary from sophisticated computer

TABLE 1.1 SUMMARY OF NEW DRUG DEVELOPMENT

Drug Discovery	NMEs discovered by design or happenstance. May be discovered elsewhere and licensed for development.	
Laboratory Screening	Screen NME for activity in specially designed tissue culture screening tests.	
Animal Testing	Screen for activity in specific animal models.	
Preclinical Testing	Toxicology, teratology, carcinogenicity, and pharmacology testing.	
Formulation Issues	Formulation, stability and synthesis both of small-scale clinical supplies and manufacturing scale-up.	
File IND Application	Before an investigational agent may be used in humans, an IND Application must be filed with the FDA.	
Clinical Trials	Phase I	"First-time-in-man" studies in normal volunteers.
	Phase II	Small-scale efficacy trials in selected patients.
	Phase III	Large-scale testing in a wider range of patients.
File NDA	Data from the clinical trials supporting safety and efficacy are assembled and submitted to the FDA with a request for permission to market the compound as a therapy or test for a specific disease.	
Marketing	The drug is marketed for the approved indication.	
Phase IV	Additional trials conducted to determine more effective dosing schedules, new formulations, new target populations, and marketing claims.	
Postmarketing Surveillance	After NDA approval, information about the drug continues to be collected.	

modeling to serendipity. The compounds are then screened for activity in vitro by established tissue culture screening panels. If activity is noted, the drug is screened for activity in animal models for pharmacology, toxicology, and effectiveness. If the compound appears to have desirable activity and is relatively safe, it will be formulated for clinical trial testing. Note that these steps are not necessarily sequential but are more likely to be concurrent. For example, while preclinical testing is being conducted in animal models, the formulation of the drug is also being designed. Additionally, the preclinical phase does not end when clinical trials begin but continues concurrently with clinical testing. Clinical testing cannot begin until an Investigational New Drug (IND) Application is filed with the FDA. Results from the clinical trials supportive of the drug's safety and effectiveness are submitted to the FDA as a New Drug Application (NDA).

Also, different types of drugs may have different developmental tracks. "Fast track" drugs typically are reviewed by the FDA prior to extensive Phase III trials. Other studies may not easily fit into a particular phase, for example, a study of a combination therapy of two approved chemotherapeutic agents.

PHASES OF CLINICAL RESEARCH

Clinical trials comprise the systematic investigation of the therapeutic effects of an investigational agent, treatment modality (e.g., surgery or radiation), methods of prevention, or detection or diagnosis of a disease state. Clinical trials are conducted under stringent conditions that are specified in the protocol.

Clinical trials involving the safety and efficacy of NMEs, new formulations, or new indications are generally conducted in phases, as indicated in Table 1.2. It is not always easy to categorize a clinical trial into a specific phase. These definitions are general categories used to describe the activity of the development of a new drug; some differences in interpretation and overlap may exist. It is important to note that there may be some overlap in development from Phase I to Phase IV. For example, Phase II trials demonstrating efficacy may focus on completing long-term evaluation, while Phase III trials are being initiated. A drug may be in Phase III testing for one indication while only in Phase II for another indication.

TABLE 1.2 PHASES OF CLINICAL DRUG TRIALS

Phase I: "First-Time-In-Man" studies

Purpose	Phase I trials are conducted to determine the **SAFETY** of an investigational agent. Also, pharmacological data may be collected to determine the absorption, distribution, metabolism, and excretion (ADME) of the compound. Generally, initial dosing of the agent is determined from **PRECLINICAL** trials in animals. Most Phase I trials are designed to begin dosing at a subtherapeutic level to avoid unexpected, catastrophic adverse events. Doses are then escalated to reach the maximum tolerated dose (MTD). Data collected from serum levels can be indicative of the effectiveness of the investigational agent in the disease, since, typically, predetermined serum levels must be reached to be effective against the disease. Usually, all subjects receive the experimental compound in single or multiple doses. The study is often conducted by a single investigator at a single site.
Length of Studies	Phase I trials are short-term studies. Individual subject participation may be from one day to several weeks, and the whole study is conducted over several months.
Subjects	Generally, normal volunteers without confounding diseases or concurrent medications are recruited to participate in Phase I trials. However, when testing antineoplastic agents or in unpredictable patient groups, it may be preferred to begin trials in a patient population. Studies involving antineoplastic agents generally utilize patients who have failed all other forms of treatment. Although the efficacy of the drug is unknown, it may provide some hope while investigating the adverse effects of the drug. Phase I trials usually enroll 20–60 subjects.

Table 1.2 continued on next page

Table 1.2 continued from previous page

Phase II: Pilot Trials

Purpose   Phase II trials are conducted to demonstrate **EFFICACY** in a particular disease. These trials are randomized, tightly controlled studies, using small numbers (60–200) of carefully selected patients. When feasible, Phase II trials employ comparisons to a placebo or an active control with known efficacy. Subjects may be receiving single or multiple doses. Phase II trials may be conducted at multiple centers.

Length of Studies   Subject participation will vary but is usually of longer duration than Phase I. Phase II trials may be completed in a few months or take up to several years.

Subjects   Subjects in Phase II trials are patients with the disease or clinical situation being examined. They should be otherwise healthy, in terms of their disease, and free of other serious medical illnesses, particularly renal or hepatic impairment. These are the subjects you would expect to do well if their disease were managed by conventional means.

Phase IIb: Pivotal Trials

Purpose   Phase IIb trials (sometimes overlapping with Phase IIIa) are conducted to gain specific efficacy and safety information for submission of an NDA and are often referred to as pivotal trials. Two "adequate and well-controlled" pivotal trials are generally required to file for an NDA with the FDA. Phase II trials also are conducted to determine dose ranging for Phase III trials.

Length of Studies   Pivotal trials may last a few months to several years. Duration of subject participation depends on the amount of time expected to demonstrate efficacy.

Subjects   Subjects are generally patients with the disease without serious complications or other concurrent diseases.

Phase IIIa: Expanded Clinical Trials

Purpose   Phase IIIa trials are designed to gain safety and efficacy information in a large number of patients. Some variables include extended dosing, dose ranging, patient populations more representative of the market situation. Phase IIIa trials are tightly controlled and are conducted with the experimental agent versus placebo or an active control. Different study designs may be employed during the drug development process. These data are often used to supplement the NDA submission.

Length of Studies   Duration of participation for individual patients varies. Phase IIIa studies tend to be of longer duration, lasting one to four years.

Subjects   Phase IIIa subjects are patients with the disease being studied and selected from a larger population of patients, although entry criteria may still be stringent. Several hundred subjects are required in Phase IIIa studies to demonstrate efficacy and to assess safety adequately.

Phase IIIb: Large-Scale Trials

Purpose   The purpose of Phase IIIb trials is to gain experience with the experimental agent in a large number of subjects reflective of the population that will use the drug. Therefore, the trials are less tightly controlled: All subjects may be receiving experimental drug, entry criteria are relaxed, and larger numbers of patients are enrolled. Trials may also be designed to specifically address special patient groups, such as geriatrics or pediatrics.

Table 1.2 continued on next page

Table 1.2 continued from previous page

Length of Studies	Phase IIIb studies last one to four years and are used to gather additional safety and long-term efficacy data about the investigational agent.
Subjects	Phase IIIb trial subjects come from a larger, heterogeneous population. The subject population may focus on specific concurrent illnesses to further delineate the drug's safety. This is especially true of trials involving geriatric and pediatric patients.

Phase IV: Postmarketing Trials

Purpose	Phase IV trials are done for a variety of reasons: to place the drug in the market, to make marketing claims, to perform pharmacoeconomic or quality-of-life studies, and to provide additional surveillance for unexpected or rare adverse events. New formulations or new indications for a marketed drug must begin with Phase I (new formulation) or Phase II clinical trial designs.
Length of Studies	The length of Phase IV trials is determined by the purpose of the study and may be indefinite, such as a continual postmarketing surveillance program.
Subjects	Subjects in Phase IV trials are drawn from the general population with the specific disease. Further conditions are defined by the purpose of the protocol.

BASIC ELEMENTS OF CLINICAL RESEARCH

The basic elements or "tools" of clinical research are the protocol, the investigational agent, data collection forms, and the study files. Understanding these elements and their purpose is critical to the clinical research process. These elements are described below.

The Protocol

The study protocol is the blueprint of the study and is required by Good Clinical Practice (GCP) guidelines [21 CFR 312.23(a)]. Generally, the protocol includes the following items:

Item	Description
Objective	The "what" of the study. A clear concise statement of the hypothesis to be tested by the clinical trial. What specific questions is the study designed to answer about the investigational agent at a particular time in its development?
Background and Rationale	The "why" of the study. The background section of the protocol discusses the known information about the class of drugs being studied. It also summarizes known information about the investi-

gational agent, specifically, side effects known to date. The discussion should lead to the rationale of using the drug(s) at this dosage in this disease for this study.

Subject Selection Criteria (Inclusion/ Exclusion Criteria)   The "who" of the study. This section defines the subject population by clearly stating the criteria for a subject to be enrolled in the trial.

Treatment Plan   The "how" of the study. The treatment course (drugs and dosages) is outlined in detail. Additional details describing the study medication include pharmaceutic information, packaging information, dose modification recommendations, concomitant medication instructions, dispensing instructions, and comprehensive instructions to patients.

Study Procedures   The "when" of the study. This section includes information concerning the study design, treatment schedule, laboratory and diagnostic tests, time intervals of subject visits and tests, and off-study evaluations. It also includes information concerning the collection of adverse events, dose modification guidelines, and the appropriate handling of study dropouts. This information is often displayed graphically in protocols in a table, commonly called the study schema or "schedule of events."

Response Evaluation Criteria   Each protocol must define a priori criteria for patient response to the test agent. These criteria are outlined in detail in this section; also included are discussions of objective responses, endpoint variables, measurement of lesions, and other evaluation criteria.

Statistical Section   The statistical section discusses the proposed plan for the evaluation of the data. It includes such items as an explanation of study design, feasibility, proposed analyses, early stopping rules, and termination of study.

Administrative Items

Toward the end of the protocol, obligations of the investigator while conducting the study are discussed. These may include adherence to FDA regulations, monitoring of the trial, data management procedures, techniques required for completion of CRFs, and other administrative details.

Bibliography

Lists references cited in the protocol.

Appendices

Some common appendices are as follows:

- Toxicity grading scales.
- Study schema and/or schedule of events charts.
- Instructions for evaluating responses.
- Instructions for collecting specific samples.
- Instructions for shipping biological supplies.
- Prescribing information (package insert) for approved active control drugs.
- Sample of patient diary.
- Staging guidelines.
- Quality of Life forms.
- Performance Status Scales (i.e., Karnofsky, Zubrod, ECOG, or WHO).
- Central Radiology Review requirements.
- Gender and Minority Target Accrual.

Investigator's Brochure

The Investigator's Brochure (IB) is a confidential document, created by the study sponsor, that summarizes all known information about the investigational drug. This very important document enables the investigator to obtain all information known to the sponsor about the investigational agent. This includes preclinical data such as chemical, pharmaceutical, toxicology data; pharmacokinetic and pharmacodynamic data in animals and humans; and the results of previously conducted clinical trials. The data should support the proposed clinical trial and contain information on expected risks or precautions. IBs are updated periodically so that data resulting from prior clinical trials and additional preclinical data can be incorporated as it becomes available.

Investigational Agent

The investigational agent is the item (drug or device) being studied. It may be an experimental drug, a combination therapy of approved drugs, or an experimental drug combined with or compared to an approved drug. Many investigational agents are studied as compared to a placebo control or active control (which are also considered investigational agents for the trial). Pacemakers, thermometers, contact lenses, and adhesive bandages are all examples of investigational agents in medical device clinical trials.

All study drugs or devices used by study subjects are considered investigational agents and must be stored, dispensed, and monitored per regulation (21 CFR 312.57, 312.59, 312.60, 312.62). More detailed information is included in Chapter 5, "Investigational Agent Management."

Case Report Forms (Data Flow Sheets)

Data in clinical trials must be identified and collected in a systematic fashion to assure that the data can be analyzed to determine the results of the trial (21 CFR 312.62). The CRF is used by most pharmaceutical sponsors to record all data pertinent to the study. The CRF is completed by the investigator or the CRC, reviewed by the sponsor monitor and data management personnel, entered into the study database by data managers, and analyzed by statisticians. In some trials, data flow sheets or data summary charts are used to collect data. The CRF or flow sheet should be designed to capture all of the data, and only that data, required by the protocol. More detailed information is included in Chapter 12, "Data Management."

Study Files

FDA regulations require that all documents pertinent to the conduct of the clinical trial be maintained (21 CFR 312.62). Items included in the study files are listed and discussed in Chapter 8, "Conducting the Study."

Summary

The development of an investigational agent is accomplished in a very systematic and controlled environment. Satisfactory participation in clinical trials requires an understanding of the process from the sponsor's point of view, as well as effective management of the individual components. These individual components will be discussed in the following chapters.

RESOURCES

Many resources are available to the clinical investigator; some are listed below. Also, check into educational opportunities offered through your institution, local universities, and regional groups.

Association of Clinical Research Professionals (ACRP)

The ACRP is an international organization of individuals involved in clinical research and other research-related professions. ACRP has a large CRC component that is very well organized and active, as well as a Clinical Research Associate (CRA) forum and investigator forum. The ACRP offers a certification exam for CRCs and CRAs. The annual meeting is held in the spring. The ACRP publishes a newsletter, *The Monitor.* Membership information can be requested as follows:

Association of Clinical Research Professionals
1012 14th Street, NW
Washington, DC 20005
202-737-8100
202-737-8101 (fax)
www.acrpnet.org

Drug Information Association (DIA)

Mission statement: "To provide a worldwide forum for the exchange and dissemination of information that is intended to advance the discovery, development, evaluation, and use of medicines and related health care technologies." The DIA holds an annual meeting and sponsors topic-specific workshops and seminars throughout the year. The DIA publishes the *Drug Information Journal* and a quarterly newsletter. A membership application form appears in each issue of the journal; information can be requested as follows:

Drug Information Association
501 Office Center Drive, Suite 450
Fort Washington, PA 19034-3211
215-628-2288
215-641-1229 (fax)
dia@diahome.org

Therapeutic Group Organizations

Therapeutic organizations such as ASCO, ASA, and ICAAC, often have a research component.

Hospital Satellite Network (HSN)

HSN offers information through videotapes and videoconferencing. Often, this service is made available through medical institutions.

National Center for Research Resources (NCRR)

NCRR publishes a bimonthly newsletter, *Reporter,* discussing current research activities. For more information:

Research Resources Information Center (NCRR)
1601 Research Blvd.
Rockville, MD 20850

Center for Clinical Research Practice

Research Practitioner (previously *Research Nurse*) is a bimonthly publication that discusses current issues and practices in clinical research and offers continuing education through self-assessment. The CCRP also offers training opportunites. Information can be requested as follows:

> Center for Clinical Research Practice
> 40 Washington Street, Suite 130
> Wellesley, MA 02481
> 781-431-7577
> www.ccrp.com

Patient Information

Taking Part in Clinical Trials is a booklet for patients with cancer [NIH publication number 98-4270 (6/98)]. It is available from

> U.S. Department of Health and Human Services
> Public Health Service
> National Institutes of Health
> Cancer Information Service: 800-4-CANCER
> www.nci.nih.gov

From Test Tube to Patient: Improving Health Through Human Drugs, FDA Consumer Special Report (9/99). This report is available free from the FDA and is a great overview of the role of the FDA in the drug approval process. It can be obtained by writing to

> U.S. Food and Drug Administration
> (HFI-40)
> 5600 Fishers Lane
> Rockville, MD 20857
> 888-INFO-FDA
> www.fda.gov/cder

BIBLIOGRAPHY

Challenges in the Design of Phase I and Early Phase II Studies. G. S. Hughes, Jr., *Drug Information Journal,* Vol. 23, pp. 693–697, 1989.

Clinical Trial Design. G. Keith Chambers and Mary Sue Fairborn, *Applied Clinical Trials,* Vol. 7 (9), p. 60, 1998.

General Considerations for the Clinical Evaluation of Drugs, U.S. Department of Health, Education and Welfare, Public Health Service, 1977 (FOI).

How Drugs Are Developed. A Practical Guide to Clinical Research. Josh Cochr, *Applied Clinical Trials,* Vol. 5 (5), p. 63, 1996.

Improving Performance in Clinical Trials. Edward Fuchs, *Research Practitioner,* Vol. 1 (3), pp. 81–86, 2000.

Learning to Conduct Research—the Hard Way. L. Witter, *RN,* February, pp. 35–40, 1990.

Lessons Learned from Coordinating a Pivotal Clinical Trial. D. Johnston, *Journal of Clinical Research and Drug Development,* Vol. 7, pp. 31–39, 1993.

Management of Clinical Trials. F. Abdellah, *Journal of Professional Nursing,* Vol. 6 (4), p. 189, 1990.

New Challenges for Nurses in Clinical Trials. J. Guy, *Seminars in Oncology Nursing,* Vol. 7 (4), pp. 297–303, 1991.

New Drugs: First Time in Man. Posnar and Sedman, *Journal of Clinical Pharmacology,* Vol. 29, pp. 961–966, 1989.

Phase I Trials: Past, Present, and Future. W. Todd, *Drug Information Journal,* Vol. 23, pp. 669–672, 1989.

The Food and Drug Administration's Fast Track Designation: Observations from the Initial Two Years of This Designation. David Cochetto, *Drug Information Journal,* Vol. 34 (3), pp. 753–760, 2000.

The Importance of the Investigator's Brochure. M. Mikhail, *Applied Clinical Trials,* Vol. 2 (6), pp. 56–58, 1993.

The Limitation of Animal Studies: What Can and Cannot Be Predicted for Man. A. D. Dayan, *Drug Information Journal,* Vol. 25, pp. 165–170, 1991.

2

THE PHYSICIAN AS INVESTIGATOR

> An investigator is responsible for ensuring that an investigation is conducted according to the signed investigator statement, the investigational plan, and applicable regulations; for protecting the rights, safety, and welfare of subjects under the investigator's care; and for the control of drugs under the investigation . . . .
>
> —21 CFR 312.60

The privilege of performing clinical research in the United States as a Principal Investigator (PI) carries with it serious responsibilities. This chapter will describe the qualifications, duties, and responsibilities of the investigator and subinvestigators.

QUALIFICATIONS OF THE PRINCIPAL INVESTIGATOR

The clinician who is considering research participation should, like Sir Gawaine, "ponder his worthiness." Whether you are interested in pharmaceutical-sponsored research or federally funded research, the single most important factor is the qualification of the investigator. First and foremost, clinician investigators must be "qualified by training and experience as appropriate experts to investigate the drug or device" (21 CFR 312.53).

In practice, this means initially that the physician was educated in an approved institution of medical training. Following successful completion of medical studies, as evidenced by a degree of Doctor of Medicine (or equivalent), the prospective investigator must have had additional training at the residency and/or fellowship level to attain the requisite training in the research field. Minimum qualifications include the possession of a valid current license to practice medicine in the state in which the prospective investigator intends to perform the research. Specialty board certification is often required as well. This should be accompanied by evidence of clinical experience, usually obtained

15

by practice in the therapeutic area of the proposed research. "Experience" is also defined as experience in clinical research, which may be obtained by initial research participation as a subinvestigator. Some sponsors are now requesting, in addition, evidence that no malpractice claims against the clinician have been settled or are pending.

An investigator who is lacking experience may consider mentoring with a known expert in the field, serving as a subinvestigator in a clinical trial, and publishing in peer-reviewed journals. The fact that you see 50 patients a day in the clinic is not as important as what research you have done. Successful experience in research provides evidence that you can think scientifically, carry out a research plan, and interpret scientific data.

Initial Investigator Information

The prospective investigator must first submit a curriculum vitae (CV) and a signed Statement of Investigator (SOI, Form FDA 1572). The CV appropriate for clinical research varies somewhat from the CV appropriate for employment searches or public relations handouts. It should list the research qualifications of the physician, including all undergraduate and postgraduate medical educational experiences. It should contain the physician's current professional address, office telephone number, medical activity (e.g., academic, private practice, sabbatical), state medical licensures, and any pertinent prior clinical research experiences.

A signed investigator statement is required by the Food and Drug Administration (FDA) before a sponsor is permitted to allow an investigator to begin study participation for an Investigational New Drug (IND) study. It must be accurate and correct in every particular, and must contain the following information:

- The name and research address of the investigator.
- The name and code number of the protocol as listed in the IND to be conducted by the investigator.
- The name and address of any other research facility where the clinical investigation will be conducted.
- The name and address of any medical laboratory facilities to be used for the study.
- The name and address of the Institutional Review Board (IRB) that is responsible for review and approval of the study.

Form FDA 1572 also includes a list of the names of the subinvestigators who will be assisting the investigator. In addition, the signature of the investigator constitutes a commitment by the investigator that he or she will

- conduct the study in accordance with the relevant protocols;
- comply with all requirements regarding the obligations of clinical investigators;
- personally conduct or supervise the described investigation;
- inform any patients that the drugs are being used for investigational purposes and ensure that the requirements relating to informed consent and IRB review are met;
- report to the sponsor adverse experiences that occur in the course of the investigation in accordance with 21 CFR 312.64;
- read and understand the information in the Investigator's Brochure; and
- ensure that all associates, colleagues, and employees assisting in the conduct of the study are informed of their obligations in meeting the above commitments.

The investigator's signature on Form FDA 1572 also constitutes a commitment by the investigator that

> for an investigation subject to an institutional review requirement under Part 56, an IRB that complies with the requirements of that part will be responsible for the initial and continuing review and approval of the clinical investigation and that the investigator will promptly report to the IRB all the changes in the research activity and all unanticipated problems involving risks to human subjects or others, and will not make any changes in the research without IRB approval except where necessary to eliminate apparent immediate hazards to the human subjects.
>
> Statement of Investigator Form (Form FDA 1572),
Section 9, Commitments.

For clinical studies of investigational devices, the investigator signs an "Investigator's Agreement" that includes the basic information on Form FDA 1572.

THE INVESTIGATIVE STAFF

The PI can rarely conduct a clinical study alone and requires the assistance of qualified staff. The subinvestigator(s) will assist in all areas of the study or specifically defined areas. The Clinical Research Coordinator (CRC) will facilitate the study procedures and assure data collection. Administrative staff may help to process the endless mounds of paperwork. Together, this study team will assist the investigator in fulfilling the obligations and responsibilities as outlined in 21 CFR 312.60–312.70.

The PI is ultimately responsible for communications regarding the clinical trial. Much of the communication may be delegated to the subinvestigator or CRC, but the success of the trial depends on the PI's ability to obtain the enthusiastic and knowledgeable cooperation of others.

SUBINVESTIGATORS

The PI will designate subinvestigators on Form FDA 1572. Only those subinvestigators named on Form 1572 should perform study-specific procedures or enter data on Case Report Forms (CRFs). These investigators may take on a great deal of the study conduct. However, the PI is ultimately responsible for all actions of the subinvestigator. Therefore, the subinvestigator should be well informed and trained in the specifics of the study and the obligations of the investigator in clinical research.

Subinvestigators may be necessary for several reasons. For example, the trial may involve several medical disciplines; an antibiotic trial involving peritonitis may require the skills of a surgeon, an infectious disease specialist, and a microbiologist. A study of peptic ulcer disease in subjects with rheumatoid arthritis may involve both a gastroscopist and a rheumatologist.

A subinvestigator may be a colleague or partner who will assist the PI in the conduct of all aspects of the trial, although the PI who has signed Form FDA 1572 is ultimately responsible. The subinvestigator may be doing as much as, or more than, the PI. There may be a number of subinvestigators involved in the trial, such as a whole department or other colleagues at a remote location. A subinvestigator may also be a specialist who will be intimately involved with the study. For example, the radiologist who reads all of the chest X-rays specifically to determine study endpoint criteria or the pathologist who reads all of the biopsies may be listed as subinvestigators.

Life is uncertain. The subinvestigator should be completely familiar with the specifics of the study protocol and with Good Clinical Practice (GCP). A prepared and involved subinvestigator will be able to perform the duties of the PI when the latter is on vacation or ill. In addition, subinvestigators can assist in the recruitment of subjects, participate in subject screening, and accelerate the enrollment and study conduct activities of the study. There must always be open communication and clear delineation of responsibilities between the PI and the subinvestigators. All study information must be disseminated to all subinvestigators to assure that the study proceeds as smoothly as possible and according to protocol.

A subinvestigator should also be qualified by training and experience to take part in the clinical trial. The responsibilities assigned to the subinvestigator should be based on this requirement. A colleague of many years experience may be able to conduct the study unsupervised after careful discussion of and agreement with the study procedures. However, a first-year fellow who is participating in his or her initial clinical research study will require close supervision.

The FDA does not give clear guidance on who should or should not be listed on Form FDA 1572 as a subinvestigator. Some sponsors or IRBs may require that only degreed physicians (MD, DO, etc.) be listed. Others may require all persons involved in the study, such as pharmacists and CRCs, be included. Minimally, physician colleagues, research fellows, residents, and associates involved in the study should be listed. As a guideline, a subinvestigator is one who

- treats the subject in lieu of the PI,
- performs critical test interpretation or endpoint determination, and
- can step into the role of the PI, if needed.

THE ROLE OF THE CLINICAL RESEARCH COORDINATOR IN CLINICAL RESEARCH

The primary responsibility of the CRC in clinical research is to ensure smooth, accurate progress of the project from the planning stage to study end (and often beyond) by acting as liaison to the investigator, the subject, the institution, the IRB, and the pharmaceutical company or government sponsor. Willing collaboration with individuals who possess the necessary clinical skills and medical expertise for the conduct of the study is necessary to ensure the quality and integrity of a clinical trial. It is imperative that there be open communication and clear delineation of responsibilities between the PI and CRC.

Qualifications of a CRC

Clinical research studies can be fun, exciting, and professionally satisfying, but research is not for everyone. During the planning phase of a clinical trial, it is most important for the PI to evaluate the coordinator's qualifications for the job and for the candidate to do some self-evaluation. Some qualifications of the ideal CRC are as follows:

- Scientific/medical background.
- Interest in research methodology.
- Detail oriented.
- Good organizational skills.
- Ability to work independently.
- Innovative and creative.
- Availability per study requirements.
- Flexibility.
- People oriented.
- Good interaction skills with patients.
- Certification in clinical research.

Responsibilities of CRCs

The responsibilities of the CRC may vary tremendously from institution to institution. The investigator should carefully evaluate the responsibilities to be assigned to the CRC. Some administrative responsibilities are as follows:

- Interact with IRB, lab staff, clinic staff, pharmacy, other departments in the institution such as radiology and nursing.
- Assist in preparation of IRB documents, including the Informed Consent Form.
- Prepare a study budget.
- Assure all study documentation is maintained.
- Interact with sponsor.
- Interact with PI and subinvestigators.
- Coordinate and participate in monitoring visits by the sponsor.
- Complete CRFs and submit to the sponsor/resolve data queries.
- Coordinate, participate in, and facilitate inspections/audits of the site.
- Document study progress.

CRC responsibilities involving patients are as follows:

- Recruit study subjects.
- Assess subjects for eligibility.

- Discuss study with subject and assist in obtaining informed consent.
- Schedule subject assessments/visits.
- Assure all study tests and visits are done at appropriate time intervals.
- Evaluate study subjects at appropriate intervals.
- Assess laboratory data and clinical signs for potential of adverse events.
- Provide information for treatments and reactions.
- Administer or dispense investigational agent, as needed, under PI supervision.
- Promote subject compliance by providing patient support and education.
- Prepare laboratory specimens; shipping biological samples and radiologic films.
- Arrange for study subject compensation.
- Comply with FDA regulations for conducting clinical trials.

Certification of CRCs

Two organizations currently certify clinical research professionals: the Association of Clinical Research Professionals (ACRP) and the Society of Clinical Research Associates (SoCRA). ACRP certifies both Clinical Research Coordinators (CCRC) and Clinical Research Associates (CCRA) as two separate test processes. Additionally, ACRP is planning for Investigator certification by the end of 2001. SoCRA certifies both CRAs and CRCs but as a single process, and the title is Certified Clinical Research Professional (CCRP). Both certifications require a minimum amount of on-the-job experience plus passing an exam. Contact each organization for additional information.

SUPPORT STAFF: ADMINISTRATIVE, PHARMACY, AND HOSPITAL

Any number of ancillary services may be involved with the study. Many large institutions have an established research pharmacy with specific operating procedures for handling investigational drug. The pharmacist who dispenses the investigational agent, even if experienced in research, must be properly supervised by the PI and instructed on dispensation of each investigational agent. The pharmacist may be responsible for the randomization of subjects and maintaining the code breaker. The research pharmacist is an integral part of the institutional study and can facilitate dispensation of the investigational agent.

The hospital or clinic medical and nursing staffs, while not directly involved in the study, will have contact with study patients and should be familiar with the expected procedures. (See the section "Preparing Hospital Staff" in Chapter 3 for further comments.)

A smooth-running study also depends on administrative support. There will always be communications with the IRB, the sponsor, and study subjects. There will be countless CRFs to copy and mail. The administrative staff can be indispensable in making this part of the study run smoothly. Communication with all individuals involved in a clinical trial is the key to success and vitally important for a smooth-running and scientifically accurate study.

THE STUDY WORK AREA

Clinical trials may be conducted in a hospital-based center for either inpatient or outpatient trials, in physician's offices or clinics, or at special research facilities that concentrate only on clinical research trials. Too often the work area allocated for clinical trial management is inadequate. The following should be considered when planning a clinical trial:

- Adequate space must be provided for the coordinator to work and store necessary documents.
- If patient exams or interviews are conducted, the room should ensure privacy. If a telephone is necessary, one should be available in the work area.
- Study information is considered confidential—files should be locked and CRFs should be secured.
- There should be a quiet and private work area reserved for the monitor during a monitoring visit (also important for an inspector during an audit).
- Exam rooms should be available for subject visits and equipped with the necessary implements for the study visit.
- If remote data entry or a computer system is used for coordinating the study, adequate equipment and hookups should be available.
- All materials required to conduct the study should be readily available for study assessments.
- A key factor in a well-run study is organization—a tidy office/work area can make all the difference.

INVESTIGATOR RESPONSIBILITIES TO THE INSTITUTIONAL REVIEW BOARD

The investigator should assure that he or she will promptly report to the IRB all changes in the research activity and all unanticipated problems involving risk to human subjects or others. The investigator is not permitted to make any changes in the research program (ignore, omit, or add any tests, procedures, or therapies) without IRB (and sponsor) approval, except where necessary to eliminate apparent immediate hazards to human subjects.

Conflict of interest and financial disclosure are important issues for investigators and IRBs and are discussed in Chapters 4 and 8.

INFORMED CONSENT

Respect for every human subject's rights and dignity requires that informed consent be obtained before a human subject participates in any clinical investigation. IRBs, clinical investigators, and sponsors all share responsibility for ensuring that the informed consent process is adequate. The investigator is responsible for ensuring that the research subject gives fully informed consent to participate in the clinical trial. The regulations for subject protection are discussed in Chapter 4, and informed consent is discussed in Chapter 7.

RECORDKEEPING RESPONSIBILITIES

The clinical investigator is responsible for keeping three basic types of records:

1.   Study drug disposition.
2.   Case histories and CRFs.
3.   Regulatory files.

Strict Control of Study Drugs

The study sponsor is responsible for shipping investigational new drugs only to investigators participating in the investigation, i.e., ones who have filed with the FDA a completed Form FDA 1572 for this IND. In practice, this means that the investigator will have submitted to the sponsor a

- completed Form FDA 1572,
- current CV containing state licensure information,
- letter of protocol approval from the IRB,
- letter of Informed Consent Form approval from the IRB, and
- copy of the protocol bearing the investigator's original signature.

The investigator is to administer the investigational drug only to subjects under the investigator's personal supervision or under the supervision of a subinvestigator responsible to the investigator, preferably as listed on the Form FDA 1572. The investigator is required to prepare and maintain adequate and current records of the dispensation and final disposition of each unit of the investigational drug. These responsibilities may be delegated to the pharmacist or subinvestigator; but the PI is ultimately responsible. This important topic is discussed in detail in Chapter 5.

Case Histories

Investigators are required to prepare and maintain adequate and accurate case histories designed to record all observations and other data pertinent to the investigation on each subject treated with the investigational drug or employed as a control in the investigation. These clinical records should contain, at a minimum, the following:

- Basic subject identification.
- Documentation of subject's study eligibility (inclusion/exclusion criteria).
- Information that supports and confirms data recorded on the CRF and submitted to the sponsor.
- Drug dispensation records documenting the subject's exposure (date, time, and duration) to the test article or control.
- Information that supports and confirms any adverse events that occurred during the study.
- Pertinent laboratory and imaging reports with written comments by the investigator pertaining to the clinical significance of any abnormalities or changes from baseline.
- Consultations or other clinical correspondence.
- Any other diagnostic test results.

Case Report Forms

The CRF is NOT a substitute for physician records (case history records), although special forms designed to capture CRF–required data may be incorporated into the physician's clinical records. Every entry or bit of information that is recorded on the CRF should be supported by matching information in the case history maintained by the physician. The investigator (and no one else) is responsible for the accuracy and completeness of the study records, including all clinical files, charts, and notes, and the investigator is responsible for any discrepancies found in these records.

According to the FDA Information Sheet "Recordkeeping in Clinical Investigations" (10/95), in certain circumstances CRFs may serve as the sole source of information.

> . . . Case report forms are a critical part of the investigation records, but in most cases they cannot serve as the complete investigation record. The case report should contain all data required by the protocol but need not duplicate all the investigator's records on the subjects' medical histories. FDA does not require that special medical records be established to meet its requirements.

This may be interpreted to mean that the CRF may be used as a source document in specific cases, for example, when a volunteer is enrolled in a short-term study in a freestanding clinical research unit or when specific data in a study is strictly study required, like logging multiple blood pressure readings over a short period of time. In these cases, logging data directly on the CRF may be permissible. However, most sponsors will still require a separate case history record for research subjects; ultimately, the sponsor should decide if the case history records kept by the PI are adequate to support data on the CRFs.

The investigator is required to retain and store all records pertaining to a clinical trial for two years following the date of marketing application approval or for two years after the investigation is discontinued. Retention of complete and accurate records is necessary to establish the validity and completeness of a clinical investigation, particularly if it is submitted in support of an application for a research or marketing permit.

Regulatory Files

Regulatory files, those documents required by federal regulation and subject to audit, include required investigator reports, progress reports to the sponsor and the IRB, safety reports (discussed in Chapter 11), a final report to the sponsor and the IRB, reports to the IRB of all changes in the research activity, and reports of unanticipated problems arising that involve risk to human subjects or others. Additionally, a copy of the signed Form FDA 1572, appropriate CVs, Informed Consent Forms, correspondences documenting major decisions, and other documents become a part of the study file. A study file checklist is included as Figure 8.7 in Chapter 8.

SUPERVISORY RESPONSIBILITIES

The investigator, like the captain of a ship, is responsible for all study-related activities of anyone participating in the clinical research project. From the first signature on Form FDA 1572, through the signature on the protocol and CRFs, to the signature on the final report, the investigator bears ultimate responsibility for the actions of subinvestigators, CRCs, office staff, and medical record staff.

BIBLIOGRAPHY

Can Investigator Certification Improve the Quality of Clinical Research? Gary Lightfoot, Sandra Sanford, and Arna Shefrin, *The Monitor*, Vol. 13 (3), pp. 17–21, 1999.

Clinical Trial Indemnifications: An Investigator's Viewpoint. G. Johnson, *The Monitor*, Vol. 9 (3), pp. 30–35, 1995.

Considerations of the Primary Care Physician in the Evaluation of New Drugs. E. Sutton and P. Sutton. In *Clinical Drug Trials and Tribulations*, edited by A. Cato, Marcel Dekker, Inc., New York, pp. 295–302, 1998.

Ethical Issues in Clinical Research: An Investigator's Perspective. G. Gillett, *Applied Clinical Trials*, Vol. 1 (5), pp. 66–68, 1992.

From MD to PI. Margaret Skelton, *Applied Clinical Trials*, Vol. 8 (12), p. 60, 2000.

GCP Responsibilities of Principal Investigators Beyond the 1572. D. Mackintosh and V. Zepp, *Applied Clinical Trials*, Vol. 5 (11), pp. 32–40, 1996.

How Clinic Coordinators Spend Their Time in a Multicenter Clinical Trial. Irene L. Goldsborough, Renee Y. Church, M. Marvin Newhouse, and Barbara S. Hawkins, *Applied Clinical Trials,* Vol. 7 (1), p. 33, 1998.

Interactions of Academic Investigators with Pharmaceutical Companies. In *Guide to Clinical Trials,* B. Spilker, Raven Press, New York, pp. 390–394, 1991.

Job Descriptions and Performance Evaluations. Carol Saunders, *Research Nurse,* Vol. 5 (6), pp. 1–3, 1999.

Measuring the Workload of Clinical Research Coordinators: Part 1—Tools to Study Workload Issues. Clement K. Gwede, Darlene Johnson, and Andy Trotti, *Applied Clinical Trials,* Vol. 9 (1), p. 40, 2000.

Measuring the Workload of Clinical Research Coordinators: Part 2—Workload Implications for Sites. Clement K. Gwede, Darlene Johnson, and Andy Trotti, *Applied Clinical Trials,* Vol. 9 (2), p. 42, 2000.

New Strategies for Academic Medical Centers (Part 3). Carolyn Peterson, *Applied Clinical Trials,* Vol. 5 (8), p. 32, 1996.

Personal Care and Randomized Clinical Trials: Resolving the Basic Conflicts. B. Miller. In *Clinical Drug Trials and Tribulations,* edited by A. Cato, Marcel Dekker, Inc., New York, pp. 303–320, 1998.

Pragmatics. A. Duggan. In *An Introduction to Clinical Research,* edited by C. DeAngelis, Oxford University Press, New York, pp. 111–132, 1990.

Roles of Private Practice Physicians in Clinical Trials. In *Guide to Clinical Trials,* B. Spilker, Raven Press, New York, pp. 395–398, 1991.

Standard Operating Procedures and the Conduct of Clinical Trials. Krishan Maggon and Daniel Brandt, *Applied Clinical Trials,* Vol. 3 (7), p. 46, 1994.

Who Is a Clinical Research Nurse? Establishing Guidelines and Standards of Practice for a Growing Profession. Karen Bowen and Linda Rice, *Research Nurse,* Vol. 4 (4), pp. 1–4, 1998.

3

PROFESSIONAL INTERACTIONS

I agree to ensure that all associates, colleagues, and employees assisting in the conduct of the study(ies) are informed about their obligations in meeting the above commitments.

—Form FDA 1572, Section 9

The Principal Investigator (PI) often undertakes research protocols that specify the co-ordination of research activities with other physicians and departments within an institution. These may include subinvestigators, clinic and in-house scheduling offices, clinical and microbiology laboratories, institutional pharmacies, the institutional nursing department, and other departments (e.g., ER, ICU, Radiology), as well as the grants administration office and the Institutional Review Board (IRB).

The PI will identify subinvestigators participating in the clinical trial on the Statement of Investigator form (Form FDA 1572) in Section 6. Only those physicians named on Form 1572 should perform study-specific procedures or determine data to be entered on Case Report Forms (CRFs). The role of subinvestigators is discussed in detail in Chapter 2.

THE INSTITUTIONAL REVIEW BOARD

The investigator has the general responsibility to protect the rights, safety, and welfare of subjects under his or her care. In addition, the investigator has a regulatory requirement to assure that an IRB that complies with Food and Drug Administration (FDA) requirements will be responsible for the initial and continuing review and approval of the proposed clinical study. IRBs are responsible for reviewing proposed research to evaluate risk to subjects and protection of subject rights.

IRB procedures will differ among institutions. The PI and the Clinical Research Coordinator (CRC) will need to become familiar with the procedures of their IRB. At some institutions, the investigator, subinvestigators, and/or CRC are requested to attend the meeting of the IRB when their project is being discussed. At others, the investigators/CRC may be asked not to attend if their research is being discussed unless needed to clarify questions. In any case, investigators and subinvestigators who sit on an IRB must abstain from voting on any research project in which they are involved (and that abstention must be recorded in the minutes of the IRB meeting).

However, all IRBs are required to adhere to FDA regulations (see Chapter 4 and Appendix A, 21 CFR 56). The protocol and the Informed Consent Form must be approved by the IRB prior to beginning the study. The IRB may request that the Investigator's Brochure be submitted for informational purposes. After initial IRB approval, the PI is required to provide periodic reports on the progress of the study. IRBs require at least an annual progress report and renewal of approval at the anniversary of the initial approval date per FDA regulation. Specific IRBs may require more frequent reports. The investigator is required to submit all protocol amendments (modifications to the research study) to the IRB and obtain approval of the amendment prior to implementation. Amendments affecting subject safety may be implemented prior to IRB notification but must be submitted in a timely matter. The IRB must be notified of Serious Adverse Events and patient deaths (see Chapters 4 and 11). The PI should make certain that documentation of all communications with the IRB is maintained in the study file.

PHARMACY

Studies conducted at larger institutions usually involve the pharmacy department. The research pharmacist should be trained in the regulations governing the storage and dispensation of study drugs, the special records required for study materials, and the procedure for monitoring study materials. In addition, the pharmacist must be familiar with the study protocol so that the study drug may be properly dispensed. In many studies, the pharmacist must perform blinding functions such as placing sleeves around bottles of medication or altering the labeling of study drug to maintain the blinding of staff physicians and nursing personnel. The PI should personally enlist the cooperation of the institutional pharmacist; the study staff should maintain that cooperation through frequent communications.

LABORATORIES AND MEDICAL IMAGING

Clinical research in institutions usually involves the collection of baseline, efficacy, and safety data from sources other than the PI's physical examination and diagnostic skills. The clinical laboratory provides hematology and clinical chemistry information. The microbiology department performs cultures and drug sensitivity testing. The medical imaging department performs X-ray, echo, and magnetic imaging studies. The electro-cardiogram (EKG) and its interpretation is often required. The incorporation of these departments into the study team and the coordination of these tests are responsibilities of the PI. Protocol training is essential to insure that the procedure is performed in the time frame and manner prescribed in the protocol and that the resultant report contains all the required research information.

Finally, the proper billing procedure for these study-specific tests should be clearly delineated. While the CRC is quite capable of maintaining the needed relationships and ongoing training, it is the duty of the PI to enlist the cooperation and enthusiasm of all ancillary department personnel.

EMERGENCY ROOM (ER) AND INTENSIVE CARE UNIT (ICU)

Depending on the nature of the study, the ER and the ICU may become fertile areas for the recruitment of study subjects. Research trials involving subjects with acute coronary ischemia, pneumonia, trauma, acute psychosis, or severe arrhythmias, for example, require the cooperation and active participation of the entire ER team if any study enrollment is to take place. The PI should personally contact the lead ER physicians and nurses prior to study onset to explain the protocol and enlist the cooperation of the entire ER staff.

The CRC should coordinate with the ER nursing staff such study procedures as subject identification and eligibility criteria, specimen collection, special testing, and most important, the avoidance of prohibited concurrent medications. Furthermore, both the PI and the CRC should visit with their ER colleagues frequently to assure continued cooperation and participation.

COMMUNICATION STRATEGIES

The following are some suggestions to facilitate communication about the study to all research staff involved.

- Meetings.
  - Schedule regular, periodic meetings with the subinvestigators and CRCs to discuss the trial.
  - Circulate a draft agenda prior to the meeting and request that participants submit items for the agenda.
  - Stick to the agenda at the meeting but also allow time for unanticipated topics.
  - Some suggested areas to discuss include subject accrual, adverse events, responses, amendments or changes to the study, information from the sponsor, and specific problem areas.
- Develop a centralized computer database that study personnel can access.
- Have multiple copies of the protocol, study procedures manual, study schema, and relevant study aids available for all study personnel.
- Develop charts, flow sheets, study schema, schedule of evaluations and visits, study pocket cards, forms, graphs, and lists that easily communicate aspects of the study.

STUDY COORDINATION

One of the major responsibilities of the PI, in conjunction with the investigative staff, is working through the logistics of study implementation, i.e., the day-to-day activities of performing a clinical trial. Working through the study logistics is probably the most critical task in planning a clinical trial. It is possible that the original protocol may require changes because it would not be feasible to do something as written. Study logistics require a great deal of organization, cooperation, ingenuity, and, sometimes, a little coaxing!

When: Study logistics should be evaluated at some point prior to the initiation site visit. The investigative staff should be prepared to walk-through the procedures at the initiation visit with the sponsor.

What: The investigative staff should critically review the protocol to ascertain that the required procedures can be performed as written and that they are reasonable considering the standard of care of the condition being studied. Table 3.1 shows some of the areas that the PI and CRC will need to coordinate within the institution. Most of these items are discussed in greater detail in other areas of this book.

PREPARING HOSPITAL STAFF

Many clinical trials are conducted in a hospital setting where other hospital staff will be involved. The nursing staff must not only be informed of the study procedures but also actively involved in the study process. Conduct in-service meetings to discuss the study. Be specific about what responsibilities the nursing staff will have—administering the test agent, observing for adverse events, and collecting vital sign data and test samples. Also be specific about the benefits to the subjects of study participation. Hospital nurses have performed heroic research tasks in addition to their voluminous daily duties when they understood that their patients would benefit. To ease hospital staff burdens, the wise investigator will

- provide standard physician's orders for study subjects;
- provide a protocol-specific checklist to attach to the subject's chart/daysheet (use a distinctive color of paper to set the list apart from all of the other chart papers);
- let subjects know what the study-specific procedures are—they may be able to save a sample by reminding the staff when necessary;
- label the subject's chart, daysheet, bed, wristband (anything and everything that makes sense!) to alert staff that this is a study subject (be sure to consider patient confidentiality);
- provide prelabeled sample collection vials and other study-specific materials; and
- distribute his or her telephone and pager numbers to all staff and be available to address their questions.

Develop a team approach with the institution staff and be sure to follow up periodically so that any errors or omissions can be caught early and corrected. Smooth interactions contribute to the success of the clinical trial.

TABLE 3.1 AREAS OF INSTITUTIONAL COORDINATION

TASK	DESCRIPTION
Administrative Functions	Includes IRB approvals, advertising, study file maintenance.
Subject Screening	How will study subjects be identified and screened? Will special screening clinics be arranged?
Subject Referrals	Notify appropriate specialty areas within the institution for subject referrals. Are subjects to be identified or enrolled through areas such as ER or ICU?
Subject Visits	How are visits to be scheduled? Is there adequate space? Plan the subject's day . . . how will the visits flow from assessment to labs to special tests and so on?
Hospitalized Subjects	If subjects are to be admitted to the hospital, what will the procedure be? Notify all staff on the floor of study-specific procedures. Place brightly colored instruction sheets in the subject's chart.
Subject Billing	What arrangements are made to assure the subject is not billed for procedures that are study related?
Obtaining Diagnostic and Laboratory Tests	Will the laboratory be available for tests during subject visits? Is it close by? Will the laboratory or diagnostic testing area create a bottleneck in the flow of the subject's visit? Can priority be given to study subjects?
Medical Records	Are medical records readily obtainable? How much lead time is needed? How long are medical records maintained by the institution (does this meet regulatory requirements for record retention)? Per FDA regulations, source documents must be kept for the same time period as study files. Will files be inadvertently altered (purged or thinned)? What happens to a subject's file after a death?
Completing Medical Records	Determine what is necessary to include in the subject's medical record or clinic chart. Is an *original* signed Informed Consent Form required (the subject may need to sign many copies)?
Dispensing Investigational Agent	Will the investigational agent be dispensed through the pharmacy? If so, is it open at the time of subject visits? Will the pharmacy be responsible for randomizing subjects? Does the pharmacy keep the code-breaker? Is a pharmacist available 24 hours in case of an emergency? Is the pharmacist qualified to dispense drug; does he/she know the study? Where and how will the drug be stored? Will the pharmacist maintain records as required?
Other Sites	Sometimes your institution will be the coordinating center for a large multicenter trial or a cooperative group effort. In this case, who will be responsible for shipping study drug, screening and enrolling subjects, and coordinating data?
Tests Performed by Other Departments	How will subjects be scheduled? How will the department be compensated? Will someone consistently be interpreting the test results? What is the report turnaround time and mechanism?
Subject Compensation	Will subjects be compensated? How will payments be made? Is compensation contingent on study adherence and completion?
Grants Management	Is there a specific department in the institution that handles all grants? What is its role? Does it make payments as well?
Legal Issues	The legal department may be involved with negotiating contracts and indemnification agreements.

BIBLIOGRAPHY

Collaborating with Colleagues (Part I): Staff Nurse Involvement. Tonya Edens and Karen Safcsak, *Research Nurse,* Vol. 4 (4), pp. 4–8, 1998.

Issues in the Review of Clinical Drug Trials by IRBs, D. Cowen. In *Clinical Drug Trials and Tribulations,* edited by A. Cato, Marcel Dekker, Inc., New York, pp. 321–345, 1988.

Learning to Conduct Research—The Hard Way. L. Witter, *Research Nurse,* February, pp. 35–40, 1990.

Negotiating Clinical Trial Agreements with Academic Institutions. G. Sanders, *Applied Clinical Trials,* Vol. 1 (6), pp. 39–45, 1992.

Preparing Nurses for Clinical Trials: The Cancer Center Approach. V. Wheeler, *Seminars in Oncology Nursing,* Vol. 7 (4), pp. 275–279, 1991.

The Interrelationships of Sponsors, Clinical Investigators, and Institutional Review Boards. T. Kirsch, *Drug Information Journal,* Vol. 21 (2), pp. 127–131, 1987.

The Role of the Coordinating Center Project Manager in a Multicenter Clinical Trial. S. Margitic, *Journal of Clinical Research and Drug Development,* Vol. 7 (4), pp. 243–252, 1993.

4

FDA REGULATIONS AND GOOD CLINICAL PRACTICE GUIDELINES

Scope.

(a) This part applies to all clinical investigations regulated by the Food and Drug Administration under sections 505(i), 507(d), and 520(g) of the Federal Food, Drug, and Cosmetic Act, as well as clinical investigations that support applications for research or marketing permits for products regulated by the Food and Drug Administration, including food and color additives, drugs for human use, medical devices for human use, biological products for human use, and electronic products. Additional specific obligations and commitments of, and standards of conduct for, persons who sponsor or monitor clinical investigations involving particular test articles may also be found in other parts (e.g., parts 312 and 812). Compliance with these parts is intended to protect the rights and safety of subjects involved in investigations filed with the Food and Drug Administration pursuant to sections 406, 409, 502, 503, 505, 506, 507, 510, 513–516, 518–520, 721, and 801 of the Federal Food, Drug, and Cosmetic Act and sections 351 and 354–360F of the Public Health Service Act.

(b) References in this part to regulatory sections of the Code of Federal Regulations are to chapter I of title 21, unless otherwise noted. [45 FR 36390, May 30, 1980; 46 FR 8979, Jan. 27, 1981]

—21 CFR 50.1

Clinical research in the United States involving investigational (nonapproved) agents is strictly regulated by the Food and Drug Administration (FDA). Historically, the regulations evolved often out of necessity and in response to tragedy (refer to reference by McCarthy for further information). A persistent theme during the early attempts at developing federal regulations was the inability of the FDA to enforce the regulations—there were no legal means to inspect manufacturers or to bring disciplinary action. Today, the FDA has the authority to inspect and bring criminal action against violators of regulatory requirements for the manufacture and sale of food, drugs, and cosmetics.

The regulations specific to the conduct of clinical research in the United States can be found in the Code of Federal Regulations (CFR), Title 21, Parts 50, 56, 312, and 314; 600 and 601 (biologics); and 812, 813, and 814 (medical devices); and Title 45, Part 46 (Subparts B, C, and D). Specific topics also may be included in other areas of the CFR. Collectively, these regulations are referred to as Good Clinical Practices (GCPs). The regulations for informed consent (Part 50) and Institutional Review Boards (IRBs) (Part 56) are aimed at protecting the subjects in clinical trials. Part 312 (Investigational New Drug [IND] regulations) enforces this policy in the context of a sponsor's submission of an application to the FDA to begin human trials with an IND. Clinical trial data are collected in support of a New Drug Application (NDA), a formal request to the FDA to approve the investigational drug for marketing for a specific indication based on data collected from clinical trials. The International Conference on Harmonisation Good Clinical Practice Guideline (ICH GCP) offers additional guidance on the conduct of clinical trials.

This chapter will focus on the regulations as they apply to investigators, sponsors, the protection of human subjects, and IRBs.

CODE OF FEDERAL REGULATIONS

The regulations governing clinical trials are collectively referred to as GCPs and are published annually in the *Code of Federal Regulations*. Anyone conducting a clinical trial should take the time to read these regulations. To better understand the regulations, it is advisable to read the preamble (Part VII) to the IND rewrite of the regulations (*Federal Register,* Vol. 52, No. 53, March 19, 1987, page 8798). The preamble explains some of the reasoning behind the regulations and provides a context for interpretation.

Relevant parts of the FDA regulations (21 CFR 50, 56, and 312 and 45 CFR 46) are included in Appendix A. These are the regulations published in 2001. THE CODE OF FEDERAL REGULATIONS IS UPDATED ANNUALLY, AND AN INVESTIGATOR SHOULD ALWAYS HAVE A COPY OF THE MOST CURRENT REGULATIONS. The information in this book is based on the regulations in effect at the date of publication. **ALWAYS** review the current regulations for changes, and always assess each situation as it applies to your particular case. Verify all processes according to institutional, local, state, and federal regulations. If you do not subscribe to the *Federal Register* or CFR, the university library or the pharmaceutical company sponsor should be able to provide you with current copies. However, when the sponsor of a clinical trial supplies the investigator with copies of appropriate regulations, verify that they are *current* regulations and not outdated copies. You can access the current regulations at www.fda.gov.

Changes to the CFR are proposed through the *Federal Register,* a document that records government business and is published daily. Generally, comments are invited to a proposed rule by a specified date. The Final Rule, which takes into consideration the comments that have been discussed in a government forum, is then published in the *Federal Register*. When appropriate, the final rule determination is incorporated in the next annual issue of the *Code of Federal Regulations*.

The FDA regulations are supplemented by Information Sheets for investigators, IRBs, and sponsors as well a series of guidelines on the development of specific classes of drugs, based on therapeutic area. The Information Sheets were published in 1989 as a set for the IRB and a set for clinical investigators. In October 1995, these Information Sheets were revised and merged into one set, "Information Sheets for Institutional Review Boards and Clinical Investigators." Individual information sheets may be revised as needed. The most current versions may be accessed at www.fda.gov/oc/guidance. Revised again in 1998, the Information Sheets* contain the following topics:

General
- Frequently Asked Questions.
  - IRB Organization
  - IRB Membership
  - IRB Procedures
  - IRB Records
  - Informed Consent Process
  - Informed Consent Document Content
  - Clinical Investigations
  - General Questions
- Cooperative Research.
- Nonlocal IRB Review.
- Continuing Review After Study Approval.
- Sponsor–Investigator–IRB Interrelationship.
- Acceptance of Foreign Clinical Studies.
- Charging for Investigational Products.
- Recruiting Study Subjects.

*See Appendix B.

- Payment to Research Subjects.
- Screening Tests Prior to Study Enrollment.
- A Guide to Informed Consent Documents.
- Informed Consent and the Clinical Investigator.
- Use of Investigational Products When Subjects Enter a Second Institution.
- Personal Importation of Unapproved Products.
- Exceptions from Informed Consent for Studies Conducted in Emergency Settings.

Drugs and Biologics

- Investigational and "Off-Label" Use of Marketed Drugs and Biologics.
- Emergency Use of an Investigational Drug or Biologic.
- Treatment Use of Investigational Drugs.
- Waiver of IRB Requirements.
- Drug Study Designs.
- Evaluation of Gender Differences.

Medical Devices

- Medical Devices.
- Frequently Asked Questions about IRB Review of Medical Devices.
- Significant Risk and Nonsignificant Risk Medical Device Studies.
- Emergency Use of Unapproved Medical Devices.

FDA Operations

- FDA Institutional Review Board Inspections.
- FDA Clinical Investigator Inspections.
- Clinical Investigator Regulatory Sanctions.

Appendices

- A List of Selected FDA Regulations.
- 21 CFR Part 50.
- 21 CFR Part 56.
- Investigations Which May Be Reviewed Through Expedited Review.

- Significant Differences in FDA and HHS Regulations.
- The Belmont Report.
- Declaration of Helsinki.
- A Self-Evaluation Checklist for IRBs.
- FDA District Offices.
- FDA Phone Numbers.
- Internet Sites of Interest for Human Subject Protection Information.

ICH GCP GUIDELINE

The International Conference on Harmonisation (ICH) of Technical Requirements for the Registration of Pharmaceuticals for Human Use was created to standardize technical guidelines and requirements for product registration across the United States, Europe, and Japan to reduce the need for duplication of clinical trials. A listing of ICH topics appears in Table 4.1.

ICH Guideline E6 is specific to GCP and provides a unified standard for designing, conducting, recording, and reporting trials including human subjects consistent with the Declaration of Helsinki. This guideline was adopted by the FDA as a guideline effective 9 May 1997. Although the ICH GCP Guideline requires more than the FDA regulations, it does not contradict the regulations.

The ICH GCP Guideline offers distinct and clear concepts to conduct clinical research trials that are a good practice in any clinical research trial. However, it is imperative to comply with ICH GCP Guidelines when the sponsor anticipates submitting a global marketing application to assure that the data will be acceptable to all participating regulatory agencies.

TABLE 4.1 ICH TOPICS AND GUIDELINE CATEGORIES

- **Quality** Relating to chemical and pharmaceutical quality assurance
  EX: Q1 Stability Testing
           Q3 Impurity Testing

- **Safety** Relating to in vitro and in vivo preclinical studies
  EX: S1 Carcinogenicity Testing
           S2 Genotoxicity Testing

- **Efficacy** Relating to clinical studies in human subjects
  EX: E3 Structure and Content of Clinical Study Reports
           E4 Dose Responses

- **Multidisciplinary**
           – M1 Medical Terminology
           – M2 Electronic Standards for Transmission of Regulatory Information
           – M3 Timing of Preclinical Studies in Relation to Clinical Trials
           – M4 The Common Technical Document

ICH Efficacy Guidelines

Code	Subject	Number of Guidelines
E1	Exposure	1
E2	Clinical Safety Data Management	3
E3	Study Reports	1
E4	Dose Response Studies	1
E5	Ethnic Factors	1
E6	**Good Clinical Practice**	**1**
E7	Populations	1
E8	Clinical Trial Design	1
E9	Statistical Considerations	1
E10	Choice of Control Group in Clinical Trials	

A reference guide to FDA regulations and ICH GCP Guidelines by topic is presented in Table 4.2. Table 4.2 also references the FDA Information Sheets that provide guidelines for investigators, IRBs, and sponsors. Table 4.3 lists clinical guidelines, by therapeutic grouping, developed by FDA Advisory Committees and consultants. These guidelines contain specific information on the conduct of clinical trials. Copies of FDA Information Sheets or clinical guidelines may be obtained by writing to

Food and Drug Administration   Executive Secretarial Staff
FOI Staff   OR   HFD-8
HFI-35, Room 12A-16   5600 Fishers Lane
5600 Fishers Lane   Rockville, MD 20857
Rockville, MD 20857   301-594-1012
301-443-6310   301-594-3302 (fax)
www.fda.gov   www.fda.gov

TABLE 4.2 REFERENCE GUIDE TO FDA REGULATIONS AND THE ICH GUIDELINE (E6)

	21 CFR	ICH (E6)	Information Sheets/References*
Abbreviated NDA	314.55	N/A	Organization of an Abbreviated NDA and an Abbreviated Antibiotic Application
Advertising	50.20 and 21 56.111	3.1	Advertising for Study Subjects (9/98) Recruiting Study Subjects (9/98) Payment to Research Subjects (9/98)
Adverse Experiences	310.305	1, 4.11, 5.16, 5.17, 6.8, 7.3.6	
Biologics	600, 601	all	Emergency Use of an Investigational Drug or Biologic (9/98) Investigational and "Off-Label" Use of Marketed Drugs and Biologics (9/98)
Children	50 45 CFR 46 Subpart D	4.8.12	
Compassionate Use IND		N/A	*In* Treatment Use of Investigational Drug (5/89)
Contract Research Organizations	312.52	5.2	
Cooperative Research Review	56.114 45 CFR 46.114	N/A	Cooperative Research (2/89), Nonlocal IRB Review (2/89)
Data Management	312.62		*see* Recordkeeping
Electronic Signature	11		
Emergency Research	50.24		Exception from Informed Consent Requirements for Emergency Research (3/2000)
Emergency Use	312.36	N/A	Emergency Use of an Investigational Drug or Biologic (10/95)

Table 4.2 continued on next page

Table 4.2 continued from previous page

	21 CFR	ICH (E6)	Information Sheets/References*
			Guidance for the Emergency Use of Unapproved Medical Devices (10/22/85)
and Informed Consent	50.23		
Expedited Review	56.110 46 CFR 8980	3.3.5	*Federal Register*, 1/27/91, Vol. 48 (17) Investigations Which May Be Reviewed Through Expedited Review
Federal Policy for the Protection of Human Subjects	46, 50, 56 49 CFR 11, 45 CFR 46 Subparts B, C, D	all	*see* Informed Consent
Fetuses	45 CFR 46 Subpart B	N/A	
Financial Disclosure	54	5.8, 5.9	Forms FDA 3454 and 3455 Financial Disclosure by Clinical Investigators (3/2001)
Food and Drug Administration	10.90		FDA District Offices (10/95), FDA Phone Numbers (10/95) Administrative Practices and Procedures; Good Guidance Practices (10/2000)
Foreign Studies (not under U.S. IND)		N/A	*Federal Register*, Vol. 56 (93) (1991) Acceptance of Foreign Clinical Studies (9/98)
Gender		N/A	Evaluation of Gender Differences (9/98)
Good Clinical Practices	50, 56, 312, 314, 812, 813	all	*Federal Register*, Vol. 52 (53), (3/19/87) Preamble and original Rewrite of regulations, Part VII, p. 8798
Good Laboratory Practices for Nonclinical Laboratory Studies	58		Good Laboratory Practice Regulations: Questions and Answers (6/81) Additional specific guidelines available
Good Manufacturing Practices	211		Specific guidelines available
Information Amendment to IND	312.31	N/A	
Informed Consent	50, 56	1.28, 2.9, 4.8	Informed Consent Regulations
	312.60		Informed Consent and the Clinical Investigator (9/98)
	45 CFR 46		A Guide to Informed Consent Documents (9/98)
Inspections	312.68, 312.58	1.29, 5.19	Compliance Program Guidance Manual (Clinical Investigators) Compliance Program Guidance Manual (Sponsors, CROs, Monitors)

Table 4.2 continued on next page

Table 4.2 continued from previous page

	21 CFR	ICH (E6)	Information Sheets/References*
			Compliance Program Guidance Manual (IRBs)
			Guide for Detecting Fraud in Bioresearch Monitoring Inspections (4/93)
			FDA Institutional Review Board Inspections (9/98)
			FDA Inspections of Clinical Investigators (9/98)
			Clinical Investigator Regulatory Sanctions (9/98)
IRB and Independent Ethics Committee (IEC)	56	1.27, 1.31, 2	IRB Regulations Institutional Review Board Guidelines
			IRB Compliance Program Guidance Manual
			Sponsor–Clinical Investigator–IRB Interrelationship (9/98)
			Recruiting Study Subjects (9/98)
			Answers to Frequently Asked Questions (9/98)
			Self-Evaluation Checklist for IRBs (9/98)
			Continuing Review (9/98)
			Cooperative Research (9/98)
			FDA Institutional Review Board Inspections (9/98)
			Use of Investigational Products When Subjects Enter a Second Institution (9/98)
			Investigational and "Off-Label" Use of Marketed Drugs and Biologics (9/98)
			IRB Frequently Asked Questions About Review of Medical Devices (9/98)
			Payment for Investigational Products (9/98)
			Payment to Research Subjects (9/98)
			Recruiting Study Subjects (9/98)
			Significant Differences in HHS and FDA Regulations for IRBs and Informed Consent (9/98)
			Waiver of IRB Requirements (9/98)
			Cooperative Research (9/98)
			Nonlocal IRB Review (9/98)

Table 4.2 continued on next page

Table 4.2 continued from previous page

	21 CFR	ICH (E6)	Information Sheets/References*
			Expedited Review, *Federal Register*, Vol. 48 (17), Jan 1991
			Investigations Which May Be Reviewed Through Expedited Review (9/98)
			FDA Institutional Review Board Inspections (9/98)
			IRB Guidebook (from OHRP)
Investigator (*see* Obligations of)			
Investigational Agent Management	312.57, 312.59, 312.6, 312.62	1.33, 4.6, 5.11, 5.12, 5.13, 5.14, 7	Preparation of Investigational New Drug Products (draft) (2/88) Charging for Investigational Products (9/98)
Investigational Devices	812, 813, 814	all	*see* Medical Devices
Investigational New Drug Application	312	N/A	Clinical Development Guidelines from the FDA (Table 2.3)
Investigator's Brochure	312.23(a.5)	7	
IND Annual Progress Report	312.33	N/A	
IND Safety Report (Report of Serious Adverse Event)	312.32, 312.64	N/A	Adverse Experience Reporting Requirements for Human Drug and Licensed Biological Products, Proposed Rule, 21 CFR 20, *Federal Register* (10/27/94)
Labeling		N/A	*see* Package Insert
Laboratory Certification	42 CFR 493	8.2.12	
Marketed Drugs		N/A	Investigational Use and "Off-Label" Use of Marketed Products and Biologics (9/98)
Medical Devices	812, 813, 814	all	Emergency Use of Unapproved Medical Devices (9/98)
			Significant and Nonsignificant Risk Device Studies (9/98)
			Medical Devices (9/98)
			IRBs and Medical Devices (9/98)
MEDWATCH Form		N/A	*see* Serious Adverse Experience, IND Safety Report
Monitoring	312.53, 312.56	1.38, 5.18	FDA Guideline for Monitoring of Clinical Investigations (1/88)
New Drug Application	314	N/A	Specific guidelines for format and content available from FDA
Obligations of Investigators	312 Subpart D	4	Regulatory Information for Investigators (9/92)
	312.60–312.70		Obligations of Investigators Required Recordkeeping in Clinical Investigations (10/95)
			Informed Consent and the Clinical Investigator (9/98)

Table 4.2 continued on next page

Table 4.2 continued from previous page

	21 CFR	ICH (E6)	Information Sheets/References*
			FDA Inspections of Clinical Investigators (9/98)
			Clinical Investigator Regulatory Sanctions (9/98)
			Treatment Use of Investigational Drugs (9/98)
			Drug Study Designs (9/98)
Obligations of Sponsors	312 Subpart D 312.50–312.58	5	Obligations of Sponsors
Package Insert (labeling)	201.57, 314.5(c.2.1)	N/A	
Parallel Track		N/A	*Federal Register*, Vol. 57 (73), (1992), Expanded Access of Investigational Drugs
Postmarketing	314.80	N/A	
Prisoners	50.40 45 CFR 46 Subpart C	N/A	
Product License Application	600, 601	N/A	*refer to* NDA
Protocol	312.23(a.6)	6	Drug Study Designs (9/98)
Protocol Amendment	312.30	6	
Recordkeeping	312.62	2.10, 4.3, 4.9, 5.5, 5.15, 6.13 8.0	Recordkeeping in Clinical Investigations (10/95)
Screening		N/A	Screening Tests Prior to Study Enrollment
Serious Adverse Experience	312.32, 312.64	1.50, 1.60	Adverse Experience Reporting Requirements for Human Drug and Licensed Biological Products, Proposed Rule, 21 CFR 20, *Federal Register* (10/27/94)
			see also IND Safety Report
			MEDWATCH form
Sponsor Investigator	312.3(b)		
Statement of Investigator	312.53	N/A	Form FDA 1572
Subjects	50	1.57	Payment to Research Subjects (9/98)
	45 CFR 46		Recruiting Study Subjects (9/98)
Treatment IND	312.34	N/A	Treatment Use of Investigational Drugs (9/98)
Veterans	41 CFR 16	N/A	
Women	45 CFR 46 Subpart B	N/A	Evaluation of Gender Differences (9/98)

*The FDA and Department of Health and Human Services (DHHS) have issued a series of guidelines or summaries of information to complement the regulations. These Information Sheets are available through Freedom of Information.

TABLE 4.3 GUIDELINES FOR THE CLINICAL EVALUATION OF DRUGS

- General Considerations for the Clinical Evaluation of Drugs
- General Considerations for the Clinical Evaluation of Drugs in Infants and Children
- General Considerations for the Clinical Evaluation of Analgesic Drugs
- General Considerations for the Clinical Evaluation of Antacid Drugs
- General Considerations for the Clinical Evaluation of Anti-Anginal Drugs
- General Considerations for the Clinical Evaluation of Anti-Anxiety Drugs
- General Considerations for the Clinical Evaluation of Anti-Inflammatory and Anti-Arrhythmic Drugs
- General Considerations for the Clinical Evaluation of Anti-Arrhythmic Drugs
- General Considerations for the Clinical Evaluation of Anticonvulsant Drugs
- General Considerations for the Clinical Evaluation of Antidepressant Drugs
- General Considerations for the Clinical Evaluation of Antidiarrheal Drugs
- General Considerations for the Clinical Evaluation of Antiepileptic Drugs
- General Considerations for the Clinical Evaluation of Antihypertensive Drugs (proposed)
- General Considerations for the Clinical Evaluation of Anti-Infective Drugs
- General Considerations for the Clinical Evaluation of Antineoplastic Drugs
- General Considerations for the Clinical Evaluation of Antiulcer Drugs
- General Considerations for the Clinical Evaluation of Bronchodilator Drugs
- General Considerations for the Clinical Evaluation of Drugs for the Treatment of Congestive Heart Failure
- General Considerations for the Clinical Evaluation of Drugs to Prevent, Control, and/or Treat Periodontal Disease
- General Considerations for the Clinical Evaluation of Drugs to Prevent Dental Caries
- General Considerations for the Clinical Evaluation of Drugs Used in the Treatment of Osteoporosis
- General Considerations for the Clinical Evaluation of Gastric Secretory Depressant Drugs
- General Considerations for the Clinical Evaluation of General Anesthetics
- General Considerations for the Clinical Evaluation of G.I. Motility-Modifying Drugs
- General Considerations for the Clinical Evaluation of Hypnotic Drugs
- General Considerations for the Clinical Evaluation of Laxative Drugs
- General Considerations for the Clinical Evaluation of Lipid-Altering Agents
- General Considerations for the Clinical Evaluation of Local Anesthetics
- General Considerations for the Clinical Evaluation of Nonsteroidal Anti-Inflammatory and Anti-Rheumatic Drugs
- General Considerations for the Clinical Evaluation of Psychoactive Drugs in Children
- General Considerations for the Clinical Evaluation of Radiopharmaceutical Drugs
- Guidelines for Abuse Liability Assessment
- Guidelines for the Evaluation of Controlled Release Drug Products
- Guideline for the Study and Evaluation of Gender Differences in the Clinical Evaluation of Drugs
- Guidelines for the Study of Drugs Likely to be Used in the Elderly

*Guidelines are available from the FDA through Freedom of Information. Additional guidelines can be obtained at www.fda.gov/cder/guidance/index.htm or www.fda.gov/foi/foia2.htm.

RESPONSIBILITIES OF THE INVESTIGATOR

In signing the Statement of Investigator form (Form FDA 1572), the investigator agrees to ensure that the investigation is conducted according to the provisions in the Statement of Investigator, the investigational plan (protocol), and applicable regulations. Specifically (adapted from Form 1572 and 21 CFR 312.60–312.70), the investigator agrees to the following:

Protocol Conduct the study according to the current protocol and make changes only after notifying the sponsor, except when necessary to protect the safety, rights, or welfare of subjects.

Study Conduct PERSONALLY conduct or supervise the investigation.

Informed Consent and IRB Requirements Inform subjects that the study is investigational and assures that informed consent regulations (21 CFR 50) and IRB regulations for review and approval (21 CFR 56) are met.

Adverse Experiences Report adverse experiences to the sponsor in accordance with 21 CFR 312.64 (see Chapter 11).

Investigator's Brochure Read and understand the information in the Investigator's Brochure, especially the potential risks and side effects of the investigational agent.

Inform Investigative Staff Ensure that all associates, colleagues, and employees assisting in the conduct of the trial are informed about their obligations in meeting these commitments.

Subject Records Maintain adequate and accurate records (21 CFR 312.62) and make the records available for inspection (21 CFR 312.68). Case histories shall be prepared and maintained to record all observations and other pertinent data for each individual treated (21 CFR 312.62).

IRB

Assure that the IRB is in compliance with 21 CFR 56. The investigator will be responsible for obtaining the initial and continuing review and approval of the study. The Principal Investigator (PI) must report to the IRB all changes in the research activity and unexpected risks to subjects or others. The PI will not make any changes in the research plan without IRB approval, except when necessary to protect subjects (21 CFR 312.66).

Regulations

Comply with pertinent requirements in 21 CFR 312.

Investigational Drug

Administer the drug only to subjects under the investigator's personal supervision or under the supervision of a subinvestigator responsible to the investigator. The PI shall not supply investigational drug to any person not authorized to receive it (21 CFR 312.61). The investigator is required to maintain adequate records of the disposition of the investigational drug, including dates, quantity, and use by subjects. Unused supplies shall be returned to the sponsor or otherwise properly disposed of (alternative disposition, 21 CFR 312.59) if the study is completed, terminated, discontinued, or suspended (21 CFR 312.62).

Record Retention

Maintain records (study files) for the study for a period of two years following the date of NDA approval for marketing for the indication being studied; or, if no application is filed or is not approved, until two years after the notification to the FDA that the investigation (IND) is discontinued (21 CFR 312.62).

Investigator Reports

Accept responsibility for submission of progress reports to the IRB, safety reports of unexpected serious adverse events (IND safety reports), and the final report summarizing the study (21 CFR 312.64).

Inspections

Permit the FDA to have access to, copy, and verify any records or reports made by the investigator. The investigator is not required to divulge subject names unless more detailed study of the cases is required or there is reason to believe that the records are not representative of actual cases or results (21 CFR 312.68).

Controlled Substance   For drugs subject to the Controlled Substances Act, take adequate measures to prevent theft or diversion into illegal channels of distribution (21 CFR 312.69).

Disqualification   May be disqualified for failure to comply with regulations (21 CFR Subpart D, Part 50, and Part 56) or submitting false information to the sponsor (21 CFR 312.70). The process of disqualification is outlined in Part 312.70.

Investigators are also responsible for subject rights protection (21 CFR 50, and 45 CFR 46, Subparts B, C, and D) and for IRB practices (21 CFR 56).

RESPONSIBILITIES OF THE SPONSOR

The sponsor also has specific obligations as outlined in the FDA regulations, one of which is to monitor the investigator to assure adherence to all obligations. The sponsor has a lot at stake here—the future of the investigational agent. Because of this, the sponsor does not want to risk the disqualification of an investigator or have any unfavorable attention from the FDA. Therefore, the sponsor will be very insistent that the investigator conduct the study as written and according to GCPs.

In addition to specific regulations regarding the conduct of clinical trials, sponsors also must adhere to regulations pertaining to the IND Application (21 CFR 312), an NDA (21 CFR 314), Good Manufacturing Practices (21 CFR 211), and Good Laboratory Practices (21 CFR 58).

The general responsibilities of the sponsor, as outlined in 21 CFR 312.50 are as follows:

- To select qualified investigators.
- To provide investigators with information needed to conduct the investigation properly.
- To ensure proper monitoring of investigations.

- To ensure the investigation is conducted according to the general investigational plan and protocols contained in the IND.
- To maintain an effective IND.
- To ensure that the FDA and investigators are promptly informed of significant new adverse events or risks.

The specific responsibilities of the sponsor (adapted from 21 CFR 312.53–312.59) are as follows:

Selecting Investigators
Investigators must be qualified by training and experience as experts to investigate the drug [21 CFR 312.53(a)].

Investigational Agent
Investigators receiving the drug must be registered with the FDA (via a Statement of Investigator form) to receive this investigational drug under this IND (21 CFR 312.53). The investigator must keep records (shipping papers, accountability logs, destruction records) to show disposition of the investigational drug (21 CFR 312.57). The sponsor shall keep in reserve any sample of a test article used in a bioequivalence or bioavailability study (21 CFR 312.57). The sponsor shall assure that adequate precautions are taken to prevent theft or diversion of a controlled substance to illegal channels (21 CFR 312.58). The sponsor shall assure the return of all unused supplies of the investigational agent from investigators whose participation is discontinued or terminated (21 CFR 312.59) or may authorize alternative disposition of unused drug (21 CFR 312.59).

Investigators
Before an investigator may begin using the investigational agent in the study, the following information must be obtained by the sponsor (21 CFR 312.53):

- Completed Statement of Investigator (Form FDA 1572).
- Curriculum Vitae.
- Protocol (authored by either the investigator or sponsor) and Case Report Forms (CRFs).

The sponsor shall keep the investigator informed by providing an Investigator's Brochure before the study begins (21 CFR 312.55). The sponsor shall keep the investigator informed of any new observations regarding adverse effects and safe use of the investigational drug and relay any important safety information resulting in an IND Safety Report to the investigator (21 CFR 312.55).

Monitoring

The sponsor must select monitors who are qualified by training and experience to monitor the progress of the investigation (21 CFR 312.53). The sponsor shall monitor the progress of all clinical investigations being conducted under the IND (21 CFR 312.56).

Investigator Compliance

If a sponsor discovers that an investigator is not complying with the signed Statement of Investigator, the investigational plan, GCPs, or other requirements, the sponsor must either secure compliance or discontinue shipment of investigational drug and end the investigator's participation in the study (21 CFR 312.56 and 312.70).

Safety

The sponsor shall review and evaluate the data relating to safety and efficacy as it is obtained from the investigator to be reported to the FDA as IND Annual Progress Reports (21 CFR 312.33) or IND Safety Reports (21 CFR 312.32 and 312.56). If the sponsor should determine that the investigational drug presents an unreasonable and significant risk to subjects, the sponsor shall discontinue those investigations, notify the FDA, notify all investigators and IRBs, assure disposition of the investigational agent, and provide the FDA with a full report. Investigations must be terminated no later than five days after the decision to discontinue is made (21 CFR 312.56).

IND Safety Reporting Requirements

According to 21 CFR 312.32, sponsors shall promptly review all information relevant to the safety of the drug. The sponsor must notify the FDA in a written report of any serious and unexpected adverse experience associated with the use of the drug within fifteen calendar days after the sponsor's initial receipt of the information. The FDA must be notified by telephone call or facsimile transmission of any unexpected fatal or life-threatening experience associated with the use of the investigational agent no later than seven calendar days after receipt of the information (21 CFR 312.32). The sponsor must relay all important safety data to all investigators in a timely manner (21 CFR 312.55). IND safety reports are discussed in greater detail in Chapter 11.

Recordkeeping and Record Retention

The sponsor must maintain adequate records of investigational agent receipt, use, and other disposition (21 CFR 312.57). The sponsor shall maintain required records of the study conduct for two years after NDA approval or until two years after notification to the FDA that shipment and delivery of the drug for investigational use has been discontinued (21 CFR 312.57).

Inspections

The sponsor shall permit the FDA to have access to, copy, and verify any records and reports relating to the investigation (21 CFR 312.58). The sponsor shall discontinue shipment of drug to any investigator who has failed to maintain or make available records or reports as required (21 CFR 312.58).

IND Annual Progress Reports

The sponsor must submit a brief report of the progress of studies under the IND within 60 days of the annual anniversary of when the IND went into effect (21 CFR 312.33). Generally, the reports include a summary of studies, summary information on all clinical data (for example, IND safety reports, tabulation of adverse experiences, dose response data), and preclinical data. Specific requirements for the content are included in 21 CFR 312.33.

Additional specific regulations apply to the sponsor and are detailed in Part 312.

The sponsor may transfer certain obligations to a Contract Research Organization (CRO). In this case, the CRO is responsible for those obligations of the sponsor that are specified in a written agreement between the CRO and the sponsor (21 CFR 312.52). The CRO may, in effect, become the "sponsor."

SPONSOR INVESTIGATOR STATUS

Sponsor investigator refers to a situation where an individual is both the sponsor, or discoverer, of a treatment modality (drug, biologic, genetically engineered product) and the investigator. In this situation, the investigator is also taking on all obligations of a sponsor as defined in 21 CFR 312, Subpart D. This includes filing for IND status and maintaining a current IND application.

FINANCIAL DISCLOSURE BY CLINICAL INVESTIGATORS

The financial disclosure regulation (21 CFR 54) was established for the FDA to address payment arrangements and financial interests of investigators that could potentially bias the outcome of the study. Sponsors must submit information concerning compensation to, and financial interests of, any clinical investigator conducting clinical trials where the FDA relies on the data to support efficacy, safety, or bioequivalency to ensure that the reliability of the data is not affected.

The financial interests of investigators, subinvestigators, and their spouses and dependent children must be disclosed if the amount meets the criteria of 21 CFR 54. This includes:

- a situation where the value of compensation for the study could affect the study outcome (i.e., compensation for a favorable outcome is higher);
- proprietary interest in the investigational agent (patent, licensing agreement, or trademark);
- equity interest in the sponsor company (greater than $50,000); and
- any other significant payment by sponsor, i.e., royalties (greater than $25,000).

The sponsor must obtain Certification of Financial Disclosure from the investigator conducting the clinical trial using one of two FDA forms:

- Form FDA 3454: Certification: Financial Interest and Arrangements of Clinical Investigators
- Form FDA 3455: Disclosure: Financial Interest and Arrangements of Clinical Investigators

When there is a significant financial interest for the investigator, the sponsor must take steps to minimize bias by that investigator.

The sponsor submits a list of all investigators (principal and subinvestigators) to the FDA and appropriate Forms 3454 and 3455. If the sponsor does not submit these records with the marketing application, the FDA may refuse to file the application.

The FDA has published a Guidance for Industry, Financial Disclosure by Clinical Investigators, effective March 20, 2001.

ELECTRONIC SIGNATURES (21 CFR 11)

The regulations pertaining to electronic signatures provide the FDA's criteria for acceptance of electronic records and signatures to ensure data integrity and validity. An electronic document is "any combination of text, graphics, data, audio, pictorial, or other information in digital form that is created, modified, maintained, archived, retrieved, or distributed by a computer system." Electronic signatures document that a file and its contents were produced or audited by a particular, authorized individual. Instead of a graphic image of a signature, the electronic signature can be a computer code, i.e., unique identification code and password, that only the originator can use. The signature code should provide identification and verification of the originator by two separate means and also create an audit trail.

The regulation requires that there are technical and procedural controls within the system to assure data integrity by using validation systems. The system must be capable of producing paper copies. The computerized system must also be maintained as systems change or have adequate and accurate means of upgrading to the newer system.

THE INSTITUTIONAL REVIEW BOARD

The IRB is responsible for reviewing clinical investigations with the intent to protect the rights and welfare of human subjects involved in such investigations. FDA regulations specific to IRBs are in 21 CFR 56. An IRB "means any board, committee, or other group formally designated by an institution to review, to approve the initiation of, and to conduct periodic review of, biomedical research involving human subjects" [21 CFR 56.102(g)]. The IRB is a generic term used to describe the committee that is responsible for review of research and protection of rights and welfare of research subjects. An institution may choose any name for this board.

Functions and Operations

IRB functions and operations are outlined in 21 CFR 56.108 and summarized below. (Note: IRBs must establish and follow WRITTEN procedures to fulfill these functions and operations.)

- Conduct initial and continuing review of research and report its findings and actions to the investigator and the institution.
- Determine which projects require review more often than annually and which projects need verification from sources other than the investigator that no material changes have occurred during the review period.
- Ensure prompt reporting to the IRB of changes in research activity.
- Ensure that changes to approved research plans are not implemented without prior IRB review and approval except when necessary to remove immediate significant hazard to protect subject safety.
- Ensure prompt reporting to the IRB, institutional officials, and the FDA of any unanticipated risks to subjects, any instance of serious or continuing noncompliance with these regulations, and any suspension or termination of IRB approval.
- Review proposed research at convened meetings where a majority of IRB members are present, including one member whose primary interests are in nonscientific areas (see membership).
- Approval by a majority of those members present is required for the research to be approved.

IRB Membership

The membership composition of IRBs is very carefully selected. According to FDA regulation (21 CFR 56.107):

- IRBs must have at least five members with varying backgrounds, qualified through experience and expertise, diversity (consideration of gender, race, cultural backgrounds), and sensitivity to community attitudes.
- Every nondiscriminatory effort should be made to ensure that an IRB does not consist entirely of all men or all women.
- Each IRB should have one member whose primary concerns are scientific and one member whose primary concerns are nonscientific.
- Each IRB shall include a member who is not personally affiliated with the institution and does not have an immediate relative affiliated with the institution.
- An IRB member may not participate in the review of any research where the member may have a conflicting interest, although he/she may provide information to the IRB as requested. Such conflicts would include reviewing studies where the member is on the investigative staff for the study, reviewing studies funded by sponsors who contribute to the member's research, or reviewing studies of sponsors in which the member has a financial stake (e.g., stock in the company) [21 CFR 56.107(e)].
- An IRB may invite experts to assist in the review of complex issues but these individuals may not participate in IRB voting.

One member may satisfy several membership requirements, for instance, one female or one male in a nonscientific field not affiliated with the institution.

Review of Research

The purpose of IRB review of research is to assure that the research is sound and that the subjects are being treated fairly and safely. Some areas that the IRB will consider during the review process are as follows:

Subjects   Are risks to subjects minimized and reasonable?

- Procedures should be consistent with sound research design and should not unnecessarily expose subjects to risk.

- Whenever possible, procedures are those that would ordinarily be performed on patients for diagnostic or treatment purposes according to standard of care regardless of study participation.
- The potential benefit of the treatment to the subject should be reasonable in relation to the risk.
- Selection of subjects must be fair and equitable.
- Informed consent will be obtained for each prospective subject as required in 21 CFR 50.
- Subjects will be adequately monitored for safety as outlined in the research plan (protocol).
- Provisions are taken to adequately protect the subject's privacy and maintain confidentiality of the data.
- Study procedures do not put subjects at risk, for example, the amount of blood collected at one visit or over a period of time is not so excessive as to put the subject's health at risk.

Study Design

Is the study well designed to answer the proposed question?

- The study is designed appropriately to obtain and collect results.
- Testing parameters are reasonable.
- Study visits are not too frequent, but subject contact is adequate to assess safety.
- Provisions for handling adverse events (dose reductions, unblinding, specific treatment plans) or subjects who fail to respond are clearly outlined.
- There is sound scientific reasoning to support the study plan.

Investigator

Is the investigator qualified?

- The investigator is qualified by experience and education to conduct the trial.
- Often the investigator has a "track record" at the institution so that the IRB is aware of his/her capabilities.

Other Duties

According to 21 CFR 56.109, the IRB shall

- Review and approve, require modification to (for approval), or disapprove research activities that are subject to FDA regulation.
- Require that information be given to subjects for obtaining informed consent as outlined in 21 CFR 50.25. The IRB will determine if that information (Informed Consent Form, video presentation, short form, etc.) is adequate and may require modifications.
- Require documentation of informed consent (21 CFR 50.27). The IRB may waive this requirement if the research presents no more than minimal risk of harm to study subjects and involves no procedures that normally require written consent outside the research context. In these situations, the IRB may require that a written summary of the research be provided to the subject.
- Notify investigators and the institution in writing of its decision to approve or disapprove a research proposal. Disapproval letters should include the reason(s) for the decision. The investigator has the opportunity to respond in writing.
- Conduct continuing review of the research at least annually but may require review more often according to the degree of risk to subjects.
- Have the authority to observe the consent process and the research or appoint a third party to do so.

Special Circumstances

Sometimes there are special circumstances that affect the review of research by the IRB:

Expedited Review   Research involving no more than minimal risk or minor changes (amendments to protocols) to approved research may be reviewed and approved without a full IRB meeting (21 CFR 56.110). Specific categories of research involving no more than minimal risk are published in the *Federal Register* (October 7, 1997, Vol. 62, No. 194). Generally, the IRB chairperson or designated member(s) will carry out the expedited review process. Expedited review is privy to all authorities of the IRB except that research may not be disapproved. A full committee must be convened to disapprove research. IRBs must establish a method to keep all members informed of expedited review activities.

Emergency Use

In certain situations, such as medical emergencies or life-threatening situations, the regulations provide for IRB review and approval for the emergency use of nonapproved drugs. In an emergency situation where there is no time for the review and approval process, the physician may treat with an investigational agent, but then must submit the following to the IRB within five working days of the emergency use of an unapproved treatment:

- Name of subject.
- Name of investigational agent.
- Statement of rationale for use of the investigational treatment.
- Copy of the signed Informed Consent Form or a statement from the investigator and a statement from a physician not involved with the clinical investigation verifying that the situation was life threatening, necessitating the use of the agent, informed consent could not be obtained, and no alternative method of approved or generally accepted therapy was available for this subject.
- Approvals for emergency situations are on an individual basis and each situation must be presented to the IRB.

Compassionate IND ("single patient use")

In a situation where a patient has failed all available treatments or there is no approved treatment available, but there is some evidence that a proposed treatment may be beneficial based on theoretical grounds, the FDA may permit the use of the proposed treatment under the sponsor's IND or under a new IND filed by the investigator for an identified patient. All outcome data must be reported to the FDA. These instances may be treated as pilot studies, and if the data look promising, controlled clinical trials should be pursued. See the FDA Information Sheet "Treatment Use of Investigational Drugs."

Treatment INDs   A current trend allowing new drugs to get to patients more quickly is for the sponsor to offer the drug (usually while the NDA is pending at the FDA) through a treatment IND. Treatment INDs may be submitted for individual patients or as a package to treat all qualified patients. Use of the drug, considered investigational since it has not received marketing approval from the FDA, is based on a substantial amount of evidence available on the drug's safety and effectiveness. Prospective IRB approval (including an Informed Consent Form) is required (21 CFR 312.34 and FDA Information Sheet "Treatment Use of Investigational Drugs").

Emergency Research   When research involves subjects who are unable to consent because of the emergency situation, such as heart attack, stroke, or motor vehicle accident, but the investigational agent or procedure is important to be studied and may benefit the patient and there are no other subject populations to answer the research question, then a sponsor may conduct the research under Emergency Research guidelines. Emergency Research is addressed in 21 CFR 50.24 but also applies to Parts 56 and 312. An Information Sheet, "Exception from Informed Consent Requirements for Research" (3/2001), clarifies some of the requirements. Some requirements are also discussed in Chapter 7.

Marketed Drugs   Research involving marketed drugs also requires IRB approval. A full protocol or study plan must be prepared and submitted to the IRB (FDA Information Sheet "Investigational and 'Off-Label' Use of Marketed Drugs, Biologics, and Medical Devices").

Medical Devices   Regulations for IRB approval apply to the investigational use of medical devices (pacemakers, IUDs, bandages, thermometers, intraocular lenses, in vitro diagnostic products are a few examples). Further information is available on the FDA information sheet "Frequently Asked Questions About IRB Review of Medical Devices."

Cooperative Research   Institutions involved in multi-institutional studies may use joint review, review by another qualified IRB, or similar arrangements to avoid duplication of effort. However, at some institutions, the IRB may still request local review for studies conducted at the institution. If there is a local IRB, notify the board of the study and approval by a separate IRB. Submittal of approval documents from other institutions may facilitate the local review (21 CFR 56.114, 45 CFR 46.114, and FDA Information Sheet "Cooperative Research").

Private Practice   Research conducted in a private practice setting must have IRB review and approval. The investigator may use the IRB at the institution where he/she has admitting privileges or an established IRB at a local institution, establish an IRB, or hire an independent IRB.

The Approval Letter

The IRB approval letter should contain the following information:

- Name and address of the IRB.
- Date of approval.
- Investigator name.
- Title of the study.
- Statement that the research protocol and informed consent have been approved.
- Identify any other materials reviewed, e.g., advertisements.
- Comment regarding updates, reapproval date.
- Signature of the IRB chairperson or appointed representative.

Ongoing Review

The IRB must conduct continuing review of the research at least annually and more frequently if the IRB determines it to be indicated by the degree of risk involved with the study. This review may be prompted by the IRB but is ultimately the responsibility of the investigator.

Suspension of Research

In accordance with 21 CFR 56.113, the IRB has the authority to suspend or terminate approval of research that is not being conducted in compliance with the IRB's requirements or that has been associated with unexpected serious risk to subjects. Suspensions and terminations must be reported with a statement for the reason(s) for the IRB's action to the investigator, institution officials, and the FDA.

Studies Exempt from IRB Approval

Certain types of studies are exempt from the IRB review and approval process, as defined in 45 CFR 46.101(b) and 21 CFR 56.104. However, it may be necessary to submit a study application to the IRB to ascertain that these conditions are met. Exempt studies may include the following:

- Research involving normal educational practices.
- Research involving the use of educational tests, survey procedures, interview procedures, or observation of public behavior.
- Research involving the review of existing data, documents, records, or specimens, if these sources are publicly available or if subjects cannot be identified, directly or through identifiers linked to the subjects.
- Research and demonstration projects designed to evaluate public benefit or service programs.
- Research involving taste and food quality evaluation and consumer acceptance.

In any of these situations, refer to specific regulations and submit the information to the IRB according to the institution's policy and governing regulations.

Advertising

IRBs are responsible for reviewing methods of recruiting study subjects to assure equitable selection of subjects [21 CFR 56.111(a)(3)]. One method used by investigators is advertising. IRBs review the information contained in the advertisement and the medium used (newspaper, TV, poster, leaflet, etc.) to determine that the rights and welfare of subjects are protected and no undue influence is exerted. Advertisements used to recruit subjects should comply with the regulations governing informed consent and subject selection processes [21 CFR 50.20, 50.25, and 56.111(a)(3)].

IRBs should assure that all information in advertisements avoids undue coercion and is not misleading to subjects. This is especially relevant to subjects with acute or severe physical or mental disabilities or subjects who are economically or educationally disadvantaged and who may need extra diligence to assure their rights [21 CFR 56.111(b)].

Generally, advertisement should be limited to

- the name and address of clinical investigator or institution,
- the purpose of the research and summary of eligibility criteria,
- a description of recruitment incentives to subject (payments or free exams),
- the location of the research,
- time commitment required of participants, and
- whom to contact for further information.

No claims of effectiveness, safety, or superiority to other drugs should be made or implied. Such information would be misleading to subjects and is a violation of FDA regulations involving promotion of investigational agents [21 CFR 312.7(a) and 812.7(d)].

When recruiting from other physicians, nurses, and so on, it is not recommended to offer a referral fee. However, you may offer to forward the subject's pertinent study data or arrange for the subject to be seen for study visits (with reimbursement) at the local physician's office (in which case you may need to include the physician as a subinvestigator on Form FDA 1572 and arrange for reimbursement).

Advertising for study subjects may be submitted to the IRB with the initial protocol or at a later date but must receive approval of the IRB prior to implementation.

IRB Records and Reports

The IRB is required to maintain documentation of IRB activities, such as the following:

- Copies of all research proposals, sample Informed Consent Forms, investigator's progress reports, and reports of injuries to subjects.
- Minutes of IRB meetings.
- Records of continuing review activities.
- Copies of all correspondence between the IRB and investigators.
- Detailed list of IRB members.
- Written operating procedures [21 CFR 56.108(a) and (b)].
- Statement of significant new findings provided to subjects (21 CFR 50.25).

The records shall be retained for at least three years after study completion. Records must be made available for inspection and copying by FDA representatives (21 CFR 56, Subpart D).

Sponsors and IRBs

Sponsors must assure that IRBs are operating in compliance with 21 CFR 56 and 50 (informed consent regulations). This is usually accomplished without direct interaction between the IRB and the sponsor.

- Documents of correspondence between the investigator and the IRB are kept in the study file and will be checked by the monitor (and by an inspector during an audit).
- The Informed Consent Form will be reviewed by the sponsor prior to study initiation. During the trial, the monitor will ensure that each subject has signed a valid consent form.
- The sponsor will usually request a list of IRB membership to ascertain that membership meets FDA requirements. Additionally, or alternatively, the sponsor may request the DHHS "general assurance" number or some other statement that the IRB conforms to 21 CFR 56.

This does not imply that the sponsor is responsible for the detailed compliance of IRBs. The sponsor must rely on the clinical investigator to assure that the IRB is in compliance, especially when the investigator and IRB are both affiliated with the same institution. When an independent IRB is used, it would be wise for both the sponsor and investigator to carefully inspect the IRB for compliance to FDA regulations. Results of

FDA IRB inspections (Establishment Inspection Reports or EIRs) may be obtained through the Freedom of Information process.

Inspection of IRBs

The FDA Bioresearch Monitoring Program includes a provision for the inspection of IRBs to ensure the protection of human subjects through well-organized and properly functioning IRBs. The FDA conducts on-site procedural reviews of the IRB to determine whether an IRB is operating in accordance with its own written procedures as well as in compliance with current FDA regulations. See the FDA Information Sheet "FDA Institutional Review Board Inspections" for further information. Administrative actions for noncompliance are included in 21 CFR 56, Subpart E.

The Office for Human Research Protections (OHRP) protects humans participating in biomedical and behavioral research under the DHHS, which includes both the National Institutes of Health (NIH) and the FDA. OHRP replaced the Office for Protection from Research Risks (OPRR) in June 2000. OPRR previously only had oversight over research sponsored by the NIH. OHRP has the authority to inspect IRBs and suspend research at institutions where there are violations of human subjects rights.

SUBJECT INFORMED CONSENT

The protection of the rights and the welfare of research subjects is the most critical aspect of the regulations. Protection of human subject rights and safety is specifically addressed in 21 CFR 50. Additional regulations under the DHHS are found in 45 CFR 46. To gain an appreciation for the intent of the regulations, it is recommended to review the Declaration of Helsinki and the Belmont Report.

Anyone involved in a research trial has a right to know what that involvement entails. To assure that research subjects are given this crucial information, the regulations require that the research subject give "informed consent" to participate in a research trial. This is most commonly accomplished by preparing an Informed Consent Form. After discussion with the subject, the subject will sign the form signifying agreement to participate in the study and documenting his/her understanding of the risks and benefits.

It is important to remember that getting the subject to sign the Informed Consent Form does not, in and of itself, constitute informed consent. The Informed Consent Form is an aid to assure that the subject is receiving adequate, consistent information about participating in the research trial. Signing the form provides documentation of the subject's consent to participate in the study. Chapter 7 suggests techniques to use when obtaining informed consent from research subjects.

A consent form should be prepared for each research trial and be specific to the protocol. *General requirements* for informed consent are contained in 21 CFR 50.20 and include the following:

- No investigator may involve a human being as a research subject unless the investigator has obtained legally effective informed consent from the subject or the subject's legally authorized representative (except as provided in 21 CFR 50.23).

- The prospective subject or representative must have sufficient opportunity to consider the study with minimal possibility of coercion or undue influence.

- The information presented to the subject must be in a language understandable to the subject.

- The informed consent shall not include exculpatory language through which the subject may waive his/her legal rights or releases the investigator, sponsor, or institution from liability for negligence.

EXCEPTION from the general requirements (21 CFR 50.23; all conditions must be met and verified in writing both by the investigator and a physician not involved in the research) would require that:

- the subject is confronted with a life-threatening situation necessitating the use of the investigational agent,

- informed consent cannot be obtained because of an inability to effectively communicate with the subject,

- consent cannot be obtained from the subject's legal representative in a timely matter, and

- there is no alternative treatment available that provides an equal or greater likelihood of saving the life of the subject.

If time is not sufficient to obtain independent determination by a noninvolved physician, the determinations of the investigator shall be made and reviewed and evaluated by a physician not participating in the study within five working days of use of the investigational agent. All documentation must be submitted to the IRB within five working days after the use of the investigational agent.

Certain exceptions may apply to the use of investigational agents under an IND sponsored by the Department of Defense. These are summarized in Part 50.23(d).

Emergency research situations may also include exception from pretreatment informed consent (21 CFR 50.24). Refer to Chapter 7.

Elements of Informed Consent

Elements of informed consent are addressed in Part 50.25. Table 4.4 summarizes the basic elements and additional elements of informed consent.

Documentation of Informed Consent

Informed consent must be documented by the use of a written form approved by the IRB except where minimal risk [as defined in 21 CFR 56.109(c)] is involved. The subject or the subject's legal representative must sign and date the form. A copy is given to the person signing the form. The consent form may be either a

- written document including all of the elements of informed consent (21 CFR 50.25, Table 4.4) or
- short form stating that the elements have been orally presented to the subject or the subject's legal representative. A witness is required for the oral presentation. Also, a written summary of what is to be presented to the subject must be approved by the IRB. The subject or his/her representative must sign the short form. The witness must sign both the short form and a copy of the summary. The person presenting the consent must sign a copy of the summary. A copy of the summary is given to the subject or his/her representative (21 CFR 50.27).

Special Groups

Specific considerations apply to certain groups and situations. Information can be found as indicated below:

Prisoners   21 CFR 50.40, 45 CFR 46, Subpart C.

Children   45 CFR 46, Subpart D; 21 CFR 50.25; "General Considerations for the Clinical Evaluation of Drugs"; and "General Considerations for the Clinical Evaluation of Drugs in Infants and Children."

TABLE 4.4 INFORMED CONSENT CHECKLIST

There are eight REQUIRED ELEMENTS of informed consent. These elements MUST be present in the Informed Consent Form or the summary for the short form:

1. The Informed Consent Form must CLEARLY state that the study involves RESEARCH.

   - State the study purpose in terms that the subject can understand.
   - Identify all experimental drugs, delivery techniques, or treatments.
   - Give a description of the experimental aspects of the study.
   - State the expected duration of participation in the study.
   - Describe briefly the procedures to be performed (e.g., lab evaluations, X-rays, office visits).
   - State the route of administration of the experimental agent.

2. Define RISKS attributable to the experimental agent and/or procedures.

3. Discuss any expected BENEFITS from participation in the trial.

4. Discuss ALTERNATIVE TREATMENTS.

5. State the policy for protection of CONFIDENTIALITY of records, noting that a qualified representative of the sponsor and the FDA may inspect subject study records.

6. Discuss whether COMPENSATION for study-related injuries is provided and if EMERGENCY TREATMENT will or will not be provided by the institution.

7. List the names and numbers of CONTACT PERSONS for research-related questions and for subject rights–related questions and questions regarding study-related injuries.

8. Clearly state that participation is VOLUNTARY and the decision to not participate or to withdraw from the study will not affect the patient's treatment plan.

Additionally, when appropriate the following items also must be included:

9. State that unexpected risks may be involved.

10. Discuss the circumstances under which the patient's participation may be terminated by the investigator or sponsor without the subject's consent.

11. Note that additional costs may be incurred by the subject due to study participation.

12. Inform the subject of the consequences of his/her decision to withdraw from the study.

13. Provide the subject with any significant new findings that relate to the subject's treatment and continued participation in the trial.

14. State the estimated number of subjects to be involved in the trial.

Other items to consider:

15. State that a copy of the Informed Consent Form shall be given to the subject.

16. The form should use terminology that the subject can understand.

In presenting the Informed Consent Form, the subject must understand what he/she is agreeing to. (See Chapter 7 for suggestions on presenting the informed consent process.)

Additional suggestions:

- Have the subject initial each page of the document.
- Keep the original in the study file or the subject's permanent record with a copy in the other file.
- Identify each version of the consent form by date or appropriate revision number.
- Note that if the informed consent is revised while the study is ongoing, subjects currently enrolled may need to sign the revised informed consent.

Elderly	"Guidelines for the Evaluation of Drugs Likely to Be Used in the Elderly."
Department of Defense	21 CFR 50.23(d).
Women	45 CFR 46, Subpart B; *Federal Register* 58, 39406; FDA Information Sheet "Evaluation of Gender Differences"; and "Women in Clinical Trials: A Change in FDA Policy."
Veterans	41 CFR 16.

Be aware of special consideration involving the use of women of child-bearing potential in clinical trials. As the practice of in vitro fertilization and the use of fetuses for research increases, protection of rights of subjects should be addressed in these cases. The Belmont Report provides some background information to help understand the principles applied to the protection of human rights and how they may apply in these special circumstances.

Protection of Human Subjects

Everyone involved in research has a primary responsibility of protection of human subjects: their safety from harm due to study participation, their right to privacy, and their overall welfare. There are several federal offices that have oversight on the protection of humans in clinical research trials.

OHRP	The Office for Human Research Protections was formerly called the Office for Protection from Research Risks. OHRP is located under the Office of Secretary of HHS. OHRP is charged with interpreting and overseeing implementation of the regulations regarding the protection of human subjects (45 CFR 46). OHRP also provides guidance on ethical issues in biomedical and behavioral research. OHRP has oversight and educational responsibilities wherever DHHS funds are used to conduct or support research involving human subjects.

OHRT   The FDA has recently created a separate office to oversee subject protection, the Office for Human Research Trials. This office oversees and coordinates all human subject protection policy for industry-sponsored studies. OHRT participates in the FDA Bioresearch Monitoring Program (inspections), international GCP, and education activities.

OHSP   The Office of Human Subject Protection is an NIH Intramural Research Program to provide oversight for the protection of subjects in clinical trials of the IRP.

Other Requirements

Note that certain state and/or local laws may also be in effect for protection of subject's rights and safety. Investigate those requirements with the IRB.

REGULATORY REFERENCES

- 21 CFR 11, 50, 54, 56, 312; 45 CFR 46
- FDA Summary Sheets:
    Informed Consent Regulations
    Informed Consent and the Clinical Investigator (9/98) (see Table 4.1)
- The Belmont Report
- The Declaration of Helsinki (Helsinki Accord)
- The ICH GCP Guideline

To obtain copies of the CFR, contact the local library or contact the FDA by calling 301-443-1382 for 21 CFR 50 and 56, and 45 CFR 46, Subparts A to D; 202-512-1800 for 21 CRF 312 and 314.50. All regulations, guidelines, and Information Sheets can be obtained through the Internet by contacting www.fda.gov. The ICH GCP Guideline can be obtained directly from the FDA or the FDA Web site.

CONTACTS

Office for Human Research Protections (OHRP)

(formerly Office for Protection from Research Risks (OPRR))

6100 Executive Boulevard

Suite 3B01 (MSC-7507)

Rockville, MD 20892-7507

301-496-7041

To obtain an IRB Guidebook, contact OHRP, or ohrp.osophs.gov/irb.

IRBs can contact the following offices to determine whether an IND or an Investigational Device Exemption (IDE) is required for the study of a drug or device:

Drugs

Document Management and Reporting Branch

Center for Drug Evaluation and Research (CDER)

301-443-4320

Fax on Demand: 800-342-2722 or 301-827-0577

www.fda.gov/cder

Biological Products

Division of Biological Investigational New Drugs

Office of Biologic Research and Review

Center for Biologic Evaluation and Research (CBER)

301-443-4864

Fax on Demand: 800-835-4709 or 301-827-3844

www.fda.gov/cber

Medical Devices

Office of Device Evaluation

Center for Devices and Radiological Health (CDRH)

301-427-8162

www.fda.gov/cdrh

To determine if a test article is a "drug" or "device":

Office of Health Affairs
301-443-1382
Fax on Demand: 800-899-0381 or 301-827-0111

BIBLIOGRAPHY

21 CFR 11—More Than Meets the Eye. Tammala Woodrum, *Applied Clinical Trials*, Vol. 9 (6), p. 86, 2000.

An IRB Primer. Celine M. Clive and Sharon Hill Price, *Applied Clinical Trials*, Vol. 6 (5), p. 62, 1997.

Biologics Development. A Regulatory Overview, *Applied Clinical Trials*, Vol. 6 (11), p. 52, 1997.

Clueless? What State Laws Do You Need to Know Before Conducting Research at Your Site? John Isidor and Sandra Kaltman, *The Monitor*, Vol. 13 (2), pp. 31–33, 1999.

Deficiencies in Ethics Committee or IRB Review. Wendy Bohaychuk, Graham Ball, Gordon Lawrence, and Katy Sotirov, *Applied Clinical Trials*, Vol. 7 (11), p. 44, 1998.

Deficiencies in Informed Consent Procedures. Wendy Bohaychuk, Graham Ball, Gordon Lawrence, and Katy Sotirov, *Applied Clinical Trials*, Vol. 7 (9), p. 32, 1998.

Document Tracking for Institutional Review Committees. Ruth Fries, Phyllis Kuhn, and Rosemarie Culmer, *Applied Clinical Trials*, Vol. 5 (4), p. 34, 1996.

Documentation Basics That Support Good Clinical Practices: The Master Plan. C. DeSain and C. Vercimak, *Applied Clinical Trials*, Vol. 2 (6), pp. 48–52, 1993.

Far Beyond the 1572. GCP Responsibilities of Principal Investigators Revisited. Douglas R. Mackintosh, Vernette J. Molloy, and G. Stephen DeCherney, *Applied Clinical Trials*, Vol. 8 (3), p. 59, 2000.

Federal Protection for Human Subjects: Historical Perspective. C. McCarthy, *Journal of Clinical Research and Drug Development*, Vol. 1, pp. 131–141, 1987.

Good Clinical Practices Made Easy. Interactive Screen Educator, *Applied Clinical Trials,* Vol. 7 (6), p. 104, 1998.

History of FDA Regulation of Clinical Research. R. Kingham, *Drug Information Journal,* Vol. 22 (2), pp. 151–155, 1988.

Introducing MEDWatch: A New Approach to Reporting Medication and Device Adverse Effects and Product Problems. D. A. Kessler for the Working Group. *JAMA,* June 2, 1993, Vol. 269 (21). Reprinted in *Journal of Clinical Research and Drug Development,* Vol. 7 (3), September 1993. (Reprint requests to Commissioner of Food and Drugs, FDA, 5600 Fishers Lane, Rockville, MD 20857.)

Issues in the Review of Clinical Drug Trials by IRBs. D. Cowen. In *Clinical Drug Trials and Tribulations,* edited by A. Cato, Marcel Dekker, Inc., New York, pp. 321–345, 1988.

Making Investigators' Responsibilities Clear. Felix Khin-Maung-Gyi and Sherry Schwarzhoff, *Applied Clinical Trials,* Vol. 6 (1), p. 60, 1997.

Placebos and Subject Protection. Jill Wechsler, *Applied Clinical Trials,* Vol. 7 (6), p. 26, 1998.

Regulatory Versus Public Health Requirements in Clinical Trials. M. Buyse, *Drug Information Journal,* Vol. 27, pp. 977–984, 1993.

Still No Single Market for Clinical Trials. Peter O'Donnell, *Applied Clinical Trials,* Vol. 9 (10), pp. 26–28, 2000.

Women in Clinical Trials of New Drugs: A Change in FDA Policy. R. Merkatz, *New England Journal of Medicine,* Vol. 329 (4), pp. 292–296, 1993.

Women in Clinical Trials: Screening and Consent Issues Revisited. Terry Vanden Bosch, *Research Practitioner,* Vol. 1 (1), pp. 17–20, 2000.

Additional articles on the subject of informed consent are listed at the end of Chapter 7.

5

INVESTIGATIONAL AGENT MANAGEMENT

An investigator shall administer the drug only to subjects under the
investigator's personal supervision or under the supervision of a
subinvestigator responsible to the investigator. The investigator shall not
supply the investigational drug to any person not authorized under this part
to receive it.

—21 CFR 312.61

The term *investigational agent* refers to the study medication, study drug, or experimental device to be studied in a research protocol. The use of investigational agents in clinical trials is strictly stipulated by Food and Drug Administration (FDA) regulations. It is imperative that the agent can be accounted for at every step of the clinical trial process—from the sponsor's shipment to the site to its use by the subject or destruction by the sponsor. This chapter will focus primarily on investigational drug substances, although the same general principles apply to devices.

INVESTIGATIONAL DRUG AGENTS IN A CLINICAL TRIAL

Most study drugs are in the form of a solid or a liquid, but you may also see investigational agents in the form of a gas, such as an inhaled anesthetic used during surgery. Whatever the investigational agent may be, the investigator must keep in mind these responsibilities:

1. The drug is to be used only as specified in the protocol and is to be handled/
    administered only under the supervision of the physician approved to do the
    study (Principal Investigator [PI]) (21 CFR 312.61).

2. The FDA requires the investigator to establish a record of receipt, use, and disposition of all investigational agents [21 CFR 312.6(a)].

3. The drug must be kept in a secure place (pharmacy or physician's office, secured clinic area) (21 CFR 312.69).

Before you receive the first shipment of an investigational agent(s), be sure you have a good understanding of the protocol and know what to expect in the way of study drugs and study design. Frequently, there will be mention of an active drug (the investigational agent) as well as a placebo ("sugar pill" or some similar inactive, innocuous substance). The study drug and the placebo are usually manufactured to look identical. In some trials, the investigational agent is being compared to an approved form of treatment (active treatment arm). In this case, the study drugs may be designed to look alike, e.g., similar appearing capsules. If this is not possible, such as when one drug is an intravenous fluid and the other a tablet, placebos for each drug may be provided. Note that all of these are considered investigational agents and must be appropriately accounted for.

A study may be double blinded, meaning that neither the investigator nor the subject knows which treatment they are receiving, the study drug or placebo or active control. Or it may be a single-blinded study in which the subject does not know which he/she is receiving.

The statistical analysis section of the protocol usually describes how the number of study subjects was determined, as well as the drug randomization schema. A computer-generated randomization pattern "assigns" the drug to the subject by subject number. Randomization by blocks is done to assure equal enrollment into each treatment group. If a study drug is randomized in groups of four, for example, this means that each group of four consecutive subject numbers was randomized separately from other subject numbers. Each group of four has the same number of active and placebo drug assigned. Separate blocks of four are randomized in different orders:

SUBJECT #	DRUG ASSIGNMENT	
1	A	
2	A	
3	P	
4	P	drug A =
5	A	active
6	P	
7	A	drug P =
8	P	placebo
9	P	
10	A	
11	A	
12	P	
etc.		

Other than for statistical significance, it is important to understand the randomization method used in the study to be sure the subject's study number is assigned properly. If the Clinical Research Coordinator (CRC) assigned the number "1" to the first subject, then "2" to the second subject, but "5" to the third subject, the randomization sequence has been broken; statistically speaking, this could make the study "biased" and ruin the study. ALWAYS ADMINISTER THE STUDY DRUG AND ASSIGN SUBJECT NUMBERS EXACTLY AS INSTRUCTED.

PACKAGING OF INVESTIGATIONAL AGENT

The contents may be packaged in several ways:

Bulk Containers   The liquid or solid form may arrive in large bulk containers from which the pharmacy may dispense study drug to multiple subjects. The drug agent name may be indicated if the study is not blinded or the container may have a code name (like "Drug A") if the pharmacist/investigator is blinded. ALL SUCH IDENTIFIED DRUGS MUST CONTAIN AN EMERGENCY CODE-BREAKING MECHANISM.

Individual Subject Containers	The study drug is divided into separate containers for each subject and is identified by subject numbers or by code name. The contents of each container are uniform—for example, there are not both active and placebo in each container. Multiple containers may be provided as a "subject kit."
Blister Cards	Study drug is packaged under clear plastic and attached to a card so that each dose (tablet or capsule) is separated under a "dome" and is clearly visible. The cards are usually set up according to dosing schedules for ease of subject compliance (see Figure 5.1).

STUDY DRUG LABELS

The labels used on study medication containers packaged for individual subjects are often "tear-off" labels. This means that one part of the label stays on the medication container, be it a bottle, vial, or blister card, and the other part is torn off at the time of dispensing and saved for drug records. An example of a tear-off label is shown in Figure 5.2.

Labels are important in study drug accountability records and should be kept in an orderly fashion in a safe place. Sometimes the identity of the drug is written and sealed inside the tear-off label and would only be revealed if the label were peeled back a certain way. In this case, the labels serve as both code breakers and drug accountability aids and must be secured once the bottle is dispensed to prevent anyone from discovering the identity of the drug, thus unblinding the study. Never dispense the study medication to the subject with the tear-off portion attached; the subject may open the label, unblind himself/herself to the study medication, and become ineligible for continued participation. Ask your sponsor representative how you should handle any tear-off labels.

FIGURE 5.1 BLISTER CARD EXAMPLE

This is a seven-day study with varying doses four times a day.

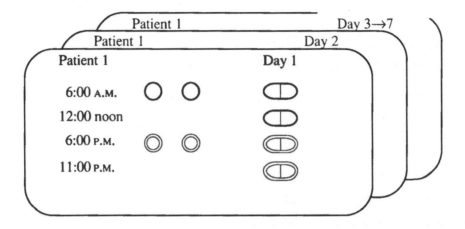

FIGURE 5.2 EXAMPLE OF A TEAR-OFF LABEL

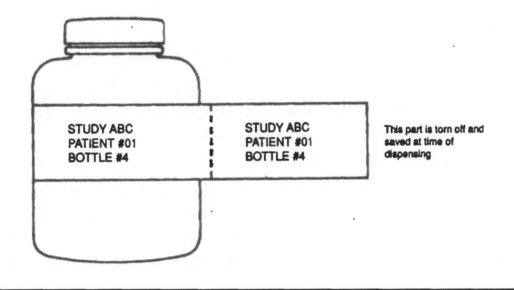

The following sections will guide you in handling the investigational drug agent from the time it is received at the study site to the time it is either administered or destroyed.

RECEIVING AND STORING THE INVESTIGATIONAL AGENT

The pharmacy is usually the best-qualified department in an institution to receive and store study drug, but in some cases, alternate storage areas may be used:

- The clinic drug storage and dispensing area.
- Private office medication storage area.
- Hospital satellite pharmacy.
- Investigator's locked private office.

The following are UNACCEPTABLE areas for storage:

- Desk drawer.
- Subject hospital room.
- Open shelf in clinic or exam room.

The following guidelines should be followed when finding a storage area for investigational drugs:

- Must meet the requirements for storage of the drug (temperature, light).
- Must be LOCKED and secured from unauthorized people.
- Space must be adequate.

Carefully review the protocol to determine the storage requirements and the QUANTITY of drug to be shipped. If each subject is dispensed 1 bottle every 2 weeks for a 3-year study with an anticipated enrollment of 100 subjects, you might have to move out of your office just to provide storage space! Work with the sponsor monitor to establish a shipment schedule that would provide enough drug to begin the study. It is then the responsibility of the investigative site to assure that there will always be enough study drug on hand for subjects. Remember to allow a lead time of **4 to 6 weeks** for packaging and shipping subsequent orders.

When the box of investigational agent arrives, immediately examine the outer box to make sure there was no damage during shipping, open it carefully, and locate the shipping document in the container, which describes the contents of the shipment. Read the document completely and compare it to the actual contents found in the shipment box. If there is any discrepancy or question, call the sponsor IMMEDIATELY. If there are no problems with the shipment, sign the appropriate portion of the shipping document, if required, and return it to the sponsor. RETAIN A COPY IN THE STUDY FILE (OR SIMILAR PHARMACY FILES) AS A RECORD OF THE SHIPMENT FOR FUTURE REFERENCE.

DISPENSING THE INVESTIGATIONAL DRUG AGENT

There are many ways in which the study drug may be dispensed:

In-House Studies
- Prepared by pharmacist in the central pharmacy or satellite pharmacy.
- Delivered to the ward or dispensed to the CRC.

Outpatient Studies
- Dispensed directly by the pharmacist to the subject.
- Dispensed by the pharmacist to the CRC who dispenses the drug to the subject.
- Stored, prepared, and dispensed in the clinic area by the PI, CRC, or other qualified individual.

Be sure to let the investigational pharmacist know before you begin screening subjects that the study will begin as soon as a subject qualifies. If the study is blinded, the pharmacist will consult the procedure guidelines for assigning subject study numbers and study medication and will prepare the study medication for the investigational staff to pick up (or to dispense to the subject for outpatient studies). Double-check the study inclusion/exclusion criteria and be sure all necessary data points have been collected before administering or dispensing the study drug.

The subject has signed the Informed Consent Form and meets all study entry criteria, and your site has been given the green light to proceed with study enrollment. Time to begin!

Before administering or dispensing any other nonstudy medication:

1. Be sure there is a written medication order by the investigator in the subject's chart.

2. Verify the dose to be given.

3. Identify the subject—if an inpatient, check the wrist band.

4. Administer the medication as ordered.

5. Record the dose (date, time, medication, dose, route, frequency, amount dispensed, container numbers) in the medication record, the subject chart, and the Case Report Form (CRF).

If the subject is to take the medication home for self-administration, be sure to do steps 1–3 first, then give the subject instructions on self-administration (see "Instructions to Study Subjects"). Record the identifiers of the study medication (date, time, medication dose, route, frequency, amount dispensed, bottle #) on both the CRF and the subject's chart. It is important to record the information in the subject record so that another healthcare provider unfamiliar with the study will be aware of the subject's study participation should the subject require treatment.

INSTRUCTIONS TO STUDY SUBJECTS

If the subject will be self-administering the study drug, it is extremely important that the drug be taken correctly. Give both oral and written instructions that are

1. as brief as possible,

2. clearly understandable to any nonmedical person, and

3. handy (either attached to the medicine container or on a small card that can fit in a wallet or pocket).

Items to include in the written instructions are as follows:

1. Exact dose in familiar dosage units, e.g., 2 pills instead of 2 mg.

2. Number of doses. It is best to relate the doses to specific times of the day rather than leaving that open to the subject. For example, indicate 8:00 A.M. and 8:00 P.M. instead of b.i.d. If possible, work with the subject's schedule to make dosing convenient without violating the protocol.

3. Highlight any restrictions, such as "take with food" or "take without food."

4. Provide instructions on storing the medication (refrigerate, avoid heat, etc.).

5. Instruct the subject to return all unused medication (including empty containers) and/or to bring the medication to EVERY appointment (for compliance checks).

6. Include the name and telephone number of someone on the investigational staff who can answer questions.

7. Provide a brief synopsis of the study drug to give to healthcare personnel not familiar with the study in case of an emergency.

8. Provide diaries for subjects to record information.

Some suggestions are illustrated in Figures 5.3–5.5.

Test the subject's understanding of the oral and written instructions by asking the subject to read them back to you and explain each step. A follow-up telephone call can be very helpful to assess subject understanding and compliance. If the subject does not have a telephone, ask for the nearest phone where you could leave a message or arrange a call-in time.

FIGURE 5.3 WRITTEN INSTRUCTIONS: ON THE APPOINTMENT CARD

Dr. B. Perfect

DERMATOLOGY

Your next appointment is scheduled for:

_____ at _____

(555) 555-5555

front

Write the instructions on the back of the subject's next appointment card.

Instructions

1. Take 1 tablet with food at breakfast, lunch, and dinner.
2. Keep refrigerated.
3. Return bottle at your next visit.

   Contact: C. Menow 555.5555

back

FIGURE 5.4 WRITTEN INSTRUCTIONS: ON A PERMANENT CARD

INSTRUCTIONS

1. Take 1 tablet with food at breakfast, lunch, and dinner.
2. Keep refrigerated.
3. Return bottle at your next visit.

Any questions? Contact C. Menow at 555-5555, beeper #315.

Write instructions on a 3 × 5 card that can be laminated to preserve it through the study.

FIGURE 5.5 WRITTEN INSTRUCTIONS: ON THE DRUG CONTAINER LABEL

Attach instructions to the label on the study drug container.

STUDY DRUG ACCOUNTABILITY

When you are working with an investigational drug, the FDA requires that you know where EVERY UNIT of drug is and how it was used from the time it arrived at your site until it was administered to the subject or returned to the pharmacy or sponsor for destruction. The key to accurate drug accountability is constant record-keeping, which becomes your "paper trail"—shipping documents, pharmacy records of drug preparation and dispensing (drug accountability logs), subject medication records, subject's drug diary (if applicable), and pharmacy final inventory log for destruction or return to the sponsor.

Some examples are demonstrated in Figures 5.6 and 5.7. In both specific examples, the study drug passes through many pairs of hands from the time it is dispensed to the time it is returned. Because records are kept at each step of the way, it should be easy to track down any unit of study drug at any time. Table 5.1 shows an example of an accountability form used for this important paper trail. Note that sometimes it is preferred to have one accountability page per patient (long-term studies with frequent visits) instead of one log for the whole study. The study sponsor will probably supply all the necessary forms for the study, but sometimes you may be able to use your own pharmacy forms if they are appropriate. ALWAYS record medication dispensing, returning, wasting, and so on, AS YOU GO. You will not be able to remember details accurately if you wait until a later time to do the paperwork.

CODE BREAKERS

A code breaker contains the identity of the drugs assigned as determined by the randomization sequence to each subject who is enrolled in a blinded study. Ask the study sponsor whether there will be a code breaker for the study drug given to the study site. A code may be "broken" in the case of a subject's serious, life-threatening event if it affects the immediate treatment of the event. Usually someone other than the investigator or the CRC should keep this document to minimize "peeking" and bias in the study. Often it is kept by the investigational pharmacist in a place that is accessible 24 hours a day. Be sure you clearly understand the circumstances under which the code may be broken. Even if done when necessary and appropriately, statistical significance is lost each time the code is broken. Any time the code is broken, it will raise a red flag during an inspection.

FIGURE 5.6 THE ROUTE OF AN INVESTIGATIONAL DRUG (EXAMPLE A)

Study patient #01 enters the study.

The pharmacist dispenses Bottle #1 for pt. #01 to the CRC.
Pharmacist records entry in drug accountability record.

The CRC dispenses Bottle #1 to pt. #01.
CRC makes an entry in the clinic chart and CRF re: dispensed medication.

Pt. #01 takes the bottle home, takes the exact amount as prescribed,
returns in one week for study visit with Bottle #1 containing 2 tablets.

The CRC verifies that 2 tablets remain and that pt. #01 has taken
all medication as prescribed and enters this information on the clinic chart and CRF.

CRC returns Bottle #1 to the pharmacy for the study sponsor
to inventory at the next site visit.

The pharmacist dispenses Bottle #2 for pt. #01 to the CRC.
Pharmacist records entry in drug accountability record.

The CRC dispenses Bottle #2 to pt. #01.
CRC makes an entry in the clinic chart and CRF re: dispensed medication.

When all bottles in the pharmacy have been inspected and
accountability records are correct, Bottle #1 for pt. #01 will be
destroyed or returned to the sponsor.

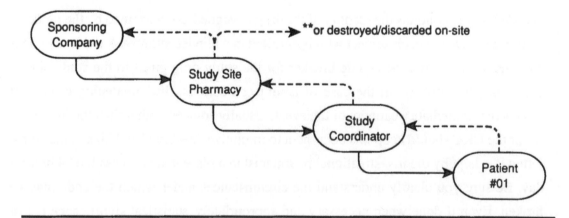

FIGURE 5.7 THE ROUTE OF AN INVESTIGATIONAL DRUG (EXAMPLE B)

Inpatient study involving study drug that is a liquid injection; the "unit" of drug in this case is a vial of liquid.

Study patient #01 enters the study.

Pharmacist draws up the dose of study drug in a syringe according to patient's weight (the used vial remains in the pharmacy) and logs out the amount in the drug accountability record.

↓

The CRC picks up the syringe to take to the patient.

↓

The dose is administered to pt. #01 and recorded in the medical chart and CRF.

↓

The CRC returns the empty syringe to the pharmacy to show that it was all used as dispensed.

↓

The pharmacist notes that none is remaining in the syringe and discards the empty syringe.

↓

The used vial for pt. #01 will remain on-site until the study sponsor inspects the accountability records.

DESTRUCTION OF THE INVESTIGATIONAL AGENT—FINAL DISPOSITION

Once the paper trail for each study subject is complete, and there are no outstanding or unaccounted for units of study drug, the remaining drug in each study container as well as the containers themselves are usually destroyed. The drug may be sent back to the manufacturer or sponsor for final counts and destruction or it may be destroyed on-site with the study sponsor's approval (include the written destruction policy in the study file). Find out what final disposition policies the company wishes you to follow and prepare accordingly *before* the study begins. If it is a large study with many vials or bottles and all containers must remain on-site until study completion, there must be

TABLE 5.1 DRUG ACCOUNTABILITY FORM

Form approved:
OMB No. 0925-0240
Expires: 1/31/2001

National Institutes of Health National Cancer Institute **Investigational Agent Accountability Record**	Division of Cancer Treatment and Diagnosis Cancer Therapy Evaluation Program	PAGE NO. CONTROL RECORD ☐ SATELLITE RECORD ☐
Name of Institution:	NCI Protocol No.:	
Agent Name:	Dose Form and Strength:	
Protocol Title:	Dispensing Area:	
Investigator Name:	NCI Investigator No.:	

Line No.	Date	Patient's Initials	Patient's ID No.	Dose	Quantity Dispensed or Received	Balance Forward / Balance	Manufacturer and Lot No.	Recorder's Initials
1.								
2.								
3.								
4.								
5.								
6.								
7.								
8.								
9.								
10.								
11.								
12.								
13.								
14.								
15.								
16.								
17.								
18.								
19.								
20.								
21.								
22.								
23.								
24.								

NIH-2564

space allocated for such bulk. The drug may be returned/destroyed before the study is completed if the sponsor checks the completed subjects' drug accountability logs at each site visit and approves disposition. Whatever the policy, do not discard any part of the investigational agent or the container without first consulting the sponsor.

On-site destruction must follow guidelines written by the Occupational Safety and Health Administration (OSHA), and most hospital pharmacies are aware of the approved methods for drug destruction. The most common methods are destruction by incineration, crushing, and autoclaving. Some liquids may be poured down the sink; capsules may be flushed IF THE EXPERIMENTAL AGENT POSES NO DANGER TO THE ENVIRONMENT SHOULD IT BECOME PART OF THE FOOD CHAIN. Containers may be recycled if clean or burned. The study drug cannot simply be thrown into the trash can for fear someone might ingest it or use it in some other way which could cause harm to him/her. The sponsor monitor may be required to witness the destruction of the investigational agent.

After study completion, the drug accountability logs must be kept in the study files along with other study information for the appropriate time period. If the pharmacy requires that records be kept in its files, maintain a copy in the study files with a memo stating that the original may be located in the pharmacy files.

COMMON QUESTIONS

Following are some common questions and answers about investigational agent management:

1. What if some of the study drug is spilled or lost, or the vial is broken?
    Your first step is to call the sponsor, especially if the study is blinded and you need to replace the drug. It may be possible to use some of the subject's existing supply to get through the crisis. Record the incident in the drug accountability record and the subject's chart.

2. As the "investigational drug manager" in a cancer clinic, I noticed during a periodic inventory of study drugs that some drug was not recorded in the accountability log. What should I do?
    There are many ways to locate the missing information (assuming it is the information that is missing and not the drug):

- Review subject charts. Sometimes study drug is not specifically identified to a subject; therefore, it may be necessary to review all charts of all subjects enrolled in the study. Determine which subjects received the drug and if it was logged out.
- Review CRFs until you can identify what had not been logged.
- Review what should have been dispensed; maybe too much or too little was dispensed.
- Determine who might have dispensed the drug and ask them if they recall who received the drug.
- You can narrow down when the drug was dispensed by the date in the log (either by what was dispensed chronologically, if the drug containers are numbered, or by the last time you did an inventory). Consult the appointment books to see which study subjects were in clinic and should have received treatment.
- When you discover the missing entry, enter it in the log on the next line with a brief explanation (or write a longer explanation in a memo to file with the logs). Remember, if the study is audited five years later, someone is going to need the information to explain the entry.

3. My subjects often forget or "lose" their drug to be returned. How important is it to return the empty containers or partially used containers?

It is very important. The FDA requires that you account for all study medication. Also, as the PI or CRC, you will want to see returned medication to measure compliance. Reinforce the importance of this to the study subjects. Written instructions help. A phone call before the visit might also help those who chronically forget. In cases where the returned container is not retrievable, write a memo to the file so that there is a record (and so that everyone else doing accountability does not waste a lot of time looking for something that does not exist).

4. The study patient is lost to follow-up; how far should the PI or CRC go to retrieve the study drug?

A valiant effort is needed here. In one study, the investigator went to the subject's funeral to retrieve study drug. Multiple phone calls are necessary. Some flexibility helps ("I'll meet you at the corner outside the clinic so you don't have to park"). You may even need to go to the subject's home (only if invited). In any case, DOCUMENT all attempts to retrieve the study drug even if unsuccessful.

5. Dr. Wan came to our clinic and requested some of the investigational drug for his laboratory studies. Can I log out the drug to him?

    NO! The investigational agent is to be used ONLY FOR THE SUBJECTS IN THE CLINICAL TRIAL. The same investigational drug supply (unless open-label and intended to be used in more than one trial) cannot be used in two different clinical trials with the same drug. NEVER switch drugs between studies. NEVER give clinical trial supplies for laboratory research use or personal use. A PI cannot authorize such use of clinical trial investigational agents.

6. Why should I record information in both the drug accountability logs and the patient CRF?

    The dispensing of study drug should be recorded in at least three places:

    •   Drug accountability log: Keeps track exclusively of all study drug. Must be maintained with the study files and is used to inventory study drug and determine usage and disposition.

    •   CRF: Records the actual amount of study drug given to the subject. This is the information that will be entered into the database for analysis.

    •   Source document (medical record/clinic chart): This is the record that will become a part of the subject's medical history and may impact his/her medical care.

REGULATORY REFERENCES

Code of Federal Regulations, Title 21, Parts 312.59, 312.61, 312.62, and 312.69

Recordkeeping in Clinical Investigations (10/95).

"Investigational Drug Accountability Record" (forms and instructions from the National Cancer Institute). Requests copies from:

Drug Management and Authorization Section
Investigational Drug Branch
Cancer Therapy Evaluation Program
Division of Cancer Treatment
National Cancer Institute
P.O. Box 30012
Bethesda, MD 20814

BIBLIOGRAPHY

Managing Clinical Trials Materials. CTM Returns Accountability. Gerald Finken, *Applied Clinical Trials,* Vol. 8 (7), p. 52, 1999.

Managing Clinical Trials Materials. Smooth the Clinical Supply Process with Clear, Open, Frequent Communication. Jim Clark, *Applied Clinical Trials,* Vol. 8 (2), p. 52, 1999.

Preparation, Packaging, and Labeling of Investigational Drug Supplies. D. Bernstein and F. Tiano, *Journal of Clinical Research and Pharmacoepidemiology,* Vol. 5, pp. 1–10, 1991.

Regulatory Authority Affecting American Drug Trials: Role of the Hospital Pharmacist. W. Gouveia and E. Decker, *Drug Information Journal,* Vol. 27 (1), pp. 129–134, 1993.

Study Medication Handling. D. J. Touw, C. D. Linseen-Schuurmans, and A. C. van Loenen, *Applied Clinical Trials,* Vol. 8 (10), p. 50, 1999.

The Use of the Three-Part Label: Contributing to Quality. P. Be'court, *Drug Information Journal,* Vol. 27 (3), pp. 921–924, 1993.

6

DESIGNING A STUDY AND PROTOCOL DEVELOPMENT

A protocol is required to contain the following, with specific elements and detail of the protocol reflecting the above distinctions depending on the phase of the study . . . .

—21 CFR 312.23(6)(iii)

Before you can conduct a study, you must have a protocol—a written, detailed PLAN for how the study will be conducted and analyzed. Protocols are developed and written in a variety of ways; it is rarely an individual effort. The author(s) must determine what information is known and what has been discovered by previous trials. In the development of a new drug by a sponsor, there is usually an overall drug development plan for the drug, and the content of the protocol is largely determined by that plan. Other elements to consider are as follows:

- What study design(s) should be used (parallel, crossover, "washout," "lead in," single blind, double blind, open label)?
- What phase is the trial?
- What is the disease, and what aspect of the disease is being examined?
- Will there be a control group (active, placebo, observation only, standard treatment)?

A variety of resources may used when preparing the protocol: research articles on similar trials, the Investigator's Brochure, similar protocols, the drug development plan. Other physicians, project leaders, Clinical Research Associates (CRAs), Clinical Research Coordinators (CRCs), statisticians, and pharmacists all make contributions to the production of a protocol, which is the study plan.

THE DRUG DEVELOPMENT PLAN

Industry sponsors develop an overall plan for how the proposed investigational agent should be studied to provide evidence to the Food and Drug Administration (FDA), in the form of a New Drug Application (NDA), of the safety and effectiveness of the investigational agent. These plans are referred to by a variety of names: Drug Development Plan, Clinical Development Plan, Project Management Plan. The purpose of an overall drug development plan is to anticipate questions raised by the FDA about the investigational agent and to design studies to answer these questions. Generally, they evaluate each trial from the Phase I studies (to establish dosing, pharmacokinetics, etc.) to the proposed Phase II and III studies (for safety and efficacy trials in patients). All of this information is pooled together to create the NDA document for submission to the FDA. Each individual study protocol fits into this overall plan.

SO YOU WANT TO DO A STUDY . . .

The first step in designing a study is to have a clear idea of what you want to prove: the hypothesis or objective. The objective of the study should define the rest of the protocol: what design to choose, which parameters to collect, and so on. Specific objectives must relate to the hypothesis. Objectives should be clear, concise, reasonable, and attainable. Industry-sponsored studies are often very focused and clear on what they want to prove—or disprove—about a drug. As an investigator, you may have some input into the study design, but don't be disappointed if a study is presented to you to conduct, and they've left out all of the "fun stuff." Rarely, a company will allow you to supplement its study with one of your own or allow you to add a few parameters to the protocol for the interest of "science," i.e., something interesting but not critical to getting the drug to market.

FDA requirements for protocol designs of Phase I, II, and III Investigational New Drug (IND) clinical trials are outlined in the 21 CFR 312.23(6). Additionally, the FDA offers "Guidelines" for the development of trials for specific therapeutic areas (see Table 4.2) and the FDA Information Sheet "Drug Study Designs." The FDA also issues Guidance for Industry documents, for example, "E10 Choice of Control Group and Related Issues in Clincal Trials" (5/01). These are invaluable resources when designing a study and writing a protocol.

STUDY DESIGNS

Chapter 1 discusses the phases of clinical research studies. Note that studies are not conducted linearly from Phase I to Phase III or IV. Generally, development proceeds along the Phase I to III track only when development of the next phase is dependent on results of studies in earlier phases, e.g., maximum tolerable dose (MTD) studies to determine the dose for efficacy trials. However, earlier phase studies may be ongoing while Phase II/III trials are being conducted. For example, a Phase I study to better determine the pharmacodynamic profile of the drug may be conducted while efficacy trials are underway. Pharmacokinetic studies are generally the first studies of an investigational agent conducted in human clinical trials to establish the ADME (absorption, distribution, metabolism, and excretion) profile of the drug. These early studies explore how the body acts on the drug. Pharmacodynamic studies, the actions of the drug on the body, also are commonly performed as Phase I trials. These types of studies lay specific groundwork for determining "how" the drug works.

Specific Designs for Clinical Trials

The way a study is designed is critical to proving your hypothesis. At least one of the studies submitted to the FDA in support of an NDA to prove safety and efficacy must be an "adequate and well-controlled" clinical trial where there is adequate comparison with a control group. FDA regulations (21 CFR 312.126) cite five types of controls:

1. Placebo concurrent control.
2. Dose-comparison concurrent control.
3. No-treatment concurrent control.
4. Active-treatment concurrent control.
5. Historical control (reserved for special circumstances).

The FDA Information Sheet "Drug Study Designs" discusses the advantages of each of these types of controls.

In addition to a well-controlled study (e.g., a comparison group to detect a difference or show no difference), clinical trials should be RANDOMIZED, when appropriate, to minimize selection bias. In a randomization procedure, subjects are equitably placed in treatment groups. Subjects also may be STRATIFIED in a study, where they are enrolled in a treatment group based on a variable that is thought to affect study outcome, e.g., stratifying subjects according to renal function to evaluate the effects of impaired renal clearance.

Another design element that adds strength to the overall study design is BLINDING or MASKING of the treatment type. Studies can be *open label*, where the subject, investigator, and other evaluators know the treatment; *single blind*, where one of the parties does not know the treatment; or *double blind*, where two or more of the parties (Principal Investigator [PI], CRC, subject, monitor) are unaware of the treatment the subject is receiving. The double-blind study is preferable since it allows for the evaluation of the effects of the drug with fewer subjective influences.

TREATMENT SEQUENCE is also an element of study design. Subjects may be randomized to one of many parallel groups where they receive the same treatment for a specified time period. In a crossover study design, subjects are randomized to receive a treatment for a specified period of time and then cross over to another treatment arm. In a factorial study design, subjects may receive multiple treatments, each for a specified time period, and may serve as their own control.

Table 6.1 displays study design features when developing a study.

TABLE 6.1 STUDY DESIGN GRID

BLINDING	TREATMENT SEQUENCE	CONTROL	SUBJECT ASSIGNMENT
Open label	Parallel Groups	Uncontrolled	Sequential enrollment
Single blind	Crossover	Placebo	Randomization
Double blind	Lead-in	Active	Stratification
	Washout	Case history	
	Multiple treatments	Matched case control	
	Survival	Untreated control	

THE PROTOCOL

By federal regulation [21 CFR 312.23(6)(iii)], a protocol for IND studies is required to contain the following, with particular detail to requirements for Phase I, II, or III studies as outlined in 21 CFR 312.23(6)(i) and (ii):

- A statement of the OBJECTIVES and PURPOSE of the study.
- The name, address, and statement of qualifications of each investigator and subinvestigator, address of the research facilities to be used, and the name and address of each reviewing Institutional Review Board (IRB). (Note: this information is commonly submitted through the Statement of Investigator form, Form FDA 1572, although the name and address of the investigator(s) are often on the cover page of the protocol.)
- The criteria for PATIENT SELECTION and for the EXCLUSION of patients and an estimate of the number of patients to be studied.
- A description of the STUDY DESIGN, including the kind of control group to be used, if any, and a description of methods to be used to minimize bias on the part of subjects, investigators, and analysts.
- The method of determining the DOSE(s) to be administered, the planned maximum dosage, and the duration of individual patient exposure to the drug.
- A description of the OBSERVATIONS and MEASUREMENTS to be made to fulfill the objectives of the study.
- A description of CLINICAL PROCEDURES, LABORATORY TESTS, or other measures to be taken to monitor the effects of the drug in human subjects and to minimize the risks.

"Basic Elements of Clinical Research" in Chapter 1 contains additional discussion of items generally included in a study protocol.

THE OBJECTIVE

The study's objective is the primary reason for conducting the study; it must be well defined, clear, and concise, and attainable in the current study. The objective should be a logical extension of the background information of the study drug and disease as explained in the study rationale. The rationale should succinctly lead the reader to the objectives of the study. The study objective should be able to be accomplished with the specific study design. Examples of study objectives are outlined in Table 6.2.

TABLE 6.2 EXAMPLES OF PROTOCOL OBJECTIVES AND PARAMETERS

PHASE/STUDY DESIGN	OBJECTIVE	PARAMETERS
I open label, escalating dose	to determine the maximum tolerated dose of a new drug	adverse events, lab values, and other specific tests
I open label, single dose or multiple dose	to establish the pharmacokinetic profile of a new drug at a specified dose level	serum/plasma levels and urine samples obtained at specified time intervals and analyzed for drug metabolites; monitor safety
II randomized, double blind, placebo controlled	to determine the efficacy of a new drug versus placebo in a well-defined patient population	efficacy data, endpoints, survival as well as safety data
III randomized, double blind, active controlled	to determine the safety and efficacy of a new drug versus an approved drug in a patient population	efficacy data, endpoints, safety data
III open label	to determine the safety of a new drug in a special population, e.g., pediatrics	safety data

NOTE: SAFETY (adverse events) is almost always a parameter in IND clinical trials.

PROTOCOL STUDY DESIGN

The study design used for a particular study must relate to the objective of the study. Design options are discussed earlier in this chapter.

A statistician needs to be involved at this stage of protocol development to determine critical components of the study, e.g., the number of subjects needed to determine a difference between treatment groups. Additionally, the statistician will be involved in the randomization procedure, if the study is randomized. The statistical section of a protocol discusses the study design in relation to the study objective and also discusses the plan for statistical evaluation of the data collected from the study. The statistician may be involved with any of the following in determining the study design:

- Total sample size (number of EVALUABLE subjects).
- Differences to be detected between (or among) treatment groups in comparative studies.

- Confidence intervals.
- Stopping rules.
- Error levels.
- Randomization.

A Data and Safety Monitoring Board (DSMB) may be used during the study to conduct interim analyses of the data to determine safety and efficacy.

TREATMENT PLAN

The treatment plan, including specific dose regimens, active controls, concurrent medications, and so on should be clearly specified so that it will be easily understood by both investigators and subjects.

INVESTIGATIONAL AGENT

There should also be a section describing the study drug(s) (chemical structure, appearance, packaging, etc.) or device, storage requirements, and how supplied. A description of all comparative agents also must be included.

PARAMETERS

The parameters to be monitored for a study should relate directly to the objective of the study. Avoid collecting extraneous information. Evaluate what is needed and at what intervals. Compare parameters to a baseline value. Are there certain restrictions or specific instructions related to the parameters, e.g., fasting before blood draws? If so, this needs to be written into the protocol.

Criteria for assessment of efficacy response should be very specific and well defined. Often, specific clinical endpoints are used and supported by diagnostic tests or measurements, such as ultrasound to measure tumor size.

Most studies also monitor patient safety. This can be accomplished by monitoring specific lab parameters, adverse events, and subject withdrawals. Generally, in early-phase studies, full chemistry and hematology panels are utilized. As the safety profile of the drug emerges, the clinical laboratory parameters are limited to only those where toxicity has appeared in earlier trials, e.g., only specific liver and kidney tests. The therapeutic guidelines from the FDA (Table 4.2) offer suggestions as to which parameters are important to collect in different phases of product development based on therapeutic objectives. Additionally, a proposed plan for treatment-emergent toxicity or adverse events may be included in the protocol. Typical adjustments include dose modifications, concomitant medications, or discontinuation of a subject from treatment.

SUBJECT SELECTION CRITERIA

The study population should be well defined; the subject group should remain consistent during the study. Uniformity of the subject population in early-phase trials allows for better control of extraneous factors so that the assessment of outcomes will be made with greater assurance, since variability is minimized. In Phase I and II studies, there is usually less subject variability. In Phase III and IV studies, variability becomes important because it is representative of the market situation; therefore, larger study populations are required to provide the needed statistical power to achieve the study objectives.

GENERAL COMMENTS

When writing a protocol, use clear, concise language. Ascertain that everyone involved in the study is interpreting the protocol in the same manner. Flowcharts and schedule of event charts (Figures 8.1 and 8.2) are useful tools. Additionally, provide specific guidelines for the determination of subjective parameters so that consistency in interpretation is maximized.

Ascertain that the various sections of the protocol contain no conflicting instructions. For example, in one section, no concomitant medications are allowed; but in another section, medications for pain may be prescribed as needed.

Provide in the protocol specific guidance on the uniform treatment and evaluation of subjects and, where possible, avoid investigator discretionary decisions; this may affect evaluability of subjects or the outcome of the study.

Finally, closely evaluate the protocol to determine its feasibility in the clinical setting. PIs should not agree to perform a protocol in which it will be impossible to collect the necessary parameters. The site's resulting poor performance will affect the sponsor's decision to choose you for future studies. HELP the sponsor recognize where the protocol can be improved to meet the objectives without compromising the quality of the study.

Remember, your objective in designing a study is to design a WELL–CONTROLLED study and to have NO DEVIATIONS from the protocol. Also, remember that protocols can be AMENDED as problems with the protocol become obvious in the application process. However, if the study was well designed and thought out, amendments should be minimal.

The Pharmacologic and Somatic Treatments Research Branch of the National Institute of Mental Health has prepared a 15-page form, "Trial Assessment Procedure Scale" (TAPS), to evaluate and critique protocols. It is available by writing to

Dr. Jerome Levine
Pharmacologic and Somatic Treatments Research Branch
NIMH
Room 10C-06
5600 Fishers Lane
Rockville, MD 20857

BIBLIOGRAPHY

An Introduction to Clinical Research. C. DeAngelis, Oxford University Press, New York, 1984.

Clinical Trial Design. G. Keith Chambers and Mary Sue Fairborn, *Applied Clinical Trials,* Vol. 7 (9), p. 60, 1998.

Clinical Trials: Design, Conduct, and Analysis. C. Meinert, Oxford University Press, New York, 1990.

Early Clinical Development: Population Pharmacokinetics and Pharmacodynamics. William Tracewell, Thomas Ludden, and Joel Owen, *Applied Clinical Trials,* Vol. 6 (10), p. 28, 1997.

Guide to Clinical Studies and Developing Protocols. B. Spilker, Raven Press, New York, 1984.

ICH Guideline "Structure and Content of Clinical Study Reports." *Federal Register,* 61 FR 37320, July 17, 1996.

Protocol Content and Management. Wendy Bohaychuk, Graham Ball, Gordon Lawrence, and Katy Sotirov, *Applied Clinical Trials,* Vol. 8 (3), p. 67, 1999.

Understanding Research Theory and Protocol Design. *Research Nurse,* Vol. 1 (4), 1995.

Note: *Applied Clinical Trials* offers many references on Early Clinical Development.

7

THE INFORMED CONSENT PROCESS

... no investigator may involve a human being as a subject in research covered by these regulations unless the investigator has obtained the legally effective informed consent of the subject or the subject's legally authorized representative ....

—21 CFR 50.20

In this century, research subjects have been cruelly, unfairly, and unethically abused in clinical trials both in the United States and abroad. To protect research subjects, the Food and Drug Administration (FDA) has developed specific regulations: 21 CFR 50 summarizes these regulations as they pertain to Investigational New Drug (IND) studies. Additionally, the Department of Health and Human Services (DHHS) has issued regulations (in 45 CFR 46 Subparts B, C, D) to protect research subjects participating in studies supported by federal grants. To fully understand the intent of the regulations protecting research rights, the following readings are highly recommended for anyone associated with enrolling human subjects in research trials:

- 21 CFR 50 and 56.*
- 45 CFR Part 46 Subparts B, C, D.*
- The Belmont Report.
- The Declaration of Helsinki.*
- FDA Information Sheets (9/98):*
  - Informed Consent Process
  - Informed Consent Document Content
  - Recruiting Study Subjects
  - Payment to Research Subjects
  - Screening Tests Prior to Study Enrollment
  - A Guide to Informed Consent Documents

*See Appendices.

- Exception from Informed Consent for Studies Conducted in Emergency Settings
- Evaluation of Gender Differences
- Significant Differences in FDA and HHS Regulations

PROPER INFORMED CONSENT

The basic premise of the FDA regulations is that respect for human subjects' rights and dignity requires that informed consent be obtained before a human subject participates in any clinical investigation. Institutional Review Boards (IRBs), clinical investigators, and sponsors all share responsibility for ensuring that the informed consent process is adequate. The investigator is responsible for ensuring that the research subject freely gives fully informed consent to participate in the clinical trial.

Proper subject consent involves much more than obtaining a subject's signature on a consent form. The consent process should be an interactive discussion with the potential subject that explains the study and the subject's expected involvement.

The following are the general requirements for informed consent:

- Informed consent must be obtained from the subject (or the subject's legally authorized representative) before a subject can be involved in research.

- The investigator must seek consent under circumstances that give a subject sufficient opportunity to consider whether to participate and that minimize possible coercion or undue influence. Timing, setting, who obtains the informed consent, and other circumstances surrounding the consent process are important to the subject's ability to comprehend the information provided.

- The information given to subjects must be understandable to them. Technical and medical terminology must be explained, and non–English-speaking subjects must have the information presented in a language that they understand.

  Subject consent interviews should be conducted in the subject's native language and with an accurately translated Informed Consent Form. The FDA Information Sheet "A Guide to Informed Consent Documents" contains more discussion on enrolling non–English-speaking subjects in a study.

- The informed consent may not include exculpatory language through which the subject is made to waive or appear to waive any of his/her legal rights or releases or appears to release the investigator, the sponsor, the institution, or their agents from liability for negligence.

Informed consent is the responsibility of the investigator, although the study sponsor often provides guidance and suggests wording for Informed Consent Forms that has been approved internally by the sponsor's advisors.

The eight basic elements of an Informed Consent Form are described in 21 CFR 50.25 and summarized in Table 4.3. The FDA Information Sheet "A Guide to Informed Consent Documents" provides an explanation of each of these elements. A summary of the basic elements follows:

1. A statement that the study involves research with
    - an explanation of the purposes of the research,
    - the expected duration of the subject's participation,
    - a description of the procedures to be followed, and
    - identification of any procedures that are experimental.
2. A description of any reasonably foreseeable risks or discomforts to the subject.
3. A description of any benefits to the subject or to others.
4. A disclosure of appropriate alternative procedures or courses of treatment, if any, that might be advantageous to the subject.
5. A statement describing the extent, if any, to which confidentiality of records identifying the subject will be maintained and that notes the possibility that the sponsor and FDA may inspect the records.
6. An explanation as to whether any compensation is available if injury occurs, and an explanation as to whether any medical treatments are available.
7. An explanation of whom to contact for answers to pertinent questions about the research and the rights of research subjects and whom to contact in the event of a research-related injury to the subject.
8. A statement that participation is voluntary, that refusal to participate will involve no penalty or loss of benefits to which the subject is otherwise entitled, and that the subject may discontinue participation at any time without penalty or loss of benefits to which the subject is otherwise entitled.

In addition, when appropriate, one or more of the following elements also must be inserted in the Informed Consent Form and provided to each subject:

- A statement that the particular treatment or procedure may involve risks to the subject, or embryo or fetus if the subject is or may become pregnant, that are currently unforeseeable.

- Anticipated circumstances under which the subject's participation may be terminated by the investigator without regard to the subject's consent.
- Any additional costs to the subject that may result from research participation.
- The consequences of a subject's decision to withdraw from the research and procedures for orderly termination of participation by the subject. This is especially important when subject safety is at issue.
- A statement that significant new findings developed during the course of the research that may relate to the subject's willingness to continue participation will be provided to the subject.
- The approximate number of subjects involved in the study.

Note that these additional elements are not optional; they must be incorporated into the informed consent process when appropriate.

EXCEPTION FROM THE GENERAL REQUIREMENTS

The regulations allow for circumstances where obtaining informed consent is not feasible. All of the conditions for exception must be met:

- Under 21 CFR 50.23, if the subject is confronted by a life-threatening situation necessitating the test article's use *and*
- informed consent cannot be obtained from the subject because of an inability to communicate with, or to obtain legally effective consent from, the subject *and*
- time is insufficient to obtain consent from the subject's legal representative *and*
- no alternative method of approved or generally recognized therapy is available that provides an equal or greater likelihood of saving the subject's life,
- *THEN*, after both the investigator and a physician who is not otherwise participating in the clinical investigation certify, in writing, that *ALL* of the above conditions exist, the clinical investigator may proceed with the use of the test article in the absence of an informed subject consent.

If the investigator believes that immediate use of the test article is required to preserve the subject's life and it is not possible to obtain timely certification from a physician who is not participating in the study, the clinical investigator may proceed with its use. However, following the test article's use, a physician who is not otherwise participating in the study must review and evaluate its use, in writing.

When test article use without informed consent has occurred, the investigator must submit the certification or the evaluation to the IRB within five working days after the test article's use. The IRB should review this documentation at its next scheduled meeting.

Emergency Research

Effective 1 November 1996, the regulations regarding informed consent and waiver of informed consent in certain emergency research were revised. In summary, the following areas of 21 CFR were affected (*Federal Register,* Vol. 61, No. 192, pp. 51498–51533, 2 Oct. 1996):

- **Part 50:** *Family member* is defined as "any one of the following legally competent persons: spouses; parents; children (including adopted children); brothers, sisters, and spouses of brothers and sisters; and any individual related by blood or affinity whose close association with the subject is equivalent of a family relationship."
- **Part 50.24, Exception from informed consent requirements for emergency use:** IRBs may approve the investigation without requiring informed consent of all research subjects be obtained if the IRB finds and documents the following:
  — The subjects are in a life-threatening situation.
  — Available treatments are unproven or unsatisfactory.
  — The collection of valid scientific evidence is necessary to determine the safety and effectiveness of a particular intervention.
  — Obtaining informed consent is not feasible because
    - subjects cannot give consent because of their medical condition;
    - intervention must be administered before consent from the subjects' legally authorized representatives is feasible; and
    - there is no reasonable way to identify prospective subjects.
  — Participation in the research offers the prospect of direct benefit to the subjects because
    - the subject is in a life-threatening situation that necessitates intervention; and
    - preclinical studies and related evidence support the potential for the intervention to provide direct benefit to the subject.

— Risks associated with the intervention are reasonable in relation to
  - what is known about the medical condition,
  - risks and benefits of standard therapy, and
  - risks and benefits of proposed therapy.
— The clinical investigation could not practically be carried out without the waiver.
— The proposed investigational plan defines the length of the potential therapeutic window based on scientific evidence.
— The investigator has committed to attempting to contact a legally authorized representative for each subject in that window of time and, if feasible, to seek consent from that representative within that window prior to proceeding without consent. Efforts made to contact legal representatives are summarized by the investigator, and this information is made available to the IRB at the time of continuing review.
— The IRB has reviewed and approved informed consent procedures and an Informed Consent Form consistent with Part 50.25.
— Additional protections of the rights and welfare of the subjects will be provided, including (at least) the following:
  - Consultation with representatives of the community.
  - Public disclosure to the communities in which the investigation will be conducted, prior to initiation of the investigation, of the investigational plan and its risks and benefits.
  - Public disclosure of information at the completion of the investigation to apprise the community and researchers of the study.
  - Establishment of an independent data monitoring committee to oversee the investigation.
  - When a legally authorized representative is not available, the investigator has committed to contacting (within the therapeutic window) the subject's family member who is not a legally authorized representative and asking if he/she objects.
— The IRB is responsible for ensuring that procedures are in place to inform the subject and/or the legally authorized representative or family member of inclusion in the research, details of the investigation, and the information contained in the Informed Consent Form, and that the subject

or representative may choose to discontinue participation at any time without penalty or loss of benefits to which the subject is otherwise entitled. *Other regulations apply as detailed in Part 50.*

- **Part 56.109, IRB review of research:** Most of the revisions to this part are discussed above as revisions in Part 50. Additionally,
  — An IRB must promptly notify in writing the investigator and the sponsor when the IRB determines that it cannot approve the research because it does not meet the criteria for the exception or because of other relevant ethical concerns and the reason for the IRB's determination.
  — The IRB shall promptly provide to the sponsor a copy of the information publicly disclosed as required by Part 50.24(a)(7)(ii) and Part 50.24(a)(7)(iii). The sponsor shall provide copies of this information to the FDA.
- **Part 312, Investigational New Drug Application:** Many revisions pertain to the IND regulations and sponsor and investigator responsibilities:
  — 312.20 Requirement for an IND: Sponsors are required to submit a *separate* IND for a clinical investigation involving an exception from informed consent (per 21 CFR 50.24). Such an investigation *may not proceed* without prior written authorization from the FDA (provided within 30 days after the FDA receives the IND).
  — 312.54 Emergency research under Part 50.24 of this chapter:
    – The sponsor shall monitor the progress of all investigations involving an exception from informed consent.
    – The sponsor shall promptly submit to the FDA the information received from IRBs regarding public disclosures.
    – The sponsor shall provide, in writing, information of IRB nonapproval of such a clinical investigation to the FDA, other investigators participating in this investigation or equivalent investigations, and other IRBs that are asked to review such investigations.
  — *Other revisions to Part 312 regulations apply:* As a consequence of these new requirements, revisions were also made to Part 314, Part 601, Part 812, and Part 814. Additionally, 45 CFR 46 regulations were modified to allow for emergency research with similar modifications as in 21 CFR 50. Anyone participating in emergency research studies should carefully review all of the comments to this proposed rule as well as the revised regulations.

This information is available in the *Federal Register* (Vol. 61, No. 192, pp. 51498–51533, 2 Oct. 1996) and has been incorporated into the current CFR. Refer to the FDA Information Sheet, "Exceptions from Informed Consent for Studies Conducted in Emergency Settings" for an interpretation of these requirements.

DOCUMENTATION OF INFORMED CONSENT

The clinical investigator is responsible for ensuring that informed consent is obtained from each subject before that subject begins participation in the research project, and a written Informed Consent Form is signed by this subject. This form must have been approved by the IRB. Before any subjects are entered into the study, the document itself is prepared by the investigator and submitted to the IRB for review and approval.

Writing the Informed Consent Form

The investigator is responsible for interpreting the study and representing the study in the Informed Consent Form in a language understandable to the typical subject, often at the sixth-grade reading level. All of the general considerations from 21 CFR 50.20, the eight basic elements of informed consent, and all appropriate additional elements must be incorporated into the Informed Consent Form. Additionally, IRBs often require that standard formats be used, especially regarding elements dealing with confidentiality, compensation, children's consent/assent, who to contact for questions about research subjects' rights, who to contact in case of a research-related injury, and the voluntary nature of the study. Many IRBs have set limits on how much blood can be drawn from subjects or standard wording to be inserted when venipuncture is required. CHECK WITH THE IRB REVIEWING THE PROPOSED RESEARCH TO DETERMINE THEIR SPECIFIC REQUIREMENTS. Also, there may be specific state or local requirements to include in the Informed Consent Form. The NCI has a template that is used for NCI–sponsored studies (and other cooperative groups).

Additionally, consider the following suggestions when writing the Informed Consent Form:

- The use of the wording "I understand . . . ." may be inappropriate.
- Use second-person—"you"—for the subject and first-person—"I/we"—for the investigator.

- Do not use "I fully understand the study" or "the study has been fully explained to me" because the subject is not ordinarily in a position to interpret the study and cannot judge if the information is complete.
- Never make claims of effectiveness or superiority or be overly optimistic. Use "experimental" or "investigational" but not "new" when referring to the investigational agent.
- Do not insinuate FDA approval or involvement, e.g., do not use "the FDA has given permission . . . ." Also, reference to IRB approval of the study can be misleading to a subject.
- Do not use "you may not participate in this study if you are a woman who could become pregnant." The use of women in clinical trials has recently been re-examined (refer to Chapter 4 for references on women in clinical trials). Women of child-bearing potential should not be excluded from participation; however, the Informed Consent Form should adequately explain the potential consequences of pregnancy while participating in the study.
- Do not say "you will be withdrawn if you do not follow study procedures." Instead, indicate that the subject may be withdrawn if he/she does not follow instructions given by the investigator.
- Payment to subjects must be fair and reasonable, not coercive. Compensation should be accrued and NOT be contingent on completion of the entire study (considered to be coercive). A payment schedule and conditions should be clearly outlined.

Refer to the FDA Information Sheet "A Guide to Informed Consent Documents" for a discussion of the FDA perspective on some of these issues.

Obtaining Consent

The granting of informed consent by the subject shall, in accordance with FDA regulations, be documented by the use of a written consent form. The subject's signature on the consent document must be dated at the time the subject signs to verify that consent was obtained before participation in the study.

After signature by the subject or the subject's legally authorized representative, a copy shall be given to the person signing the form, and a copy shall be retained in the investigator's files. Informed consent documentation is a primary topic for monitor review and FDA audits. Customarily, two "originals" (each copy bearing original signatures) are generated—one for the study file and one for the subject. It is not necessary, per regulation, that the subject sign his/her copy; but it is *recommended* that the subject sign his/her copy, especially if other signatures are required on the document.

The IRB may waive the requirements for a written consent form if it finds that the research "represents no more than minimal risk of harm to subjects" [defined in 21 CFR 56.102(I)] and "involves no procedures for which written consent is normally required outside the research context" (21 CFR 56.109). This does not permit a waiver or alteration of any elements of informed consent; the subject still must be properly informed—except that a signed document is not required. A written statement regarding the research may be required by the IRB to be provided to the subject. Additional information may be found in the FDA Information Sheet entitled "Informed Consent Process."

OBTAINING INFORMED CONSENT FROM SUBJECTS

The process of presenting informed consent can affect a subject's decision to enroll in the study. When all of the elements are included, the Informed Consent Form can sound very scary. Be honest with the subject at all times, but be careful not to frighten the subject away. For example, to a subject who raises a concern over the adverse events listed, the investigator may respond: "Yes, you may be concerned about the adverse events, but we will be monitoring you very closely to avoid or treat any serious event. You would also expect those same symptoms if you were taking Brand X, the approved treatment for this disease." The investigator or person obtaining consent must be knowledgeable about the study, the investigational agent, similar treatments for the disease, and the disease itself.

When: Timing of consent can be variable. The IRB should be aware of the timing of consent. The subject should be presented with the Informed Consent Form prior to any assessment or screening for study participation. While it is imperative that informed consent be obtained prior to performing any study-specific procedures or prior to administering a test agent, it is preferable to begin the informed consent process with the subject at the outset of consideration so that the subject is aware that the process involves research and is informed of all options. Note that this may occur after a lab panel or a medical history has been taken for clinical purposes. Such data may be used both in the treatment of the condition and for study eligibility or baseline data. Testing that is required for the study, outside normal clinical practice, is study related, and informed consent is required prior to obtaining such tests.

Why: It is a regulatory and ethical requirement to inform subjects about participation in clinical trials.

Who: There is no FDA regulation that requires the investigator to personally obtain consent from the subject. However, the investigator must ensure that the individual obtaining consent is fully knowledgeable about the research and the study so that he/she is able to answer any questions the subject may have. The IRB should be aware of who will be discussing study participation with the subjects. Generally, the Principal Investigator (PI), a subinvestigator, or the Clinical Research Coordinator (CRC) discusses the Informed Consent Form with the subject. Additionally, he/she should be able to ascertain that the subject fully understands what is being agreed to. If the investigator/subinvestigator is not obtaining the consent, he/she should be physically present at some point to answer questions of the subject.

How: The way in which the informed consent process is conducted is critical to subject participation and may affect study recruitment. Some steps to follow are given here:

- Discuss the study with the subject in general terms. Outline the purpose, procedures, and time commitment.
- Give the subject a copy of the Informed Consent Form to read. Also, with the subject's approval, give a copy to any family member accompanying the subject. Allow the subject ample time to read through the document.
- *Read through* the form with the subject. Stop at each section to ask if the subject understands or has any questions.
- At the end, again allow the subject to ask questions or raise concerns.
- If the investigator is not the person presenting the informed consent process, then he/she should ideally make time to talk to the subject to see if there are any concerns or, at the very least, be available to discuss the study with the subject if the subject so requests.
- When the subject indicates that he/she fully understands the Informed Consent Form, ask him/her to sign and date the document. Note that it may be necessary for the subject to sign multiple copies of the Informed Consent Form. Other signatures, such as those of the PI, the person presenting the informed consent process, or a witness may be required. These signatures should be obtained at this time on all copies.

- Give the subject a copy of the Informed Consent Form to keep for his/her own information.

It is important to keep the following factors in mind when presenting the informed consent process to the subject (see also Table 7.1):

- The form must be APPROVED by the IRB. Only the most CURRENT approved form may be used.
- The form should be CLEAR, DIRECT, AND SIMPLE to understand. Avoid research or technical terms.
- The form should be given to the subject to read, and then it must be EXPLAINED AND DISCUSSED THOROUGHLY with the subject. All questions must be answered to the satisfaction of the subject.
- Make sure the consent form is in a LANGUAGE that the subject understands.
- In some instances, e.g., illiterate subjects, it may be necessary to READ the Informed Consent Form to the subject and then obtain his/her signature on the form. These types of informed consents *must* be witnessed.
- "SHORT FORM" Informed Consent Forms generally only indicate that the trial was discussed with the subject and that the subject agrees to participate. A written summary of what is said to the subject must be provided. A witness must be present and sign both the summary and the short form. The subject must receive a copy of the summary and the short form.
- A VIDEOTAPE showing the procedures and other study-related items may facilitate the process.
- Explain the RISKS AND BENEFITS completely. You may need to respond to questions regarding the probability of any specific risk to the subject. KNOW THE DATA based on information in the protocol, the Investigator's Brochure, and any other available source. Explain that since this is research, part of the objective of the study is to better determine the likelihood of these risks as well as unidentified risks. Explain how risks will be minimized by medical screening and careful monitoring.
- CHOOSE YOUR WORDS CAREFULLY. Although it is necessary to inform subjects of the risks of study participation, measure your responses so that you are not unnecessarily alarming or inappropriately reassuring subjects.

TABLE 7.1 COMMON PROBLEMS WITH CONSENT FORMS

Failure to include all required elements specified in 21 CFR 50.25 and additional elements when appropriate.

The eight required elements of informed consent are required in ALL research. The additional elements of informed consent are NOT OPTIONAL but are required when they apply.

Cannot guarantee confidentiality.

Because disclosure to third parties may be required, complete confidentiality cannot be guaranteed. Additionally, inform subjects that sponsor representatives will review their records, and FDA representatives may inspect their records.

Failure to use lay language.

Explain scientific language in detail. Write to a sixth-grade reading level.

Failure to state experimental nature.

Explain that this is research, experimental or investigational. Do not insinuate approval/permission by the FDA or IRB since this may be misleading to the subject.

Failure to state the purpose of the research.

Do not limit discussion to only those purposes considered "most beneficial" to the subject.

Failure to state the expected duration of the subject's participation.

Overly optimistic in tone.

State the known facts without insinuating safety or effectiveness. This should not be at all promotional in tone. All consents should state that the drug is being studied for "safety" and "effectiveness" (when appropriate, i.e., Phase II–IV). Do not say "...has been proven safe and effective in early clinical trials," although you may say "has been tested in early clinical trials in X number of patients."

Failure to completely describe all study-related procedures to be followed.

Describe for the subject which procedures are "standard care" and which are additional for the purposes of the study.

Failure to adequately describe alternatives.

All treatment alternatives should be presented to the subject in a fair and detailed manner. The option of choosing no treatment may also be included when appropriate.

Failure to adequately describe the risks or benefits of the alternative treatments.

Risks or benefits of alternative treatment as it relates to the experimental treatment should be described fairly.

Failure to adequately describe foreseeable risks.

The risks of all study-related procedures, not only of the experimental treatment or drug itself, should be described. For example, if venipuncture is required that is outside the standard care of the patient, risks of venipuncture (discomfort, bruising, etc.) should be described.

Failure to describe compensation/Compensation coercive.

The manner of payment should be outlined in the consent document. Compensation must not be coercive. Payment cannot be contingent on completion of the entire study nor are "bonuses" for completion appropriate.

Table 7.1 continued on next page

Table 7.1 continued from previous page

Failure to clearly state subject's right to withdraw at any time [21 CFR 50.25(a)(8)].

Subjects may withdraw at any time without being forced into study termination procedures. However, where subject safety is concerned (e.g., patient must receive alternative treatment), subjects should be encouraged to complete appropriate termination procedures. Distinguish between research procedures and patient treatment.

Failure to provide contact for answers to questions.

Contacts, complete with phone numbers, should be written in the consent document. Contacts for questions regarding the research (usually the investigator), research subjects' rights (usually an independent office within the institution), and who to contact in case of a research-related injury (possibly the same independent office) should be provided. It is preferable that the contacts for subjects' rights and research-related injuries be someone other than the investigators, i.e., an independent third party.

Uses exculpatory language.

Informed Consent Forms should not include statements where subjects are giving up their rights, e.g., "I understand that I am not entitled to monetary compensation in the case of a research-related injury."

Failure to describe conditions of premature termination by the investigator or sponsor.

Specific reasons for subject termination (e.g., failure to follow instructions, adverse event or injury, or cancellation of the study) should be included in the Informed Consent Form.

Omits the written summary for short Informed Consent Form.

The short-form process requires that a written summary be reviewed and approved by the IRB and that summary should be provided to the subject as part of the consent process.

Not study specific.

When using boilerplate forms, you must insert specific details about the research study.

Failure to obtain IRB review and approval.

Uses wrong form.

The most recent (revised) Informed Consent Form must be used. It is suggested to put the date of approval and/or version # on the front of the form.

Inadequate provisions for vulnerable subjects:

Non-English speaking	Should have an accurately translated Informed Consent Form in the native language. Including the occasional non–English-speaking subject may be questionable since it will be difficult to ascertain that the investigator can relate all study-specific information to the subject in an understandable way. To exclude non–English-speaking patients may be considered bias.
Children	Children over age 6 (7 and older, per the Belmont Report) should require documentation of the child's ASSENT to be in the study.
Illiterate subjects	Illiterate subjects should be consented per short form procedures with a witness to the subject's mark.
HIV patients	Subjects must be informed of the possible consequences of the HIV screening and their right to confidentiality and/or anonymity.

- Explain OTHER BENEFITS of participating in a clinical trial:
  - Benefit mankind, benefit to others.
  - Contribution to science, on the cutting edge of research, being a part of history.
  - Free examinations and health monitoring.
  - Priority in office visits.
  - Improved personal health.
  - Opportunity to learn more about the disease.
  - Compensation, if applicable.
- At least TWO COPIES of the Informed Consent Form must be presented to the subject—one for the study files that is signed by the subject/witness/guardian and one for the subject to keep as a record and reference. A third signed copy may be required for the medical record. Alternatively, the subject may be given a photocopy of a signed form.
- Emphasize to subjects that participation is VOLUNTARY and that they may withdraw at any time. Assure them that you are their ally, regardless of their decision to participate in the study.

SPECIAL CIRCUMSTANCES

Keep in mind that certain groups, such as children, women of child-bearing potential, HIV-infected (human immunodeficiency virus–infected) individuals, and prisoners are considered "special groups" that may require diligence in assuring that their rights are protected. Refer to Chapter 4 for regulatory references regarding these special circumstances.

The investigator must always assure that the rights and welfare of study subjects are protected. The FDA regulations and IRBs provide guidance on how to accomplish this goal; but investigator and staff interaction with the subject, fully informing subjects about study participation, is the single most critical aspect of informed consent.

BIBLIOGRAPHY

Acting Without Asking: An Ethical Analysis of the Food and Drug Administration Waiver of Informed Consent for Emergency Research. James Adams and Joel Wegener, *Annals of Emergency Medicine,* Vol. 33 (2), 1999.

Back on the Soapbox Again. Jane Ganter, *Applied Clinical Trials,* Vol. 8 (5), p. 10, 1999.

Benefit/Risk Assessment: Perspective of a Patient Advocate. A. Bowen, *Drug Information Journal,* Vol. 27, pp. 1031–1035, 1993.

Consent Documents. Erica Heath, *Applied Clinical Trials,* Vol. 8 (6), p. 100, 1999.

Essential Information to Be Given to Volunteers and Recorded in a Protocol. D. Jackson and G. Richardson, *Journal of Pharmaceutical Medicine,* Vol. 2, pp. 99–101, 1991.

Glossary of Lay Language Synonyms for Common Terms Used in Informed Consent Documents for Clinical Studies. D. Norris, Pharmaceutical Information Associates, Ltd., Levittown, Penn., 1996.

Guidelines for Writing an Informed Consent Document, Sandra Sanford and B. Tilman Jolly, *The Monitor,* Vol. 13 (2), pp. 17–23, 1999.

Informed Consent Forms. Erica Heath, *Applied Clinical Trials,* Vol. 7 (9), p. 12, 1998.

Informed Consent Glossary. Bruce Steinert, *Applied Clinical Trials,* Vol. 6 (5), p. 71, 1997.

Human Subjects: Winning the Cold War. Deb Jolda, *Applied Clinical Trials,* Vol. 9 (7), pp. 48–50, 2000.

Protecting Research Subjects—What Must Be Done. Donna Shalala, *New England Journal of Medicine,* Vol. 343 (11), 2000.

Safeguarding Subjects Also Protects Data. Pamela McGahee, *Applied Clinical Trials,* Vol. 6 (5), p. 77, 1997.

Two Models of Implementing Informed Consent. C. Lidz, P. Applebaum, and A. Meisel, *Archives of Internal Medicine,* Vol. 148, pp. 1385–1389, 1988.

Using Instructive Videotapes to Increase Patient Comprehension of Informed Consent. D. Norris and M. Phillips, *Journal of Clinical Research and Pharmacoepidemiology,* Vol. 4, pp. 263–268, 1990.

8

CONDUCTING THE STUDY

I agree to conduct the study(ies) in accordance with the relevant, current protocol(s) and will only make changes in a protocol after notifying the sponsor, except when necessary to protect the safety, rights, or welfare of subjects. . . .

—Form FDA 1572, Section 9, Commitments

A clinical trial requires a great deal of work before the first subject is enrolled. The key to a successful study is careful planning. The protocol and study design need to be concise, clear, and feasible as well as being able to provide the data to prove (or disprove) the objectives of the study. Each of the details of conducting the study must be delegated to a member of the study team before the first subject is enrolled, not in the midst of subject visits. The more problems that can be anticipated and solved prior to the initiation of a trial, the more likely it is that the trial will proceed (relatively) smoothly. Once the trial has started, stay organized and continue solving those situations that inevitably pop up. Termination of the trial is a time to put all things in order to store away for future reference.

THE EVALUATION STAGE OF A STUDY

Before an investigator agrees to conduct a clinical trial, it is crucial that the study be evaluated in collaboration with the study team. Consider the following aspects of the study.

Protocol

Critically review the protocol. The protocol, discussed in Chapter 6, is the blueprint for the study. You are being considered as an investigator by the sponsor, and you want to tell the sponsor that you CAN do all of the study-required procedures as written in the protocol. BUT, be honest with the sponsor, and do not agree to something that will be impossible to deliver. You are being considered for your expertise in the area, so offer advice to the sponsor on how to improve the study and make it more doable while still being scientifically sound; this will contribute to a better study. A sponsor should be willing to listen to your comments. There are some situations where changes are not feasible, such as participating in a large, already ongoing multicenter trial. But don't let this stop you from making a comment. And if something in a protocol is absolutely objectionable and the sponsor is unwilling to make a change, then you may want to reconsider your participation: Do you want your professional reputation tainted by association with a poorly designed study? Questions to ask include the following:

- Is it a good protocol?
- Is it logistically feasible to conduct at your study site?
- Do you have adequate space, time, and staff?
- Do you have an adequate subject population?

Determine whether the study can be done at your institution per the protocol and discuss with the study team the potential "trouble spots." It is important to resolve problems before the study commences. Use the protocol to consult with the pharmacy, inpatient/outpatient units, clinical laboratory, ECG, Radiology, and any other institutional department mentioned in the protocol. An especially useful part of the protocol is a graphic schedule of visits and evaluations (Figure 8.1) and a study schema (Figure 8.2). If they are not included in the protocol, create the table and flowchart for your own use.

Evaluate the available subject population (see Chapter 10) and establish accrual goals. Evaluate referral services to determine additional sources of subjects.

Contracts

Many components of conducting the clinical trial are contractual in nature. Your signature on the Statement of Investigator form is a commitment by you to conduct the trial according to Food and Drug Administration (FDA) regulations. Your signature on the protocol signifies agreement with and your commitment to compliance with each of the protocol-specific procedures. You will also likely sign a Confidentiality Agreement prior to reading the protocol to protect the proprietary rights of the sponsor.

FIGURE 8.1 SAMPLE SCHEDULE OF STUDY VISITS AND EVALUATIONS

	Screen	Baseline	Day 1	Day 2	Week 1	Week 2	Week 4	Off-Study
Informed Consent	x							
Entry Criteria	x							
Medical History	x							
Physical Exam		x						x
Abbreviated Exam					x	x	x	
Chest X-ray		x					x	x
CBC/differential		x						x
WBC	x		x	x	x	x	x	
Hemoglobin	x		x	x	x	x	x	
Platelet Count	x		x	x	x	x	x	
Chemistry Panel		x						x
SGPT	x			x	x	x	x	
Alkaline Phosphatase	x			x	x	x	x	
BUN		x		x	x	x	x	x
Creatinine		x		x	x	x	x	x
Urinalysis	x	x		x	x	x	x	x
Adverse Event			x	x	x	x	x	x
Dispense Drug			x	x	x	x	x	
Returned Drug				x	x	x	x	x

FIGURE 8.2 SCHEMATIC DIAGRAM OF STUDY ENTRY AND GUIDELINES

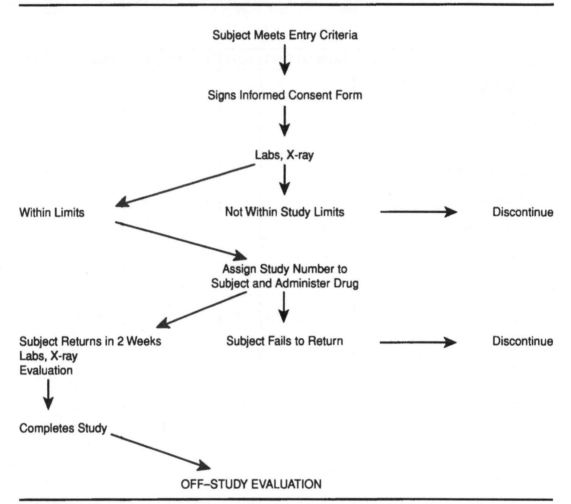

Additionally, you, your institution, or the sponsor will want a Clinical Trial Agreement (CTA). A relatively standard type of agreement, the CTA outlines who will do what and what will be paid (see "Budget"). It is very important to clearly determine what is considered an "evaluable" patient if this term is used as a criterion for reimbursement. Issues regarding completion dates, payment schedules, and publication rights might also be included. You may also have a separate consulting role with the sponsor. This may be treated as a separate contract or as part of the CTA.

Many investigators and/or institutions also require an Indemnification Agreement with the sponsor. This agreement serves to aid in protecting the investigator or institution from litigation should a subject be harmed by participation in the study. Indemnification is often covered in the CTA. In rare instances, the institution may request a

Certificate of Insurance from the sponsor to provide evidence of insurance to cover a claim. A sponsor does not generally offer the Indemnification Agreement, so determine if you will need one and request it early in your discussions with the sponsor.

Additional information that you may need to provide is as follows: a tax identification number (for yourself and/or the institution, depending on to whom the check is issued) and/or your social security number if any payment is made to you (including reimbursement for travel expenses). Also, know who in your institution is responsible for signature on the above agreements. If you know of any particular language or components required by your institution, notify the sponsor immediately. It is best to put any representing lawyers together and let them iron out the nuances of the contract, but be sure to read the final version before you sign it to make sure it is something with which you can comply. Determine if any of these contractual agreements are required by your institution as early as possible and begin this process immediately. Many studies have been delayed waiting for the lawyers to decide on just the right wording. Note that it is not spelled legal-ease!

Financial Disclosure

The financial arrangements of clinical research investigators have come under marked scrutiny in the past few years. The public is increasingly dismayed at physicians who have financial interests in the very same drugs that they are supposed to be objectively testing. While any entrepreneur has the right to profit from one's labor, the thought of an investigator simultaneously being both an objective scientist and a partisan entrepreneur is somehow disconcerting.

Therefore, the FDA has recently required financial disclosure for all investigators involved in clinical research. As stated in 21 CFR 54, the FDA is addressing payment arrangements and financial interests of investigators that could potentially bias the outcome of the study. In an effort to ensure that the reliability of research data is not affected, sponsors are now required to submit to the FDA information concerning any financial arrangements of each investigator who is participating in any clinical trial in which the FDA relies on the data to support efficacy, safety, or bioequivalency claims.

- Form FDA 3454: Certification of absence of financial interest.
- Form FDA 3455: Disclosure statement which reveals the presence of financial interests.

These forms are intended to capture the financial interests of investigators, subinvestigators, and their spouses and dependent children under the following circumstances:

- Equity interest of greater than $50,000 in the sponsor company.
- Proprietary interest (licensing agreement, trademark, or patent) in the investigational agent.
- Situations in which the investigator's compensation is greater if a favorable outcome is reported.
- Any other significant payment of greater than $25,000 by the sponsor.

The FDA may refuse to file the application of any sponsor who fails to submit a list of all investigators and subinvestigators, together with all appropriate Forms FDA 3454 and 3455, with the marketing application.

Budget

Can the study be done at your institution with the money allocated? Make a list of all costs after each group (pharmacy, lab, etc.) has read the protocol and submitted its budget to you:

- Can you get discounts for a certain volume of tests done?
- Are outside agencies cheaper or more reliable for certain study services (labs, X-ray, etc.)?
- What is your institution's fee for administrative overhead?
- Do staffing costs include benefits?

Be sure you understand where all costs come from—do you or the company sponsor pay for photocopying, mailing, duplication of X-rays, and so on? Be sure to include these figures in your budget proposal.

Prepare a sample budget on a per-patient cost (see Figure 8.3) and compare the sum total to the money allocated by the sponsor. You should at least break even. Reevaluate the study budget after the study begins in order to identify any extra costs incurred.

Usually, patients who participate in studies pay, at most, only for the care they would normally receive if they were not on a study. Any study-related tests or procedures should be covered by the study budget. Be sure you understand what is "customary care" for these patients so that the budget will be reasonable. Assure that billing is

FIGURE 8.3 BUDGET PLANNING WORKSHEET

PROTOCOL _____

ANTICIPATED NUMBER OF SUBJECTS _____

		Cost/Subj	(x #subjs)	Total
Screening				
Clinical Laboratory:	CBC	_____		_____
	Blood Chemistries	_____		_____
	Urinalysis	_____		_____
OTHER, specify:	_____	_____		_____
	_____	_____		_____
	_____	_____		_____
ECG		_____		_____
X-rays		_____		_____
OTHER evaluation:	_____	_____		_____
	_____	_____		_____
	_____	_____		_____
Professional time (exams, interviews)		_____		_____
Consultation, specify: _____		_____		_____
Screening Total		_____		_____

Interim Visits
(begin with enrollment visit)

		Required # Visits	Cost/Subj	(x # subjs)	Total
Clinical Laboratory:	CBC	_____	_____		_____
	Blood Chemistries	_____	_____		_____
	Urinalysis	_____	_____		_____
OTHER, specify:	_____	_____	_____		_____
	_____	_____	_____		_____
	_____	_____	_____		_____
ECG	_____	_____	_____		_____
X-rays	_____	_____	_____		_____
OTHER evaluation:	_____	_____	_____		_____
	_____	_____	_____		_____
	_____	_____	_____		_____
Professional time (exams, interviews)		_____	_____		_____
Consultation, specify _____		_____	_____		_____
Hospitalization (# days)		_____	_____		_____
Interim Visits Total		_____	_____		_____

Figure 8.3 continued on next page

Figure 8.3 continued from previous page

	Required # Visits	Cost/Subj	(# subjs)	Total
Other Costs				
Special Equipment, specify:				
_____				_____
_____				_____
_____				_____
_____				_____
_____				_____
Supplies, specify:				
_____	_____	_____		_____
_____	_____	_____		_____
_____	_____	_____		_____
_____	_____	_____		_____
_____	_____	_____		_____
_____	_____	_____		_____
Pharmacy	_____	_____		_____
Subject Compensation	_____	_____		_____
Secretarial/Clerical Support				_____
Advertising				_____
Clinical Research Coordinator (estimated # hours times base salary)				_____
Copying, Mailing, etc.				_____
Institutional Overhead (_____% of total)				_____
Other Costs Total				_____

Totals

    Screening Total _____

    Interim Visit Total _____

    Other Costs _____

    Total Cost of Study _____

        (÷ # subjs) = _____ cost per subject

handled in such a way that patients are not billed for study-specific procedures. Remember that the Institutional Review Board (IRB) may want this information included in the Informed Consent Form.

Clarify the basis of reimbursement with the sponsor:

- Are screening costs covered for subjects not enrolled?
- Are only evaluable subjects reimbursed?
- Precisely how is an evaluable subject defined?
- Who pays for miscellaneous costs such as photocopying and mailings?

Determine the payment schedule with the sponsor. Usually, payments are made in increments depending on subject enrollment and completion. The following is an example:

- 25 percent total amount as an initial payment.
- 25 percent when half of the subjects are enrolled.
- 25 percent when all subjects are enrolled and at least half have completed.
- 25 percent at the conclusion of the trial after the final study report is submitted.

In determining a payment schedule, make sure you can cover your costs, especially if it involves subject reimbursement. Work closely with the grants administrator at your institution. This is the person who actually handles the money.

THE PLANNING STAGE OF A STUDY

Evaluation of the protocol and the contract and budget preparation is the beginning stage of planning and preparing to conduct the study. Once you and the study sponsor have come to agreement on the protocol, budget, and contract, you must plan for the beginning of the study.

Study Conduct

Review the protocol again and determine how and by whom each of the requirements will be accomplished. Discuss the protocol with the study staff and delegate responsibilities accordingly. Walk through the procedures to anticipate problem areas. Develop study aids, e.g., flowcharts, patient calendars, schedule of events, to make the study proceed easily and minimize violations to the protocol. Begin a list of potential subjects.

Pharmacy

At the planning stages of the study, you will need to evaluate the facilities for storing and dispensing the investigational agent. Some questions to ask include the following:

- Will the institution pharmacy be used?
- Are there study-specific pharmacy procedures? Who will develop these?
- Who will be the pharmacist in charge of the study? Establish good communications early in study planning.
- Is the study drug dispensed as a unit dose or by infusion, bottle, carton, card, or other form?
- Will the pharmacy submit its budget (usually standardized by the institution) to you?
- Will the drug volume be large? Will the drug need to be mixed, requiring additional supplies and cost? What are the storage requirements (out of light, refrigerated)? Check for drug storage capabilities per the protocol.
- Are the pharmacy services available on weekends, holidays, or during non-business hours?
- What is the drug destruction policy?
- Does the pharmacy have its own forms, which it prefers to use for drug accountability? How does this compare with the sponsor's form?
- Consider the study drug's shelf life, shipping and receiving requirements, and satellite pharmacy interactions.
- How will the pharmacy staff be trained and supervised for this study?
- Is there a randomization code? Will the pharmacist be responsible for randomizing patients? In blinded studies, will the pharmacist need to unblind the code for preparation of the study material?
- Will there be a code breaker? If so, who will have it, and under what conditions should the code be broken? Is there someone available 24 hours a day to break the code in an emergency?
- If the pharmacy is not being used for the study drug, evaluate the alternative storage area:
  - Is it locked and secured?
  - Does it meet the storage requirements of the drug?
  - Is the temperature monitored?

    – Is it accessible by study personnel?

    – Who will have access?

    – Who will have responsibility for the study drug (under the supervision of the Principal Investigator [PI])?

Figure 8.4 is a worksheet to assist in planning with the pharmacy.

Clinical Laboratory, ECG, X-ray, and Other Clinical Tests

You will be working with many other departments who provide services for the study. Some suggestions are as follows:

- Establish one individual from each department as your contact person. Determine the cost of studies performed and try to find the most economical way of ordering tests (for example, it is often less expensive for a lab to run an automated blood panel than it is for them to test for only certain parameters, e.g., a chemistry panel vs. just sodium and potassium).

- Will there be a code or special requisition forms to assure the test is billed to the study account and not the patient? You may need to establish an account for the study.

- What are the weekend/holiday/daily hours for availability and/or sample pickup? You may need to arrange a different lab setup when patients are seen on days when the regular lab is closed (e.g., using the ER lab on weekends).

- Obtain laboratory licenses and/or certifications, normal value ranges, and the curriculum vitae (CV) of the clinical laboratory director prior to study initiation. Collect this documentation on EACH lab you will be using. Periodically assess the lab for license renewals and changes in normal value ranges.

Medical Records

Find out the policy for obtaining, reviewing, and duplicating patient records (hospital medical records as well as clinic charts). These may require a signed permission form from the patient ahead of time. Additionally, determine the policy for retention of subject records. These are considered source documents for the study and must be maintained as long as the study files.

FIGURE 8.4 CLINICAL INVESTIGATION INFORMATION FORM FOR THE PHARMACY

Title of Study _____

Principal Investigator _____ Telephone _____ Pager _____

Department _____

Subinvestigators _____

Clinical Research Coordinator _____ Telephone _____ Pager _____

Sponsor_____ Contact _____ Telephone _____

Name of Investigational Agent

	Dose	Frequency	Route
#1_____	_____	_____	_____
#2_____	_____	_____	_____
#3_____	_____	_____	_____
#4_____	_____	_____	_____

How supplied: _____

_____

Storage requirements:_____

_____

Study design (check all that apply):

- ☐ Randomized
- ☐ Single blinded
- ☐ Double blinded
- ☐ Open label
- ☐ Crossover

Randomization Code: _____

Code Breaker ☐ Yes ☐ No

Subjects: ☐ Inpatient ☐ Outpatient ☐ Day Hospital

Number anticipated: _____

Sites receiving investigational drug other than hospital/clinic: _____
(supply mailing addresses and contact persons on separate page)

To Be Completed by Pharmacy:

Responsible pharmacist _____ Telephone _____ Pager _____

Who will be responsible for the following:

	Pharmacy	Investigator	Other	N/A
Drug storage	☐	☐	☐	☐
Randomization	☐	☐	☐	☐
Drug preparation	☐	☐	☐	☐
Drug dispensing	☐	☐	☐	☐
Accountability records	☐	☐	☐	☐
Inventory control	☐	☐	☐	☐
Shipping drug to other sites	☐	☐	☐	☐

Administrative

Arrange for backup persons to be available for yourself and the lead Clinical Research Coordinator (CRC). Instruct these persons on the basics of the trial and orient them to all of the study information materials (protocol, procedure manual, special forms, etc.). Meet with all subinvestigators and CRCs to organize your efforts and to have everyone working from the same plan. Arrange an in-service meeting with hospital staff. Designate one person as the liaison with the IRB.

Regulatory

Write the Informed Consent Form, and obtain sponsor review and approval. Submit protocol, informed consent, and any proposed advertising to the IRB. Complete, sign (by the PI), and submit to the sponsor the Statement of Investigator form (Form FDA 1572) and appropriate CV. The sponsor will also require the following documents: CVs of subinvestigators, clinical laboratory certifications and normal value ranges, IRB membership list, and/or Department of Health and Human Services (DHHS) general assurance number and IRB approval letters.

STUDY COMMENCEMENT

After what may seem like endless preparation, the time has come for entering the first patient. Review the study start-up checklist (Figure 8.5) and make sure the following are completed:

- Protocol is finalized (at least for now).
- Regulatory paperwork is completed.
- Ancillary services are ready (labs, X-ray, pharmacy, etc.).
- The investigational agent is on hand, securely stored, and ready to be dispensed.
- Nursing or other personnel on whose ward or clinic you may be conducting the study have been trained in all study procedures and data collection techniques.
- You have a payment "advance" or initial grant payment or at least have adequate funding to get started.
- Case Report Forms (CRFs) or flow sheets are ready.
- Your study work areas have been reserved for your use and are equipped as necessary.

FIGURE 8.5 STUDY START–UP CHECKLIST

Completed   **Sent to Sponsor**

☐   Signed confidentiality agreement   ☐
☐   Contract agreement with sponsor   ☐
☐   Indemnification agreement   ☐
☐   Statement of Investigator completed   ☐
☐   Curriculum Vitae prepared   ☐
☐   Financial Disclosure Forms completed   ☐
☐   Protocol reviewed and signed   ☐
☐   Informed Consent Form prepared   ☐
☐   Protocol and Informed Consent Form to IRB
☐   List of IRB Membership   ☐
☐   IRB meeting date _____
☐   IRB approvals received   ☐
☐   IRB renewal date _____
☐   Clinical laboratory certification on file   ☐
☐   Clinical laboratory normal values on file   ☐
☐   List of key site personnel with signatures   ☐
☐   Receive and review study procedure manual

       Arrangements made with:
☐   Pharmacy
☐   Clinical laboratory
☐   In-house ward
☐   Clinic scheduling
☐   Other departments   _____
☐   _____
☐   _____

       Evaluate subject population:
☐   Develop a recruitment strategy
☐   Prepare advertising
☐   Submit advertising to IRB   ☐
☐   Notify subjects
☐   Prescreening clinic

       Budget:
☐   Prepare
☐   Internal approval
☐   Submitted to sponsor   ☐
☐   Sponsor approval
☐   Initial payment received

       Supplies received:
☐   Case Report Forms
☐   Investigational agent   ☐
☐   Clinical lab supplies
☐   Special equipment
☐   Shipping materials
☐   _____
☐   _____

Here are some helpful hints for the study staff:

- *Stay organized.*
- Keep a copy of the protocol available at all times.
- *Highlight areas* of importance on a reference copy of the protocol for quick reference.
- Prepare a set of *cue cards* for your pocket. Write out *3 X 5 cards* to help get started and to keep vital information and phone numbers at your fingertips (see Figure 8.6).
- Use a schematic of the study (Figure 8.2) or schedule of events calendar (Figure 8.1).
- Keep telephone numbers of everyone you may need on a card in your pocket.
- Color code your cards, group them, laminate them—whatever makes your efforts more organized.

FIGURE 8.6 3 × 5 CARD EXAMPLES

Inclusion/Exclusion Criteria (list)

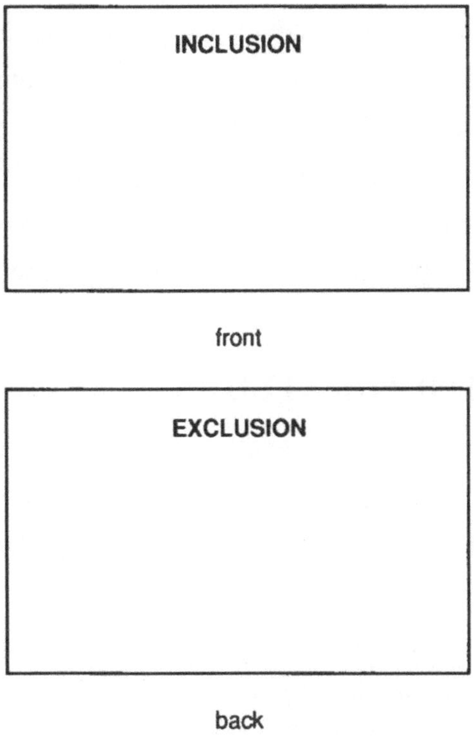

INCLUSION

front

EXCLUSION

back

The Protocol Violation

Despite careful planning, things may still go wrong when it comes to actual implementation. You may make a mistake in protocol requirements (referred to as a "protocol violation") or miss a data point on your first subject or two. This is a new experience for all involved, and some bumps along the path to the protocol's success are expected. But, CORRECT your problems as soon as possible and PREVENT the same mistakes for the next study subject. If the protocol is very complicated, it may be constructive to pause after entering one or two subjects to meet and discuss the problems and progress with the study team. If problems are still not getting resolved, enlist the assistance of the sponsor representative.

Subjects and Families

Be sure to obtain a truly *informed* consent. Does the subject really understand what he/she is signing? Is he/she aware of all the tests/procedures involved? A subject who feels misled may not comply with the study or may drop out. Is there a subject who looks at you suspiciously and tape records everything you say (it has happened!) for fear you are using him/her as a test rat? BEWARE! Perhaps he/she is not an appropriate candidate for the study.

Be sure subjects have a way of contacting you during the study if they have questions. Give clear directions, written and illustrated, if appropriate, and guide them step-by-step through the study. Point out to subjects not only the risks of participating in the study but also the potential and real benefits. Telephone calls as reminders before study appointments can be helpful. That little extra attention can keep a subject interested and cooperative. Refer to Chapter 10, "Interactions with the Study Subject."

KEEPING UP WITH THE STUDY

Now that the study is started, it must continue to run smoothly. There is so much to keep organized—all of that paperwork to attend to, and, of course, the subjects. First, get your priorities straight: The subject's well-being should always come first, as long as you stay within the guidelines of the protocol.

Next, designate a backup person to assist you or to handle the study in your absence. Keep him/her informed on all developments in the study.

Also, you need to tackle all of that paperwork on a regular basis or it gets out of control. Here are some suggestions:

- SCHEDULE time for paperwork and don't let anything interfere.
- Keep CRFs and data flow sheets current, both for data accuracy and data management.
- Don't let the "to be filed" pile get out of control. File things as you go along.
- Set up a filing system that is easily accessible (desktop) to automatically deposit items where they need to be (e.g., lab results, X-rays, IRB, sponsor) instead of one big "to be filed" pile where it becomes impossible to locate anything.

Other suggestions for staying organized:

- Keep a calendar just for the study; write in dates for scheduled and projected subject visits, monitoring site visits, due dates (IRB progress reports, annual renewal), investigational drug reorder dates, study staff meetings, and so on.
- Make and use checklists, charts, notecards, or whatever system works for you to stay ahead.
- Keep current versions of the Study Procedures Manual, Protocol, and other study aids close by for easy reference.
- Keep patient materials (lab requisitions, appointment cards, etc.) nearby for easy access.
- Keep each individual subject's paperwork in a separate file.

Starting the study is only the beginning. There are a number of activities that need to be done on a regular basis:

Study Files   The study files must be set up and kept current. The sponsor monitor should review the study files periodically during monitoring visits. A study documentation checklist is given in Figure 8.7. Section 8 of the ICH GCP Guideline provides a concise outline of essential documents for the conduct of clinical trials.

FIGURE 8.7 STUDY DOCUMENTATION CHECKLIST

☐ Form FDA 1572
☐ Financial Disclosure Forms
☐ Curriculum vitae
☐ IRB approvals and correspondence
      Submittal package
      Initial approval letter
      Progress reports
      Annual renewals
      Protocol amendments
      IND Safety Reports
      Final Report
☐ Correspondence between investigator and sponsor and other study-related correspondence
☐ Informed Consent Forms
      Blank copy of all IRB–approved versions
      Original signed copies of study participants' Informed Consent Forms
☐ Protocol
      Original signed version
      All amended versions
☐ Investigator's Brochure
      All versions applicable to the study
☐ Investigational Agent Shipping Records
☐ Investigational Agent Dispensing/Accountability Records
☐ Investigational Agent Final Disposition Records (documenting return of drug to sponsor, destroyed, or all used in study)
☐ Case Report Forms
      Blank copy of all versions
      Completed and signed copies for each study participant
☐ Serious Adverse Event Reporting Forms (blank)
☐ Reports of serious adverse events (IND Safety Reports)
☐ Study Progress Reports and Final Report
☐ Clinical laboratory certification
☐ Clinical laboratory normal value ranges
☐ Study Procedures Manual
☐ Telephone monitoring records
☐ Monitoring log

Maintain separately:
☐ Budget
☐ Indemnification and study contracts

IRB Correspondence

In addition to the initial contact with the IRB, the investigator must communicate with the IRB during the study in specific cases:

- Annual renewals.
- Progress Reports (frequency determined by the IRB).
- Submission of Protocol Amendments.
- Revised Informed Consent Forms resulting from amendments or safety reports.
- Reports of serious adverse events occurring at the site.
- Reports of IND safety reports at other sites.
- Final Study Report.

Investigational Agent

Documentation for the investigational agent must be kept current (refer to Chapter 5, "Investigational Agent Management"). The supply of investigational agent must be monitored to assure that an adequate stock is available for subjects enrolled in the trial as well as new subjects. DON'T RUN OUT! In some studies, it may be necessary to collect and store the tear-off labels from the containers. You may also have to receive, organize, and store investigational agent returned by subjects in the trial. It will help if you create a study-specific tracking chart of investigational drug for each subject and attach it to the subject's record.

Case Report Forms

CRFs should be completed immediately after a subject's visit. Some pages may be completed during the study assessment visit; in this case, make sure there is adequate documentation in the subject's medical record/clinic chart to use as source documentation. Don't let completion of the CRFs become backlogged. Also, the CRFs must be current before each monitoring visit by the sponsor so that the monitor can review all of the data. Your staff may also have to make copies of the CRFs or separate NCR papers to submit the data to the sponsor. The sponsor may require that data on CRFs be faxed directly to the sponsor to speed up data entry and analysis.

If you are using remote data entry, the same concept applies; keep up with data entry. Also, be sure to generate hard copies of the data and file them immediately.

Budget   Follow the budget closely. Make sure the sponsor is initiating payments as agreed (note there will probably be some lag time for administrative preparation and mailing of payments—allow four to eight weeks). Work with the grants administrator. Ensure that the bills are getting paid. You may find that your original arrangement is not working. You may have to be creative and borrow from other resources or renegotiate your arrangement with the sponsor.

STUDY TERMINATION

Have heart; most studies do end (although they never go away). At that time, there is much cleanup work to do.

Subjects   It's time to say good-bye to your subjects, at least in terms of the trial. The subjects may be part of your patient population, and you may see them again for regular clinic visits. If they have been recruited solely for participation in this study, arrange for records to be transferred to the referring physician. Make sure all final assessments are done. Resolve all compensation issues with each subject. Instruct subjects to return all study-specific materials, such as investigational agent, monitoring apparatus, and so on. Make sure you have a way to contact the subject (maintain a confidential patient log of subject names, addresses, and phone numbers) should anything about the study arise that requires notification. If the subject was compliant, consider using him/her again or recommending him/her to your colleagues for other trials.

Investigational Agent   All study drug must be returned to the sponsor or properly disposed of as indicated by the sponsor (see Chapter 5, "Investigational Agent Management"). This includes all unused drug, all returned drug and containers, and tear-off labels, if applicable. Additionally, ensure that the paperwork is complete. Place the original shipping records, invoices, and subject drug accountability (dispensing) logs in the permanent study file. If the pharmacy requests to keep the originals, a copy should be placed in the study files with a memo explaining the location of the original records. Code breakers should also be filed in the study files.

Informed Consent Forms

The original, signed Informed Consent Form for each subject should be stored in the study files. If it is not possible to keep the original in the file (e.g., the institution requires it to be stored in the patient's medical record), then a copy can be placed in the file with a memo describing the reason and where the original may be located. A blank copy of each version of the Informed Consent Form used in the study should be kept in the study files.

Data

All data must be completed and submitted for data analysis. This should occur as quickly as possible. Copies of subject CRFs or data flow sheets should be maintained and stored with the study files. Data queries may continue long after the study has ended, and you will need to provide the requested information.

Study Files

Now is the time to go through the study files AGAIN and assure that everything is there and all unnecessary articles are removed. A study documentation checklist is in Figure 8.7. The files should be boxed for storage. Locate the final storage place for these files and show the location to the monitor during the termination visit. Also, record the location somewhere so that, years later, anyone with the need (such as an FDA inspector) will be able to locate the files. CLEARLY mark the box with the study number, title, investigator, and contact person.

Final Report to the Sponsor and IRB

A final report must be written and submitted to the IRB and the sponsor (generally, the sponsor will withhold the final payment until the final report is filed). Different IRBs require different elements to include in the report. Some general items are as follows:

- Number of subjects enrolled.
- Number of subjects completing.
- Number of subjects dropping out and reasons.
- Frequently noted adverse events.
- Serious adverse events.
- Subject deaths, if any.
- Investigator's opinion of the results.

IRB

In addition to the final report, the IRB should also receive formal notification of completion of the study. This might be incorporated into the submission of the Final Report.

Study Supplies

All study supplies (CRFs, study-specific lab requisitions, sample collection materials) must be returned to the sponsor or destroyed, if so designated by the sponsor. Equipment purchased or supplied by the sponsor may be returned to the sponsor or may be made available for the investigator's clinical use at the sponsor's discretion.

Budget

Prepare information for the final payment. Ensure that all bills are paid.

Figure 8.8 is a checklist for study termination activities. Figure 9.4, "The Termination (Study Close-Out) Site Visit" (in Chapter 9), will also help in preparing to close the study.

FIGURE 8.8 STUDY TERMINATION CHECKLIST

Subjects:
- ☐ Complete final assessments
- ☐ Return and inventory investigational agent
- ☐ Return study materials
- ☐ Obtain address
- ☐ Resolve compensation

Investigational Agent:
- ☐ Inventory all drug, used and unused
- ☐ Resolve all accountability issues
- ☐ Assemble files, "paper trail"
- ☐ Prepare investigational agent for inventory by sponsor
- ☐ Prepare tear-off labels, randomization cards, code-breakers for inventory
- ☐ Complete accountability logs to indicate return/destruction of drug
- ☐ Package and return investigational agent
- ☐ File paperwork relating to final disposition (returned to sponsor, destroyed, used in trial)

Data:
- ☐ Record all data on CRFs or data flow sheets
- ☐ Submit to investigator for review and signature
- ☐ Assure subject's medical record/clinic charts are available and accurate for source document verification
- ☐ Maintain copies of all CRFs or data flow sheets in study files
- ☐ Generate and file hard copies of final data reports if remote data entry is used
- ☐ Return or destroy all unused CRFs
- ☐ Resolve queries related to data entry

Regulatory:
- ☐ Inventory and file all signed Informed Consent Forms in the master study file
- ☐ Prepare and submit Final Study Report to IRB and sponsor
- ☐ Notify IRB of study termination

Study Files:
- ☐ Organize study files (see study file checklist)
- ☐ Consolidate all files into a master study file to be stored
- ☐ Box master study file
- ☐ CLEARLY label the storage box
- ☐ Make a notation of the storage location of the master study files/notify sponsor

Study Supplies:
- ☐ Return or destroy unused CRFs
- ☐ Destroy study-specific lab requisitions and labels
- ☐ Return or release study specific supplies (e.g., collection tubes, packaging materials)
- ☐ Return special equipment, if required
- ☐ Return remote data entry equipment

Biological Samples:
- ☐ Inventory all samples
- ☐ Ship for analysis as indicated by the sponsor

Budget:
- ☐ Assure final payment is received from sponsor
- ☐ Assure all bills are paid

Sponsor:
- ☐ Schedule and prepare for termination site visit (see Figure 9.4)
- ☐ Resolve all outstanding issues

REGULATORY REFERENCE

21 CFR 54.

BIBLIOGRAPHY

7 Steps to Assessing a Potential Clinical Research Study. G. Keith Chambers, *Applied Clinical Trials,* Vol. 8 (3), p. 66, 2000.

Archiving Original Trial Records Data and Documents. Cristina Pintus, *Applied Clinical Trials,* Vol. 4 (6), p. 60, 1995.

Financial Disclosure Is Needed at the Lectern, as Well as in Print. Lawrence Landow, letter to the editor, *Critical Care Medicine,* Vol. 26 (12), 1998.

Interactions of Academic Investigators with Pharmaceutical Companies. In *Guide to Clinical Trials,* B. Spilker, Raven Press, New York, pp. 390–394, 1991.

Interesting Conflicts and Conflicting Interests. Peter Whitehouse, *Journal of the American Geriatrics Society,* Vol. 47 (6), 1999.

Is Academic Medicine for Sale? Marcia Angell, *New England Journal of Medicine,* Vol. 342 (20), 2000.

Issues in Clinical Trials Management: Planning and Budgeting. Felicia Favorito, *Research Nurse,* Vol. 3 (5), pp. 6–15, 1997.

Learning to Conduct Research—the Hard Way. L. Witter, *RN,* February, pp. 35–40, 1990.

Lessons Learned from Coordinating a Pivotal Clinical Trial. D. Johnston, *Journal of Clinical Research and Drug Development,* Vol. 7, pp. 31–39, 1993.

Management of Clinical Trials. F. Abdellah, *Journal of Professional Nursing,* Vol. 6 (4), p. 189, 1990.

Negotiating Clinical Trial Agreements with Academic Institutions. Garrett Sanders, *Applied Clinical Trials,* Vol. 1 (11), p. 39, 1992.

Roles of Private Practice Physicians in Clinical Trials. In *Guide to Clinical Trials,* B. Spilker, Raven Press, New York, pp. 395–398, 1991.

Source Documentation in Clinical Research. Celine M. Clive and Alyson Hall, *Applied Clinical Trials,* Vol. 9 (5), p. 73, 2000.

Sponsor Education of Clinical Investigators in the Clinical Research Process. T. Kirsch, *Drug Information Journal,* Vol. 22 (2), pp. 181–186, 1988.

The Interrelationships of Sponsors, Clinical Investigators, and Institutional Review Boards. T. Kirsch, *Drug Information Journal,* Vol. 21 (2), pp. 127–131, 1987.

Uneasy Alliance—Clinical Investigators and the Pharmaceutical Industry. Thomas Bodenheimer, *New England Journal of Medicine,* Vol. 342 (20), 2000.

Working with the Clinical Research Site: The Clinic's Perspective. K. Drennan, *Applied Clinical Trials,* Vol. 1 (5), pp. 62–65, 1992.

9

MONITORING OF CLINICAL TRIALS

The sponsor shall monitor the progress of all clinical investigations being conducted under its IND.

—21 CFR 312.56(a)

Many clinical trials are conducted by "sponsors." The sponsor is typically a pharmaceutical company but may also be a Contract Research Organization (CRO) working for the sponsor, or a government agency (e.g., NCI) or a cooperative research group (e.g., NSABP, SWOG). The pharmaceutical industry sponsor's objective generally is to get a new drug to market or to establish new applications for a marketed product. These pharmaceutical research activities are closely observed by the Food and Drug Administration (FDA), and approvals are contingent on studies that are appropriately designed and completed. Additionally, FDA regulation states that the sponsor will monitor the clinical trial [21 CFR 312.56(a)]. The sponsor, therefore, will closely monitor the progress of a clinical trial.

The frequency of monitoring will vary depending on the phase of the study, type of study, study progress, amount of data, deadlines, and so on. Communication between the investigator and the sponsor is critical to the success of the trial. Typically, the contact person for the sponsor is the Clinical Research Associate (CRA). The most effective means for the sponsor to assure that a clinical trial is conducted properly is to conduct periodic site visits. Telephone communication and written correspondence are other common methods of communication between the sponsor and the clinical site.

Studies are also funded by grants from specific research organizations. The funding organization may be considered the sponsor in these situations. Their requirements for Good Clinical Practice (GCP) and adherence to FDA regulations are similar to those of the pharmaceutical industry. However, the monitoring process may not be as intense.

This chapter focuses on the pharmaceutical sponsor, but the general concepts apply to all studies.

SITE MONITORING VISITS

The most significant interactions between the investigative site and the sponsor occur during site monitoring visits. There are four basic types of site visits:

1. **Prestudy:** The sponsor and investigator evaluate the possibility of doing the clinical trial together. This stage may also involve protocol development.

2. **Initiation:** The contract and protocol have been signed, the study is ready to begin, and both parties are ascertaining the readiness of the site to start enrolling patients. During the visit, instructions are given concerning the procedures of the study. Sometimes, it may be planned to enroll the first subject in conjunction with the initiation visit.

3. **Periodic:** The study has started, and the sponsor visits the site regularly to ensure that the study is being conducted according to the protocol and FDA regulations, there are no unanticipated problems at the study site, Case Report Forms (CRFs) are being completed appropriately, and there are no unreported serious adverse events.

4. **Termination or study closeout:** The study is over; during this visit, the sponsor ascertains that all documentation is in place, all data have been submitted, all investigational supplies returned to the sponsor, and all outstanding questions are addressed.

A less common type of site visit is a *preinspection* visit. Often, when an investigator is notified of an inspection by the FDA, the sponsor will return to the site to review the study materials with the investigator and help prepare for the inspection.

Quality assurance audits may be conducted by the sponsor to assess not only the site's performance but also that of the sponsor's monitoring staff. See Chapter 14 for a discussion of quality assurance audits conducted by the sponsor.

Careful preparation is required by both parties to assure that site visits are accomplished efficiently and are an effective use of time for both the site and the sponsor. Figures 9.1 to 9.4 summarize the preparation steps for each different type of site visit. The ICH GCP Guideline discusses specific responsibilities of monitoring the clinical trial in Section 5.18.

GRANT–SPONSORED VISITS (AUDITS AND INSPECTIONS)

Cooperative Research Bases, as well as other sponsors, must monitor their clinical trials for quality assurance based on appropriate execution of the trial and quality of the data. Ancillary service areas important in the conduct of the trial, such as pathology or radiotherapy, may also be monitored. Often grant approval, or reapproval, is contingent on a site inspection. Monitoring practices may differ from one group to another. It is best to prepare for these visits by following the guidelines set forth by the grant-issuing body, as well as those given in Figures 9.1 to 9.4.

RESOLUTION OF PROBLEMS IDENTIFIED AT SITE VISITS

At the end of each site visit, the investigator should meet with the monitor to discuss the findings. The study is a team effort with everyone working toward the same goal—a successful study. The monitor has the experience of working with other sites, often on the same study, and may have valuable suggestions as to how to improve the way the study is conducted. Additionally, it is a regulatory responsibility of the monitor to assure that

- the study is being conducted according to protocol,
- FDA regulations are followed, and
- subject safety is monitored.

The monitor and the investigator should try to resolve problems or identify solutions to problems prior to the monitor's leaving the site. Outstanding problems are likely to be documented in a follow-up letter to the investigator. THE SITE SHOULD DOCUMENT RESOLUTIONS TO ANY PROBLEMS IN WRITING.

It is desirable to resolve all data queries on the CRFs at the time of the site visit. However, because of time constraints, the monitor may leave questions noted in the CRFs requiring resolution or clarification. These may be indicated on a data audit checklist or with yellow Post-it® notes. Reconcile these questions as soon as possible and forward the corrected CRFs to the sponsor.

FIGURE 9.1 THE PRESTUDY SITE VISIT

OBJECTIVE

Sponsor   To assess the study site and the investigator's ability to perform the clinical trial. During this evaluation visit, the sponsor will be reviewing the investigator's previous research experience, the availability of patients/volunteers, the staff's ability to perform the study at the facility, and personnel support.

Investigator   To learn about the study, investigational drug, and sponsor; assess the logistics of conducting the trial at the facility; and propose possible changes to the protocol.

PREPARATION

Sponsor   Review curriculum vitaes (CVs), Form FDA 1572; prepare contract.

Investigator
- REVIEW the protocol, Investigator's Brochure (IB), and CRFs by all pertinent staff with comments returned to the Principal Investigator (PI) prior to the visit.
- Present the PI's CREDENTIALS (and those of key staff) to the sponsor—generally through a current CV.
- Evaluate training and experience of other participants (Clinical Research Coordinator [CRC], pharmacy, laboratory technician, etc.).
- Present the ORGANIZATIONAL STRUCTURE of the personnel at the facility to the sponsor.
- Assess the SUBJECT POPULATION and provide information on where and how subjects will be recruited as well as an estimate of the number of study participants available.
- Arrange for a tour of the FACILITY and schedule with other participants. Include all areas key to the conduct of the trial.
- Prepare an AGENDA for the meeting with the sponsor (the sponsor will often suggest an agenda).
- Reserve a MEETING ROOM and assure that all relevant staff can attend.
- Prepare a BUDGET for conducting the clinical trial.

DURING THE VISIT

The monitor will
- tour
    - exam rooms/areas where subjects will be treated/evaluated.
    - laboratory facilities.
    - special testing facilities.
    - pharmacy (main pharmacy and applicable satellite pharmacies).
    - in-hospital areas, if hospitalization required.
    - work areas for staff.

Figure 9.1 continued on next page

Figure 9.1 continued from previous page

      –   administrative areas.

      –   storage areas for study supplies.

- Discuss the study protocol and Investigator's Brochure.

- Discuss the contract (Clinical Contract Agreement [CTA]).

- Collect completed documents.

DOCUMENTATION

The following documents may be presented or collected prior to or during the prestudy site visit:

- CONFIDENTIALITY AGREEMENT: A statement signed by the investigator on behalf of his/her study staff agreeing to hold all study information confidential. This form is customarily executed prior to shipment of the protocol, IB, and other documents to the site.

- CLINIC VISIT RECORDS: The PI may substantiate that he/she has access to the needed patient population by producing clinic records, hospital census, or other appropriate information.

- PREVIOUS STUDIES: The PI may establish his/her experience by discussing previous trials done by him/her at the facility without breaching confidentiality.

- CURRICULUM VITAE: The current CV is used to demonstrate adequate training and to establish the PI as an expert in the field.

- BUDGET PROPOSAL: A budget proposal may be requested by the sponsor either before the site visit or as a follow-up to the site visit.

- STATEMENT OF INVESTIGATOR (FORM FDA 1572): This form should be reviewed prior to or during a prestudy visit and information completed so that the investigator understands the obligations set forth by the FDA regulations and GCP guidelines.

FOLLOW–UP

If both the sponsor and the investigator agree to proceed with the clinical trial, the following items need to be done:

- Read and sign the Statement of Investigator form (Form FDA 1572) and send to sponsor with CV of each person named therein.

- Draft Informed Consent Form; submit to sponsor for review.

- Submit final protocol and Informed Consent Form to the Institutional Review Board (IRB) for approval.

- Send IRB approval and copy of Informed Consent Form to sponsor.

- Sign indemnification and study contract agreements, if needed.

- Obtain clinical laboratory certifications and normal ranges and send to the sponsor.

- Prepare for arrival of study supplies.

- Finalize budget.

- Set up initiation meeting. Plan to attend any investigator's meeting/workshops.

- Make arrangements within institution to begin screening study subjects.

FIGURE 9.2 THE SITE INITIATION VISIT

OBJECTIVE

Sponsor   To prepare the site personnel to perform the clinical trial. May include study-specific training, review of materials, attendance at an investigator's workshop. Assure all regulatory documentation has been completed, all details of study procedures have been addressed, drug has been properly received at site and securely stored, and potential subjects have been identified.

Investigator   To review all aspects of the study to assure proper implementation. All required regulatory documentation completed, all details of study procedures have been addressed, drug has been received at site and securely stored, and potential subjects have been screened and identified.

NOTE: At times, the first subject may be enrolled in conjunction with the initiation visit (or another site visit) while the sponsor is at the site.

PREPARATION

Sponsor
- Finalize and sign contract.
- Finalize and sign protocol by PI.
- Develop Study Procedures Manual or other aids for study conduct.
- Finalize CRFs; ship to site.
- Ship investigational agent.
- Ship other pertinent study supplies.

Investigator
- The FINAL PROTOCOL should have been reviewed and study logistics determined. Any outstanding questions should be resolved during the initiation visit. Note that any major changes to the protocol by the sponsor at this point are likely to delay the start of the trial.
- All preparatory STUDY DOCUMENTS should have been completed/obtained and filed in the study files. Additional items that may be required prior to study initiation are the clinical study agreement (the contract between the sponsor and investigator/institution) and the indemnification agreement. Generally, it is up to the investigator to request these agreements from the sponsor, and they are often handled through the corresponding legal departments. Note that even minor disagreements in these legal contracts can greatly delay the beginning of the clinical trial.
- The INVESTIGATIONAL AGENT should have arrived at the site and be ready for inspection by the monitor.
- Review STUDY PROCEDURES MANUAL, if one is being used.
- Review CASE REPORT FORMS so that you can ask questions regarding the completion of the CRFs.
- Screening of subjects may have already occurred. If so, present a list of POTENTIAL SUBJECTS (identified by initials) to the monitor.

Figure 9.2 continued on next page

Figure 9.2 continued from previous page

- Be prepared to show the FACILITIES to the monitor, including work area, patient visit areas, phlebotomy areas. Make appointments with appropriate ancillary personnel, e.g., the pharmacist, to meet with the monitor.

- If the first subject is to be treated during this visit, be sure to schedule this as part of the visit.

Figure 8.5 in Chapter 8, "Study Start-Up Checklist," will assist in preparing for the initiation site visit.

DURING THE VISIT

The monitor will conduct a(n)

- detailed review of the final protocol, preferably with the entire study staff present;

- detailed review of the CRF;

- review of Study Procedures Manual, if applicable; and

- inventory and inspection of investigational agent.

DOCUMENTATION

The following documents should have been submitted to the sponsor PRIOR TO the initiation site visit:

- Signed Form FDA 1572 and CVs for investigator and subinvestigators.

- Financial Disclosure Forms.

- IRB approval letter.

- IRB approved Informed Consent Form.

- Signed copy of the final protocol.

- Clinical laboratory certification/laboratory normal ranges.

- Indemnification agreement/study contract agreement.

- Final budget.

- List of key site personnel with signatures.

FOLLOW–UP

If there are no outstanding issues between the sponsor and the site, subject enrollment may begin. If there are outstanding issues between the sponsor and the site, these need to be resolved (and the resolution documented, generally by a letter) prior to subject enrollment. Then

- notify sponsor of first subject enrolled,

- inform the sponsor of any problems AS THEY OCCUR, and

- establish the date for the first periodic site visit.

FIGURE 9.3 THE PERIODIC SITE VISIT

OBJECTIVE

Sponsor   To assure that the study is being conducted according to protocol and FDA regulations and that the clinical trial is progressing smoothly. Specifically, this includes the following:

- FDA regulations and GCP guidelines are being followed.

- Subject safety is being assessed.

- The protocol is being strictly followed.

- Entry criteria are met.

- Drug dispensed appropriately/blind maintained/accountability records completed and current/drug inventory agrees with log.

- CRFs completed/verify data per original source documents (medical record, charts, lab reports)/CRFs signed by investigator after thorough scrutiny.

- Informed Consent Form signed for each subject prior to screening.

- Protocol violations identified and discussed.

- Subject accrual rate is adequate.

- Subjects compliant.

- Facilities remain adequate.

- Changes in site staff noted and replacement personnel well trained.

Investigator   To cooperate with the sponsor to meet the above objectives. To discuss or resolve any study-related problems.

PREPARATION

Sponsor
- Review previous monitoring reports and other communications to identify issues to be discussed/resolved.

- Review records for reports of serious adverse events (SAEs) or Investigational New Drug (IND) safety reports to discuss with the investigator.

- Review protocol/amendments.

- Review the study budget and payment schedule.

Investigator
- Keep CASE REPORT FORMS current. Have the investigator review and sign all completed CRFs prior to the site visit.

- Have all CRFs, MEDICAL RECORDS, and CLINIC CHARTS available in a designated work area for the monitor to review.

- Have all INFORMED CONSENT FORMS available for review and verification.

- If a SERIOUS ADVERSE EVENT was reported, have all information available for review and verification.

Figure 9.3 continued on next page

Figure 9.3 continued from previous page

- Keep DRUG ACCOUNTABILITY RECORDS current.

- Make an appointment with the PHARMACIST to review drug accountability records and shipping records and to inventory the investigational agent(s). If required, have all tear-off labels, randomization envelopes, and/or code breakers available for the monitor to check to assure that study blinding is maintained.

- Have STUDY FILE DOCUMENTS available and filed appropriately.

- Set up an appointment, generally toward the end of the visit, for the monitor to meet with the INVESTIGATOR to discuss findings of the site visit.

- Adjust your CALENDAR and that of the CRC so that you both have time to meet with and work with the monitor to resolve discrepancies.

DURING THE SITE VISIT

The monitor will

- review CRFs and source documents; the CRC or PI should be available to answer questions;

- verify informed consent for each subject enrolled;

- review the study files for completeness and assess IRB status;

- review reports of SAEs, if any;

- inventory investigational agent;

- discuss findings with the CRC and PI; and

- meet briefly with the PI at the conclusion of every periodic site visit.

DOCUMENTATION

- Any NEW documentation—such as IRB reapprovals, lab recertifications, updates to CVs, or changes to the Statement of Investigator—will be collected by the monitor.

- Signed INFORMED CONSENT FORMS will be reviewed.

- Information pertaining to an IND SAFETY REPORT may be collected.

- Completed and signed CRFs will be taken to the sponsor offices by the monitor.

NOTE: In many cases it will be necessary to photocopy the CRFs or separate NCR (no carbon required) forms; allot time accordingly. Some sponsors may take all CRFs back to the company offices and have copies made and returned to the site. The site must retain a copy of every CRF submitted.

FOLLOW-UP

- All discrepancies identified during the periodic site visit must be resolved and the resolution documented (usually by a letter to the sponsor or in the monitor's site visit report).

- Often the monitor will leave comments for clarification on the CRFs. These issues need to be resolved as soon as possible by the study team.

FIGURE 9.4 THE TERMINATION (STUDY CLOSEOUT) SITE VISIT

OBJECTIVE

Sponsor

To assure that the study is completed and all study supplies and investigational agents returned to the sponsor, all documentation is in place and a storage location has been identified, all data (CRFs) are accurate and returned to the sponsor; to discuss the requirements for retention of study materials.

Investigator

In addition to the above objectives of the sponsor, the investigator may want to know the results of the study, what the plans for publication may be, what future trials he/she may be involved in, and a final resolution of compensation issues.

PREPARATION

Sponsor

- Review study files for accuracy and reconcile with the site's files.

- Review previous monitoring reports and other communications to identify issues to be discussed/resolved.

- Review records for reports of SAEs or IND safety reports to discuss with the investigator.

- Review protocol.

- Review the study budget and payment schedule.

- Determine deadlines for final study report and publication plans.

Investigator:

- Review the STUDY FILES for completeness and accurate filing.

- Consolidate all records into one set of study files. (Refer to Figure 8.7, "Study Documentation Checklist.") Prepare to box the files for storage.

- Complete, review, and sign all CRFs. There should be no outstanding data queries at the end of the termination site visit, although queries may continue when the data are processed by the sponsor. Copies of the CRFs must be maintained with the study documents.

- Assure that the subjects' MEDICAL RECORDS and clinic charts are available for the monitor to verify data and address queries.

- Have all subjects' signed INFORMED CONSENT FORMS available for review and storage with the study records.

- Resolve any outstanding DISCREPANCIES identified on previous periodic site visits.

- Complete the FINAL STUDY REPORT for submission to the IRB and sponsor.

- Notify the IRB of study completion or termination. Submit a final study report to the IRB.

- Establish an AGENDA for the monitor's site visit. Make appropriate appointments.

- Review BUDGET agreement to determine outstanding compensation.

Figure 9.4 continued on next page

Figure 9.4 continued from previous page

- Review INVESTIGATIONAL AGENT accountability records. Have all shipping forms available for review. Have access to all study drug, randomization cards, and tear-off labels, if used for the study. Organize study drug according to

  - unassigned drug,

  - assigned but undispensed study drug, and

  - study drug returned by subjects.

- Organize the study drug and tear-off labels by subject number to facilitate inventory of the investigational agent. Have material available to pack the investigational agent for return to the sponsor. Alternatively, if the sponsor permits, arrange to have the investigational agent destroyed on-site by an acceptable means. The monitor may be required to witness the destruction of the investigational agent.

- Arrange for the return of STUDY SUPPLIES (unused CRFs, clinical laboratory supplies, etc.). Remove any preprinted clinical laboratory requisition forms that are specific for this trial.

- Inventory any BIOLOGICAL SAMPLES collected for the study and determine their disposition.

- Arrange for the monitor to meet at length with the INVESTIGATOR to discuss study termination activities, the final study report, outstanding discrepancies or problems, arrangements for compensation, requirements for retention of study records, publication policies, and instructions for responding to a request for an audit by the FDA.

- Arrange for a site CONTACT PERSON in order to resolve any additional discrepancies and data clarifications of CRFs.

Figure 8.8 in Chapter 8, "Study Termination Checklist," will assist in preparing for the study termination site visit.

DURING THE SITE VISIT

The monitor will

- conduct standard periodic site visit activities;

- inventory investigational agent and determine final disposition;

- inventory all study supplies and biological samples and determine disposition;

- review study files for completeness and prepare for storage; and

- meet with the investigator to discuss outstanding issues, budget, record retention requirements, and publication.

DOCUMENTATION

The following items may be collected during the termination site visit:

- Updates to regulatory documents.

- IND Safety Reports (adverse drug experience reports), if any.

- Final Study Report.

- New or missing correspondence.

Figure 9.4 continued on next page

Figure 9.4 continued from previous page

- Drug accountability records (one copy for the sponsor, originals to the study files unless the pharmacy requests to keep the originals, in which case a copy should be placed in the study files with a memo explaining the location of the original records).

- CRFs.

Other items collected:

- Biological samples.

- Unused and returned investigational agent.

- Unused study supplies and CRFs.

FOLLOW–UP

- Resolve any outstanding discrepancies identified during the termination site visit and document the resolution.

- Assure that all clinical trial supplies and investigational agent are returned to the sponsor or appropriately disposed of.

- Store study documents in an acceptable area as discussed with the monitor.

TELEPHONE MONITORING

The investigative site and the sponsor also communicate by telephone. It is important that the investigative staff feel comfortable in contacting the sponsor of the trial to resolve questions or report new findings.

Telephone discussions, particularly when they involve decisions regarding the conduct of the trial or the handling of subjects, should be documented. This can be accomplished in one of several ways:

Telephone Monitoring Reports
A detailed report is written and filed in the study files. It may be initiated by the monitor, investigative staff, or both. If each party writes a report, the content should be consistent. The other party to the discussion may request a copy of the report. An example of a report form is in Figure 9.5.

Telephone Logs
A log is maintained with entries of each telephone contact. Key information may be noted in the log. An example is in Figure 9.6.

Follow-Up Letters
A letter is written to confirm decisions made during the telephone conversation. They may be initiated by either the monitor or the investigative staff.

FIGURE 9.5 SAMPLE TELEPHONE MONITORING REPORT

NAME OF SITE REPRESENTATIVE _____

NAME OF SPONSOR REPRESENTATIVE _____

CALL INITIATED BY   ☐ SITE   ☐ SPONSOR

DATE _____ TIME _____

PROTOCOL TITLE _____

PROTOCOL NO. _____

INVESTIGATOR _____

REASON FOR CALL (check all that apply):

☐ subject accrual	☐ protocol	☐ order supplies
☐ subject dropout	☐ administrative	☐ shipment of samples
☐ adverse event	☐ investigational agent	☐ data clarification
☐ IND safety report	☐ blind broken	☐ other _____
☐ subject death	☐ personnel/facility change	☐ _____
☐ protocol violation	☐ problem with lab	☐ _____

DISCUSSION:

SIGNATURE _____   DATE _____

cc: Principal Investigator
     Sponsor representative
     original to study files

FIGURE 9.6 SAMPLE TELEPHONE MONITORING LOG

PROTOCOL TITLE _____

PROTOCOL NUMBER _____

INVESTIGATOR _____

Date	Site Representative	Sponsor Representative	Discussion

Store with study files.

WRITTEN CORRESPONDENCE

Many items are documented through written correspondence between the investigator and the sponsor. All of this correspondence must be retained as part of the study file records. Additionally, any other correspondence regarding the clinical trial, such as internal memos and correspondence to other sites and labs, also must be retained in the study files.

What Should Be Documented by Written Correspondence or Telephone Contact?

The discussion of specific issues relating to the administration and conduct of the clinical trial should be documented for the study records. These items may include, but are not limited to, the following:

Study Administration
IRB approval status.
Informed consent process.
Budget.
Study files and documentation.

Protocol
Entry criteria.
Drug dosages/modifications.
Randomization and blinding procedures.
Protocol violations (notify sponsor immediately).
Protocol amendments.

Accrual
Randomization to study groups.
Accrual rates/goals.
Withdrawals/dropouts.
Subject registration.

Case Report Forms
Status of completion.
Clarifications/answers to queries.
Shipment to sponsor.
Requests for copies.
Storage of CRFs.

Adverse Events

Clinical management.

Reporting abnormal lab values.

Reporting trends.

Reporting Serious Adverse Events.

Subject deaths.

Data

Reporting of efficacy data.

Interim reports of subject data.

Data clarification forms.

Investigational Agent

Ordering/shipping/disposition.

Storage.

Coding/randomization.

Recall or retesting for potency.

Breaking the study blind (revealing the code).

Clinical Supplies

Ordering/storage/shipping.

Biological Samples

Collection.

Storage/shipping.

Labeling.

Analysis/results.

Facility Changes

Clinic.

Hospital.

Pharmacy.

Clinical or analytical laboratories.

Pharmacy/study supply/study files storage areas.

New satellite sites.

Personnel Changes

New investigator/subinvestigators (especially around June/July in major medical centers).

Change in Clinical Research Coordinator.

Change in pharmacists.

Change in Clinical Research Associate.

INVESTIGATOR'S MEETINGS

An investigator's meeting is often conducted at the beginning of a clinical trial to discuss the study and drug development plan or to review current data as a group. Both investigators and CRCs usually participate in these meetings. When the meeting is study specific, the format is similar to a prestudy site visit where the protocol and study design are discussed; study logistics may be addressed as a group, and CRFs may be reviewed.

Additionally, an investigator's meeting may be scheduled during the middle of a large trial for many reasons: to discuss findings and work through logistics, to assure consistency across sites, and to increase enthusiasm about the study.

STUDY PROCEDURES MANUAL

Many sponsors or coordinating CRCs will prepare a Study Procedures Manual delineating all of the specific procedures for a clinical trial. Generally, the procedures manual will include additional, more detailed information on conducting particular aspects of a study. If the sponsor does not provide a manual and you are coordinating a large trial with many interacting people, it may be a good idea to prepare a brief manual for all involved. Some topics to include are as follows:

Regulatory Requirements	A summary of regulations and/or a current copy of 21 CFR 50, 56, 312.
Entry Criteria	Inclusion/exclusion criteria (as a checklist or directly from the protocol or CRF).
Schedule of Visits	A graphic representation of scheduled study visits and evaluations (see Figure 8.1).
Study Schema	A schematic presentation of the study process (see Figure 8.2).
Protocol	A current copy of the protocol. Be sure to update this copy as amendments are received.
Special Tests	Instructions for conducting special tests or assessments.

Rating Scales	Any special rating scales, lesion scoring guidelines, and so on should be included.
Toxicity Grading Scale	The toxicity grading scale from the protocol should be used.
Completion of CRFs	Guidelines for completing CRFs or flow sheets with examples.
Investigational Agent	Guidelines on handling the investigational agent, completing accountability records, instructions to subjects.
Collection of Biological Samples	Special instructions for the collection, storage, and shipping of biological samples.
Serious Adverse Events	Information and instructions for reporting an SAE.
Subject Death	Measures to take in case of a subject death.

The study sponsor is charged by the FDA with the responsibility of monitoring the study's progress and the site's study activities. Monitoring occurs through personal site visits, telephone communication, and correspondence. Personal monitoring visits are usually performed by the sponsor's representative, the Clinical Research Associate. Site visits have specific purposes and agendas, depending on the stage of the study. The study site's participation in this process determines the success of the clinical trial.

BIBLIOGRAPHY

Bioresearch Monitoring: Regulation and Reality. *Applied Clinical Trials,* Vol. 4 (1), p. 36, 1995.

Collaborating with Colleagues (Part II): Collaboration Between the Monitor and the Research Nurse. Arna Shefrin, *Research Nurse,* Vol. 4 (4), pp. 9–10, 1998.

Have You Noticed That There Aren't Many Old Monitors Around? D. Cocchetto, *Journal of Clinical Research and Drug Development,* Vol. 1, pp. 87–89, 1987.

Interactions of Academic Investigators with Pharmaceutical Companies. In *Guide to Clinical Trials,* B. Spilker, Raven Press, New York, pp. 390–394, 1991.

Roles of Private Practice Physicians in Clinical Trials. In *Guide to Clinical Trials,* B. Spilker, Raven Press, New York, pp. 395–398, 1991.

Sponsor Education of Clinical Investigators in the Clinical Research Process. T. Kirsch, *Drug Information Association,* Vol. 22 (2), pp. 181–186, 1988.

The Interrelationships of Sponsors, Clinical Investigators, and Institutional Review Boards. T. Kirsch, *Drug Information Journal,* Vol. 21 (2), pp. 127–131, 1987.

Working with the Clinical Research Site: The Clinic's Perspective. K. Drennan, *Applied Clinical Trials,* Vol. 1 (5), pp. 62–65, 1992.

10

INTERACTIONS WITH THE STUDY SUBJECT

The purpose of conducting clinical investigations of a drug is to distinguish the effect of a drug from other influences, such as spontaneous change in the course of the disease, placebo effect, or biased observation.

—21 CFR 314.126(a)

The one indispensable element of any clinical trial is the study subject. This chapter will describe techniques of subject recruitment (identifying, screening, and enrolling eligible subjects), subject retention, and subject recognition and rewards. You must recruit only appropriate subjects into the study. Once enrolled, subjects must remain motivated to participate throughout the entire study.

Money, in the opinion of most researchers, is the main motivation for subject participation in clinical trials. In a survey done at Virginia Commonwealth University (*DIJ*, 1991), the main reason cited by healthy adults who volunteered to participate in Phase I clinical trials was money. (The main reason individuals declined participation in subsequent studies was schedule conflicts.) There are other reasons why people enroll in a clinical trial: free medical care, free medicine, no other treatment options, altruism, and education. Whatever the reason, it is important to keep the subject motivated throughout the entire process, from the recruiting phase through study completion.

SUBJECT RECRUITING

Subject Recruitment Plan

The first step in study subject recruitment is the creation of a Subject Recruitment Plan. With this plan, you will have a specific "road map" outlining the steps to be taken to enroll the number of subjects for whom you have contracted (and possibly a second contract for even more) in the shortest possible time.

Rapid enrollment into the clinical trial is important for scientific, statistical, and financial reasons. Scientifically, the research is designed to answer a specific question; delay may permit others to answer the research question and make the clinical trial irrelevant. The sponsor is often racing against other pharmaceutical companies to obtain approval for its drug and increasing market share and is counting on your ability to keep your contractual commitment to achieve your subject enrollment goals.

Statistically, in multicenter trials, subject recruitment should be equitable among sites across the differing geographical areas and across time. Failure to enroll as scheduled may jeopardize this subject uniformity. The longer you take to enroll subjects, the more likely it is that the baseline characteristics of your study subjects will differ from those of other sites or from those whom you enrolled at the beginning of the study.

Financially, since reimbursement is usually key to subject enrollment, recruitment delays result in negative cash flow to you, increased study costs, and increased time to complete the study. Inevitably, your site personnel will change over time. If your study enrollment is delayed, a change in research staff may occur. Replacement and training of research staff are expensive and time-consuming, increasing your study costs.

Furthermore, your compensation is usually based on the number of evaluable subjects enrolled. If enrollment is competitive, other sites may enroll sufficient subjects to complete the study before you have enrolled enough subjects to earn a sum equal to your expenses. Conversely, the more rapidly you enroll study subjects, the more income your site will earn.

The Subject Recruitment Plan should cover the following topics:

- Subject identification strategies.
- Submission of advertising copy and Informed Consent Forms to the Institutional Review Board (IRB).
- Communication systems.
- Subject data collection systems.
- Advertising programs.
- Speaking engagements.
- Office personnel assigned to subject screening.
- Subject screening procedures.
- Subject enrollment procedures.
- Subject incentive plans.

Subject Identification

Subjects must be contacted before they can be screened for eligibility. You can identify subjects in three areas: within your practice, within your medical community, and from the community at large.

Evaluate your practice. You should already have available, from your office computer, lists of every patient in your practice with a particular diagnosis, sorted by sex and age. (If you can't do this, then you'd better upgrade if you plan to participate in clinical research.) Have your office personnel call every patient in your practice who might be eligible, using a prepared script to avoid misunderstandings and protect patient confidentiality.

If your practice is not capable of providing all the needed study subjects, then you must market your study to the medical community and perhaps to the community at large. Solicit referrals from the local medical community by personally visiting physicians in your office or clinic complex and those primary-care physicians who have referred patients to you in the past (your referral base). Perhaps some of your colleagues would be interested in becoming subinvestigators and further help you with the study. As subinvestigators, they gain experience in the research process, making them eligible for studies of their own some day. Also as subinvestigators, they may become coauthors in any publications that you might generate as a result of study participation. They might also receive a portion of the investigator's budget, depending on the level of their participation. In any event, provide posters and leaflets for the office waiting areas of your colleagues. Provide study pocket cards for their use.

Be sensitive to the fact that not all of your colleagues will initially be convinced that you are acting purely out of zeal for the enhancement of medical knowledge. You can best obtain the cooperation and local referral support of these colleagues by assuring each physician that you will not "steal" patients; you will refuse to accept study subjects referred by colleagues into your practice. You should reiterate that, at the end of the study, each subject will be returned to the care of the original referring physician and that all subject information and laboratory data obtained during the trial will be shared with that referring physician.

Then, periodically throughout the study, call the local physicians personally either to thank them for subject referrals or to remind them about the study; send reminder postcards; follow up referrals with a thank-you letter and progress reports. Finally, at the conclusion of the study, refer the subject back to the referring physician and send a copy of all laboratory data and other subject information to that physician together with a final summary of the subject's progress during the study.

Advertising

The community at large is reached by advertising. Clinical research marketing techniques are identical to practice marketing techniques. The following common practice marketing techniques are also applicable to clinical research subject recruiting:

- Periodic practice newsletter and announcements in hospital and clinic newsletters.
- Office signs encouraging referrals from subjects.
- Speaking engagements and presentations to medical groups, lay groups, and, most rewarding, subject support groups.
- Presentations at your local hospital and medical societies.
- Public advertisements in newspapers, television, and radio.

You can describe the study in your practice newsletter. If you do not produce a newsletter, then a notice about the study may be inserted into your monthly billing notices. Posters should be installed in your offices and examining rooms to inform your patients about the opportunity to participate in the clinical trial. Provide "ready-to-use," study-specific copy to your local clinic and hospital newsletters. While they are marketing themselves, they are also providing advertising for your study that costs you nothing.

Speaking engagements will cause many potential subjects to contact your office. You can start by attending meetings of local support groups affected by the disease entity associated with your study. Often, merely showing interest is enough. Offer to give presentations on study-related topics to these groups, as well as to church groups, senior citizen centers, fraternal organizations, and community service organizations. Try to target groups that will contain potential subjects. These groups are often most eager to provide a forum for any speakers who will make the effort to prepare and present an informative talk.

A presentation to your hospital staff and to the local and county medical societies can be rewarding. Your colleagues will be interested in the "inside information" about the promising new therapy or procedure, and you will be recognized as a local leader in this field. Study subject referrals will follow.

Mass advertising for study subjects is often successful. Advertisements can take many forms: posters, leaflets or pamphlets, newspaper ads, study aids (e.g., a laminated study pocket card with information about the study eligibility criteria), educational material, and others (see the following). Here are some formats and locations that have been effective for other investigators:

- Posters in your waiting room.
- Posters or leaflets throughout your institution in elevators and at nursing stations.
- Posters in local pharmacies, medical supply stores, and medical equipment rental shops.
- A booth at health fairs.
- A pocket-sized card containing eligibility criteria and other study information to distribute to house staff, nurses, and other medical personnel who regularly come into contact with the target subject population.
- Ads in newspapers that target the study population.
- A public service announcement (free to you) that describes the study in your local paper, television, or radio.
- Local television or radio advertisements.

IRB Submission

While advertising will increase the number of potential study subjects that your staff will screen, care must be taken not to be coercive. REMEMBER, ALL ADVERTISING MATERIALS AND TEXT MUST BE SUBMITTED TO THE IRB FOR REVIEW AND APPROVAL. IRB approval must be obtained and documented *before* any advertising materials can be used. In addition to approving all advertising materials, the IRB also must approve the *Informed Consent Form*, a form that is written according to federal regulations to insure that all relevant risks and benefits are explained to the subject. Elements required for Informed Consent Forms are discussed in Chapter 4, and the informed consent process is discussed in Chapter 7.

Communication Systems

If your advertising activities are successful, you will have many potential study volunteers contacting your office. Is your current telephone system adequate to handle the extra load? Even two calls per hour will tie up one phone line continually, leaving one less line open for your current patients. Unless you already have a great deal of excess telephone capacity, you should consider obtaining a special "800" number during the period of study recruitment. Nothing discourages a potential study subject (or a potential practice patient) more than a constantly busy telephone line or dealing with an answering machine.

Study Volunteer Data Collection Systems

You want to capture and retain basic information about every volunteer who calls your study team. If the caller is not eligible for this study, he/she may be perfect for the next study, and you want to be able to call him/her back when that next study begins. The best system is a special computer database, with the study team member entering the basic data directly into the database while the caller is still on the line. Several established vendors will be delighted to demonstrate their software packages to you, allowing you to select the one that is best for you. A most valuable and inexpensive device is the telephone headset, which leaves both hands of the operator free to write or enter data on a keyboard.

Paper forms may be used, but they are much less efficient and are extremely expensive to search and sort by hand at a future date (hands are attached to very expensive humans).

If you do create a paper assessment system, medical history forms or special questionnaires that you develop for the study are also useful as screening tools. It may even be desirable to perform standardized personality assessments (especially if the investigational agent is a narcotic).

When screening subjects, it is important to get accurate information. Enrolling ineligible subjects is to no one's advantage and may even endanger the subject. Review the inclusion and exclusion criteria for the study with each subject. The potential subject should meet all of these criteria without exception. Often, a subject is "close" to meeting the criteria, and you may be tempted to request the sponsor to "relax" the criteria to enroll the subject. Remember to think of all the possible ramifications of enrolling a "borderline eligible" subject before doing so. Are you putting the subject in danger? At the end of the trial, after all your hard work, will the subject be unevaluable (and you unreimbursed)? A telephone consultation with the Project Medical Officer prior to beginning any enrollment procedures will be well worth the effort.

Office Personnel

Do not expect your office nurse to handle telephone responses to your campaign "in her spare time." How many minutes does a staff person need to screen a potential study subject during a typical telephone call? Many experienced research teams plan for 30 minutes per call, for establishing rapport and obtaining demographic information, concurrent conditions, concurrent medications, allergies, and subject availability throughout the trial. Less time per call is needed if you do not obtain all information, cutting off the subject as soon as an exclusion point is reached, but then you are wasting much of the investment you have already made in the advertising process. Realistically, 15–20 volunteer interactions per 8-hour day is about the best you can hope to achieve. Many experienced researchers will hire additional temporary staff to assist the permanent research staff during the screening process.

SUBJECT SCREENING

The collection of volunteer data information has already been discussed. Your screening staff must compare the protocol requirements against that data. The basic information needed to define an appropriate study subject is found in the protocol inclusion and exclusion criteria. By far, it is best to accomplish this screening while the subject is still on the telephone, to avoid another time-consuming telephone call. As an alternative, you might consider establishing a "prescreening visit," keying in on select criteria (e.g., key variable determinant of disease) that were difficult to assess over the telephone (such as standing blood pressure readings), to assess eligibility quickly before investing more time and money in the subject. Of course, potential study subjects are not charged for such visits.

SUBJECT ENROLLMENT

After you ascertain that the subject is eligible to participate, you are then ready to enroll the subject into the trial. You are required both by regulation and by medical ethics to fully inform subjects about participation in clinical trials. While it is *imperative* that informed consent be granted by the subject prior to any study-specific procedures or prior to receiving a test agent, it may be *preferable* to present the informed consent to the subject at the very outset of consideration so that the subject is aware that the process involves research and is informed of all available diagnostic and therapeutic options. You must explain the risks and benefits of participation and obtain the subject's informed consent to participate.

SUBJECT RETENTION

Now that you have identified eligible subjects, screened them efficiently and appropriately, and enrolled them into the study, your next task is to retain their interest to minimize premature withdrawal from the study. You also want to assure their compliance with the study requirements. Potential problem areas include the study visit requirements of the protocol, your office study milieu, adverse events, and boredom. The site research team, headed by the investigator, must anticipate these problems and minimize or avoid them.

Unpleasantries

The protocol may require procedures that by their very nature are unpleasant, such as gastroscopy or colonoscopy. You can minimize this drawback by your subject selection process; enroll only those subjects who seem willing to undergo the required procedures as often as necessary. With your IRB's permission, some financial compensation may be awarded to the subjects for their participation in particularly unpleasant or time-consuming study activities. One approach is to emphasize, to the subject, the benefit of these procedures, such as reassurance, evaluation of lack of disease progression, and optimal titration of therapy.

Adverse events are a part of life. Each of us gets headaches, body aches, bowel irregularities, and skin itching at one time or another. When we are sick, we can have lots of symptoms. Each symptom may become more frightening to the subject who is enrolled in a study. Additionally, nobody wants to feel "like a guinea pig." Take the time to discuss these feelings with your study subjects; put these feelings in their proper perspective, and the subjects will be more likely to stay the course.

Office Appearance and Amenities

Your office setting plays a big role in study subject retention (and in practice subject retention, as well). "Walk in the study subject's moccasins" for awhile by simulating the role of a study subject and observing your staff's interactions with you. Also observe their interactions with study subjects. Did you have adequate parking and an attractive, well lit, and adequately labeled entrance? Are study subjects greeted warmly and seen promptly? Do you have special waiting room chairs prominently reserved for study subjects? Are they given special preference in the lab, medical imaging, and pharmacy departments? Did your staff give each subject an attractive printed schedule of the day's study activities well in advance of the visit? Does each study subject know how long the study visit will take and how to best prepare for the visit procedures? If you answer "no" to any of these questions, then your subject retention rate is in jeopardy.

Boredom

Boredom is another inevitable consequence of modern life, both for the study subjects and for your study staff. You, the investigator, are the best source of enthusiasm. The study staff will emulate your attitudes and your behavior; remain enthusiastic!

You can keep the study excitement alive and increase study completion rates by increasing the contacts between your study staff and the subject. Newsletters, postcards, phone calls, and subject diaries are techniques for keeping the study and its importance fresh in the minds of your study subjects. Try rotating the tasks of study personnel; a fresh approach to the study may renew enthusiasm.

Recognition and Rewards

We all want to be recognized for the good work that we do and to be rewarded for behavior that fosters the public good. Study subjects are no exception to this generalization of human nature, and you should insure that their need for recognition and reward is well satisfied.

Subject rewards begin at the moment of study enrollment; show your pleasure and appreciation by smiling and saying "thanks for participating." If the subject is taking time off from work, or has to hire a baby-sitter, or takes public transportation to the study site, you should consider reimbursing the subject up to the amount permitted by your IRB (that sum above which is considered undue coercion). You can also provide free drinks, or, better yet, snacks or a light meal, to the study subjects who participate in clinical studies, particularly to those who must fast for blood chemistry testing.

Consider giving simple mementos to your study subjects, such as imprinted pens, carrying bags, shirts, or caps. Such gifts can be used both as motivating items to increase retention and as gifts to express gratitude for successful participation. Many sponsors will obtain modest gift items from their marketing departments for you to share with the study subjects.

No clinical trial will succeed without those courageous souls, the study subjects, who, on your recommendation, take the study drug or use the study device. The subjects must be imaginatively contacted, accurately screened, and rapidly enrolled in complete accordance with FDA regulatory requirements. The study subject must then be motivated to continue study participation until the study is completed. By performing these activities efficiently and competently, you will help assure a successful clinical trial.

BIBLIOGRAPHY

A Research-Based Approach to Patient-Focused Subject Recruitment. James A. Weinrebe, *Applied Clinical Trials,* Vol. 7 (11), p. 56, 1998.

Benefit/Risk Assessment: Perspective of a Patient Advocate. A. Bowen, *Drug Information Journal,* Vol. 27, pp. 1031–1035, 1993.

Effective Subject Recruitment. Jennifer Westrick, *Applied Clinical Trials,* Vol. 6 (7), p. 41, 1997.

Essential Information to Be Given to Volunteers and Recorded in a Protocol. D. Jackson and G. Richardson, *Journal of Pharmaceutical Medicine,* Vol. 2, pp. 99–101, 1991.

Ethical Issues in Clinical Research: An Investigator's Perspective. G. Gillett, *Applied Clinical Trials,* Vol. 1 (5), pp. 66–68, 1992.

Factors That Motivate Healthy Adults to Participate in Phase I Trials. M. A. Kirkpatrick, *Drug Information Journal,* Vol. 25 (1), pp. 109–113, 1991.

HHS Recommends Subject Recruitment Changes. Jill Wechsler, *Applied Clinical Trials,* Vol. 9 (8), p. 20, 2000.

How Patients Become Subjects. John R. Wilson, Jr., *Applied Clinical Trials,* Vol. 7 (1), p. 10, 1998.

Impact of Risk Communication on Accrual, Regimen, and Follow-Up Compliance. L. Morris and I. Barofsky. In *Patient Compliance in Medical Practice and Clinical Trials,* edited by J. A. Cramer and B. Spilker, Raven Press, New York, 1991.

Industry Reimbursement for Entering Patients into Clinical Trials: Legal and Ethical Issues. D. Shimm and R. Spece, *Annals of Internal Medicine,* Vol. 115, pp. 148–151, 1991.

Internet Subject Recruitment. Ann T. Ken, *Applied Clinical Trials,* Vol. 7 (2), p. 52, 1998.

Improving Patient Compliance in Clinical Trials: A Practical Approach. Joanne E. Karvonen, Meri Hauge, Julie Levin, and Hanna Bloomfield Rubins, *Research Nurse,* Vol. 2 (4), pp. 9–12, 1996.

Methods of Assessing and Improving Patient Compliance in Clinical Trials. B. Spilker. In *Patient Compliance in Medical Practice and Clinical Trials,* edited by J. A. Cramer and B. Spilker, Raven Press, New York, 1991.

Patient Compliance in Clinical Trials. John T. Fowler and Robert E. Hauser, *Applied Clinical Trials,* Vol. 4 (3), p. 62, 1995.

Patient Recruitment and Enrollment into Clinical Trials. J. Swinehart, *Journal of Clinical Research and Pharmacoepidemiology,* Vol. 5, pp. 35–47, 1991.

Part 1: Reimbursement for Patient Costs. Caroly Petersen, *Applied Clinical Trials,* Vol. 5 (6), p. 72, 1996.

Professional Phase I Clinical Trial Participants. Joe E. Scarborough, Philip M. Brown, and Joseph M. Scavone, *Applied Clinical Trials,* Vol. 4 (10), p. 38, 1995.

Recruiting Patients for Clinical Trials. Jill Wechsler, *Applied Clinical Trials,* Vol. 6 (6), p. 20, 1997.

Recruitment for Volunteers for Phase I and II Drug Development Trials. B. DeVries, G. Hughes, and S. Francom, *Drug Information Journal,* Vol. 23 (4), pp. 699–703, 1989.

Recruitment of Subjects in Clinical Trials. Kathleen Steger, *Research Nurse,* Vol. 3 (6), pp. 1–12, 1997.

Screening for Illicit Drug Use in Drug Development Studies. B. DeVries, G. Hughes, and L. Huyser, *Drug Information Journal,* Vol. 25 (1), pp. 49–53, 1991.

The Rights of Cancer Patients. B. Atwell, *Arizona Counseling Journal,* Vol. 10, pp. 67–76, 1985.

11

ADVERSE EVENTS AND DRUG SAFETY

An investigator shall promptly report to the sponsor any adverse effect that may reasonably be regarded as caused by, or probably caused by, the drug. If the adverse effect is alarming, the investigator shall report the adverse effect immediately.

—21 CFR 312.64(b)

A major objective of most clinical trials is the assessment of the safety of investigational agents. Safety is assessed by the reporting of serious adverse events (SAEs; adverse experiences or side effects) as well as the assessment of changes in clinical laboratory results, electrocardiogram (ECG) patterns, subject physical examinations, and vital signs. Serious and unexpected adverse events are subject to prompt Food and Drug Administration (FDA) reporting requirements.

ADVERSE EVENTS

The FDA requires the reporting of adverse events of investigational agents to assure subject safety. The pharmaceutical sponsor who is dependent on reports from the investigative site generates these reports. Most common adverse events are collected on the Case Report Form (CRF) and submitted to the sponsor for inclusion in the New Drug Application (NDA) and ultimately the package labeling. For those adverse events defined as "serious and unexpected," the FDA accepts narrative reports but prefers reports submitted on a MedWatch form. They will also accept reports on a CIOMS form, which is the form used in many international studies and recognized by the International Conference on Harmonisation (ICH). These reporting requirements are discussed later in this chapter.

The NCI has adopted a system called AdEERS (Adverse Event Expedited Reporting System) as a Web-based system designed to allow cooperative groups, cancer centers, and single institutions to submit expedited reports for serious and or unexpected events to the NCI for all trials using NCI–sponsored investigational agents.

Table 11.1 provides definitions of adverse events/adverse experiences as defined by different regulatory agencies.

TABLE 11.1 DEFINITIONS OF ADVERSE EVENTS

Source	Terminology	Definition
21 CFR 310.305 (21 CFR 312.32 for discussion of serious adverse events)	Adverse Drug Experience	"Any adverse event associated with the use of a drug in humans, whether or not considered drug related, including the following: An adverse event occurring in the course of the use of a drug product in professional practice; an adverse event occurring from drug overdose whether accidental or intentional; an adverse event occurring from drug abuse; an adverse event occurring from drug withdrawal; and any failure of expected pharmacological action." The FDA defines both serious and life-threatening adverse drug experiences in 21 CFR 312.32.
ICH GCP	Adverse Drug Reaction	". . . all noxious and unintended responses to a medicinal product related to any dose should be considered adverse drug reactions. . . ."
ICH GCP	Adverse Event	"An AE is any untoward medical occurrence in a patient or clinical investigation subject administered a pharmaceutical product and that does not necessarily have a causal relationship with this treatment. An AE can therefore be any unfavorable and unintended sign (including an abnormal laboratory finding), symptom, or disease temporally associated with the use of a medicinal (investigational) product, whether or not related to the medicinal (investigational) product."
NIH Guidelines, 1/01	Adverse Event	Any unfavorable and unintended sign (including abnormal laboratory findings), symptom, or disease temporally associated with the use of a medical treatment or procedure regardless of whether it is considered related to the medical treatment procedure (attribution of unrelated, unlikely, possible, probable, or definite).
OHSR Information Sheet #17	Adverse Event	Any unfavorable and unintended diagnosis, symptom, sign (including an abnormal laboratory finding), syndrome, or disease that either occurs during the study, having been absent at baseline, or, if present at baseline, appears to worsen.

DEFINITIONS

The pharmaceutical industry uses specific terms to describe adverse events, and the definition of these terms in this context may vary somewhat from common usage. Since you can communicate effectively in clinical research only by using terms in their appropriate context, the specific definitions of terms commonly used in drug safety assessment follow.

Event	A single sign or symptom, a set of related signs or symptoms, a disease entity, an exacerbation of a preexisting condition, or a recurrence of an intermittent condition or disease. For example, an inadvertent ingestion of an extra dose of medication that resulted in no adverse signs or symptoms is not an "event." A surgical procedure is not an "event"; the medical condition that caused the procedure to be necessary is the "event."
Adverse	A noxious or pathological change as compared to preexisting conditions, any undesirable medical experience, an untoward event, whether or not related to the investigational agent. Therapeutic failures are usually recorded elsewhere in the CRF as "lack of efficacy" information, rather than as information related to drug safety.
Concurrent Conditions	Other, nonprotocol-related medical conditions that are present during the study period.
Concurrent Medications	All medications that were administered at the onset, or within 30 days (depending on the design of the protocol) of onset, of the adverse event that are not investigational agents.
Concurrent Therapy	All other therapies administered at the onset, or within 30 days of onset, of the adverse event that are not investigational agents.
Concomitant Therapy	All therapies taken as part of the study that are not investigational agents or therapies.

Test Article or Treatment Therapy

The investigational agent or study drug. Any drug (including a biological product for human use), medical device for human use, human food additive, color additive, electronic product, or any other article subject to regulation under the Federal Food, Drug, and Cosmetic Act or under sections 351 and 354–360F of the Public Health Service Act.

Action Taken

Treatments administered to ameliorate an adverse event, including investigational agent withdrawal and administration of other therapies.

Treatment-Emergent

An acute event that first manifests itself only after the initial administration of investigational agent, or a chronic event that was quiescent at baseline and has recurred since the initial administration of investigational agent, or a chronic condition that was present at baseline that has now worsened since the first administration of the investigational agent.

Serious

A *serious adverse drug experience* is defined in the Code of Federal Regulations (CFR), Title 21, Part 312.32 (see Appendix A), as "any adverse drug experience occurring at any dose that results in any of the following outcomes:

• death,
• life-threatening (i.e., the subject was at immediate risk of death as the event actually occurred, not as it might have occurred in a more serious form),
• inpatient hospitalization or prolongation of existing hospitalization,
• persistent or significant disability/incapacity, or
• congenital anomaly/birth defect."

Expected

An adverse event that has been previously identified in the Investigator's Brochure or package insert; an adverse event that has been identified on the basis of prior experience with the drug under investigation or with related drugs as associated with the investigational agent.

Unexpected   Any adverse drug experience, the specificity or severity of which is not consistent with the current Investigator's Brochure or other risk information.

Causality   The assessment of a possible causal relationship between the adverse event and the investigational agent. In this assessment, factors to consider include are as follows:

- Clinical condition: perhaps the adverse event is known to occur in subjects with the study's clinical disease state or condition.
- Known relationship: perhaps the event is "expected," as previously defined.
- Temporal relationship: perhaps the event began only after administration of the investigational agent and/or ceased after withdrawal of the agent.
- Concomitant condition: perhaps another condition provides a more conclusive and reasonable explanation of causality or is known to produce the adverse event.
- Concurrent medication: perhaps the event is known to be associated with a concurrent medication.
- Rechallenge: perhaps the adverse event ceased upon withdrawal of the investigational agent, only to recur with rechallenge of the investigational agent.

Associated   There is a reasonable possibility (or greater) that the adverse event may have been caused by the investigational agent.

Severity   The degree of interference with the subject's activities of daily living, usually divided into *mild*, *moderate*, or *severe* or defined by a toxicity grading scale. A detailed grading scale, Common Toxicity Criteria (CTC), is available at http://ctep.info.nih.gov.

Onset Date   The date on which the event signs or symptoms first manifested themselves, even though the diagnosis may have been made only in retrospect on some later date.

Resolution Date   The date on which the adverse event resolved, or, if the adverse event concerns a chronic condition, the date on which the adverse event was determined to be stable or to have returned to the baseline state.

ASSESSMENT OF ADVERSE EVENTS

Most protocols require the investigator or Clinical Research Coordinator (CRC) to inquire about adverse events at each subject visit. In some studies, periodic telephone communications with the subject also serve as opportunities to inquire about adverse events. Every effort should be made to avoid any suggestion that any particular event is expected. To avoid introducing bias, the inquiry should be open-ended and unfocused, such as "Have you had any medical problems since your last visit?"

Most protocols also require the investigator or CRC to inquire about new concurrent medications since the last visit. Such an inquiry will not only protect the subject by possibly ascertaining the inadvertent prescription by a nonstudy physician of a drug that is incompatible with the study medications, but it also serves to reinforce the inquiry into unreported adverse events. For example, if a subject reports taking an analgesic, the study personnel should ask the subject about pains that might indicate a previously unreported adverse event.

Clinical laboratory reports should be compared with the baseline reports at each interval throughout the trial. The emergence of an abnormal result might indicate the presence of a previously unreported adverse event.

Interim and final study physical examinations should be performed with the intent of comparing all findings to the comparable baseline state. The appearance of a mass, or petechiae, or a paralysis, or degradation of auditory or visual acuity should be reported as an adverse event, even if the subject was unaware of the change.

RECORDING ADVERSE EVENTS

Since the primary duty of the investigator is to care for the subject in accordance with Good Clinical Practices (GCPs), each adverse event should be recorded on the subject chart, together with an assessment of causality and a course of treatment or action, if appropriate. Each adverse event should also be recorded on the appropriate CRF. Each adverse event should be recorded IN THE APPROPRIATE MEDICAL TERMINOLOGY.

Some sponsors provide a dictionary of acceptable terms. Other sponsors expect the investigator to use terms that can be found in a medical dictionary. Avoid the use of such phrases as "locked bowels," "pain in chest," or "bad back"; such use is indicative of both unprofessional clinical practice and inadequate research work. Much more appropriate is the use of such diagnostic terms as "constipation" (or "diarrhea," if the bowels were "locked open"), "angina" (or "musculoskeletal chest pain"), or "osteoarthritis of the lumbar spine."

SERIOUS ADVERSE EVENT REPORTING

By regulation, any adverse event that suggests a significant hazard, contraindication, side effect, or precaution, particularly any that meet the definition given above, must be reported by the study sponsor to the FDA. If the event is "serious," associated with the use of the investigational agent, and "unexpected" as defined above, then the event must be reported to the FDA by the study sponsor within 15 calendar days. Fatal or life-threatening events that are also "unexpected" and "associated" must be reported to the FDA by telephone or fax within 7 calendar days, followed by a written report within 15 days. Most sponsors require the investigator to report SAEs immediately upon discovery, so that evaluations of the event can take place prior to FDA notification and still remain within the regulatory time limits. The report may be submitted to the sponsor using Form FDA 3500A (MedWatch Form at fda.gov/medwatch) or a CIOMS I form. SAEs in NIH studies can be reported under the AdEERs system electronically at http://webapps.ctep.nai.nih.gov. Additionally, investigators must report SAEs to the Institutional Review Board (IRB).

SAE CHARACTERISTICS	FDA REPORTING REQUIREMENTS
Serious Unexpected Associated with study agent	Report in writing to FDA within 15 calendar days
Serious Unexpected Associated with study agent FATAL OR LIFE–THREATENING	Report to FDA by telephone or fax within 7 calendar days

Table 11.2 summarizes the reporting forms for adverse events.

TABLE 11.2 REPORTING FORMS FOR ADVERSE EVENTS AND CONTACT INFORMATION

Reporting Form	Use	Source
MedWatch Voluntary FDA Form 3500 Mandatory FDA Form 3500A	FDA Used for reporting adverse events in clinical trials as well as voluntary reporting of approved products	http://www.fda.gov/medwatch/
VAERS Vaccine Adverse Event Reporting System	FDA Reporting adverse events occurring with the use of vaccines	1-800-822-7967
CIOMS Council for International Organizations of Medical Sciences	International filings	
AdEERS	NCI Reporting of adverse events occurring in NCI and other cooperative group studies	http://cancertrials.nci.nih.gov

MEDICAL MANAGEMENT OF AN ADVERSE EVENT

The primary duty of a physician is to provide the best care for the subject. Therefore, the investigator should treat the adverse event in accordance with best clinical judgment, regardless of the effect that either the event or the treatment will have on the integrity of the protocol. Some investigational agents have a known toxicity, and the protocol may specify the appropriate treatment. In other cases, where the investigational agent is implicated in the causality of the event, the sponsor should be contacted to inquire about any previous reports of this event or newly discovered therapies. If further diagnostic tests are necessary for the treatment of the study subject, they should be performed; if specialist consultations are indicated, they should be obtained.

If, on the other hand, further diagnostic tests or consultations are needed solely to assess causality more accurately, then the study sponsor should be contacted in advance and full agreement should be reached on the extent of such testing and the assumption of test costs, prior to the performance of such tests or consultations. If the investigational agent is implicated in the etiology of the adverse event, and the subject's safety and well-being would be enhanced by cessation of study therapy, then the study drug should be withdrawn.

UNBLINDING—IDENTIFYING THE TREATMENT REGIMEN IN A BLINDED STUDY

Occasionally, in studies in which the investigational agent is unknown to the investigator (double-blind studies), the appropriate treatment for an adverse event cannot be discerned in the absence of knowledge of the identity of the ingested drug. Such events are rare but do occur. For example, if a subject in a study of mild pain is taking either acetaminophen or placebo, and a child ingests many study tablets, the appropriate treatment of that child will vary, depending on whether the child had swallowed acetaminophen or a harmless placebo. In such cases, the Project Medical Officer should be consulted immediately, and the identity of the investigational agent ascertained. On the other hand, in cases where the treatment will be unchanged regardless of the identity of the specific treatment agent, unblinding the treatment is unnecessary. For example, if the investigational agent is either of two macrolide antibiotics, then the precise identity is of little importance in selecting the appropriate therapy for the adverse event of diarrhea, and the investigator should remain blinded to the precise identity of the investigational agent until the study has been completed. Subjects are usually considered to be unevaluable if the identity of the investigational agent has been ascertained during a blinded study, and the subject is customarily withdrawn from the study. In addition, FDA inspectors have been known to pay particular attention to sites where the blind has been prematurely broken.

ADVERSE EVENTS AND
CONTINUED STUDY PARTICIPATION

Once the study subject has been appropriately diagnosed and treated, the question then arises: May the subject be retained in the study? As a general rule, a study subject is retained in the study unless

- further participation would put the study subject at risk,
- retention would expressly violate the study protocol procedures,
- the adverse event (or its treatment) renders the subject unable to further participate,
- further participation would confound study results, or
- the subject wishes to withdraw from the study.

Consultation with the Project Medical Officer is appropriate to clarify the issues and resolve this question.

ADVERSE EVENT TREATMENT COSTS

Nobody wants to pay for medical care these days. Costs related to adverse events associated with the investigational agent are often treated differently from costs related to adverse events that occur independent of the study (automobile accidents, for example). Many sponsors will willingly reimburse the investigator for costs incurred during the evaluation and treatment of adverse events that are considered to be related to the investigational agent, but much less willing to reimburse the investigator for costs incurred as a result of adverse events considered to be unrelated to the investigational agent. The payment for such tests, consultations, and treatments of adverse events is a matter of contract, and complete agreement between the sponsor and the investigator should occur before the study contract is signed. The study sponsor usually addresses the topic of reimbursement for study-related adverse event treatment in the Clinical Trial Agreement (CTA). The IRB usually insists on a clear statement of cost allocation in the informed consent, since such costs are included among the risks of study participation. Because the IRB is charged with the protection of human subjects, the IRB will attempt to protect the subject from incurring added costs by volunteering to participate in a clinical trial. The investigator will try vigorously to avoid incurring these costs, or, at least, negotiate an increase in the investigator's grant sufficient to cover most anticipated adverse event treatment costs.

The skilled investigator who is aware of the physician's responsibilities in the area of adverse event reporting will be able to protect the best interests of the study subject while at the same time protecting the integrity of the study.

REGULATORY REFERENCES

Serious Adverse Event Reporting

21 CFR 312.32, 312.64, and 310.305.

FDA Guideline for Reporting ABRs (21 CFR 600.80) (3/90).

Compliance Program Guidance Manual: Enforcement of the Adverse Drug Experience Reporting Regulations (5/91).

NIH Guidelines, January 2001 (www.nih.gov).

Postmarketing Reporting

21 CFR 314.80.

Guideline for Postmarketing Reporting of Adverse Drug Experiences (3/92).

BIBLIOGRAPHY

A Systematic Approach for Handling Adverse Events. H. Nowak, *Drug Information Journal,* Vol. 27, pp. 1001–1007, 1993.

Adverse Drug Events: Identification and Attribution. A. Smith Rogers, *Drug Intelligence and Clinical Pharmacology,* Vol. 21, pp. 915–920, 1987.

ASHP Guidelines on Adverse Drug Reaction Monitoring and Reporting. *American Journal of Hospital Pharmacy,* Vol. 46, pp. 336–337, 1989.

Clinical Trial Adverse Events: The Case for Descriptive Techniques. W. Huster, *Drug Information Journal,* Vol. 25 (3), pp. 457–459, 1991.

Guidelines for the Management of Adverse Events Occurring During Clinical Trials. C. Benichou and G. Danan, *Drug Information Journal,* Vol. 25 (4), pp. 565–571, 1991.

Informing Subjects of Adverse Effects. Slobodan M. Jankovic, *Applied Clinical Trials,* Vol. 6 (3), p. 58, 1997.

Introducing MEDWatch: A New Approach to Reporting Medication and Device Adverse Effects and Product Problems. D. A. Kessler for the Working Group. *JAMA,* Vol. 269 (21), 1993. Reprinted in *Journal of Clinical Research and Drug Development,* Vol. 7 (3), 1993. (Reprint requests to Commissioner of Food and Drugs, FDA, 5600 Fishers Lane, Rockville, MD 20857.)

Reasonable Possibility: Causality and Postmarketing Surveillance. J. Johnson, *Drug Information Journal,* Vol. 26 (4), pp. 553–558, 1992.

Recognizing and Reporting Adverse Events. Alan Sugar, *Research Nurse,* Vol. 4 (3), pp. 1–7, 1998.

Some Observations on the Collection of Medical Event Data. N. Mohberg, *Drug Information Journal,* Vol. 21, pp. 55–62, 1987.

12

DATA MANAGEMENT

An investigator is required to prepare and maintain adequate and accurate case histories designed to record all observations and other data pertinent to the investigation on each individual treated with the investigational drug or employed as a control in the investigation.

—21 CFR 312.62(b)

Clinical trials in human subjects are performed *in addition to* the physician-patient interaction, *not* as a substitute. The underlying premise is that the patient is treated, and information is recorded on the office charts, as if the study did not exist. The patient's case history should contain adequate and accurate observations and other data pertinent to the study, and it remains the property of the practitioner or the hospital. Case Report Forms (CRFs) are used to record the data required by the protocol and to submit the data to the sponsor for data entry and analysis.

CASE HISTORY RECORDS

The case history records should contain the following:

- Basic subject identification information.
- Information showing that the subject met the study selection criteria.
- Sufficient information to support data on the CRF.
- Information describing the subject's exposure to the test article, including the date and time of each administration and the quantity administered.
- Copies of CRFs submitted to the sponsor.

The sum total of a practitioner's (or hospital's) case history records, including observations and consultation reports and laboratory, electrocardiogram (ECG), and radiology reports concerning a patient are called the *source documents* because they are the source of all information extracted for clinical research purposes. In addition, any document where information concerning a subject is initially recorded also becomes a source document. These records should remain in the patient's medical file under the supervision of the physician investigator. (See the discussion on case histories in Chapter 2.)

In theory, when a patient volunteers to become a subject in a clinical trial, the desired study-specific data are then extracted from the patient's medical record (the "source" of research data) and recorded on separate forms, called Case Report Forms. In contrast to the patient's case history records, the CRFs are part of the clinical study and remain the property of the study sponsor, although copies must be retained at the research site.

In practice, the process is often more complicated. When a patient appears for treatment or evaluation at a research site without any prior relationship with the physician investigator, both a physician medical record and a CRF should be created and maintained as two separate and distinct files. When the patient initiates a professional relationship solely for the purpose of study participation, and maintains a professional relationship with another physician for the delivery of nonstudy-related medical care, the patient medical record and the CRF may contain identical information (the investigator may choose to collect only the information required by the study protocol). Nevertheless, the investigator must always remember that, for the duration of the clinical trial and for the treatment or evaluation of the clinical entity under study, the investigator has accepted a full physician-patient relationship with that research subject, with all the responsibilities that the relationship implies. This topic has been delineated in the FDA Information Sheet "Record Keeping for Clinical Investigations" (10/95).

The site monitor, or Clinical Research Associate (CRA), is employed by the sponsor to verify that the information contained in the CRFs is recorded accurately and supported by the source documents. In addition, the monitor is charged with the responsibility of ascertaining that no information appears in the source documents that would cast any doubt on study eligibility or participation of the study subject. The Food and Drug Administration (FDA) inspector also will review this information to assure validity of the data and to ascertain that the patient truly existed and was available for the study.

PREDESIGNED CASE REPORT FORMS

Guidelines

In sponsored studies, the protocol and data collection tools (CRFs) are created either by the sponsor or under the direction of the sponsor, and the research site usually has little control over the appearance and content of the CRFs. The site should therefore make every effort to ascertain the rationale and data entry expectations of the sponsor for every item (data field) on the CRF. In a multicenter study (when uniform data collection among sites is paramount), the experienced sponsor will prepare an instruction manual, or completion guideline manual, for use at each site. This manual will contain the definition, with examples, of every line and every data entry field in the entire CRF. In the absence of such a manual, the site study team should make every effort to obtain this information from the study monitor.

Source Document Design

The investigator is required by regulation [21 CFR 312.62(b)] to prepare and maintain adequate and accurate case histories designed to record all observations and other data pertinent to the investigation on each individual treated with the investigational agent or employed as a control in the investigation.

Once the site has a clear understanding of the data to be collected for the study, the site may create study-specific source documents to insure the accurate and complete capture of this information at every study subject interaction. Physicians have been using preprinted forms to collect patient historical and current medical information for generations, as well as preprinted checklists for the performance of physical examinations. As long as the CRF is not merely copied for source document use, preprinted forms are a perfectly acceptable technique for recording patient care information, having successfully withstood many FDA audits. Such source document data collection forms, of course, must be designed to collect ALL the information needed for completion of the CRFs; anything less is not worth the time it takes to create them.

Corrections

"To err is human; to correct divine." Even the most conscientious and precise Clinical Research Coordinator (CRC) will need to make corrections to the CRF, either because of error or because erroneous information was initially provided. We all have met patients who cannot remember the precise name of their medications, or the year of their hysterectomy, or even their birthday. It is normal and customary to correct the CRF when doing so will improve the accuracy of the data collected.

Since the CRF is part of the study record, remains the property of the sponsor, and is subject to audit, the correction must be performed in a manner that will permit an inspector to understand the change made and to identify the prior erroneous entry. The inspector will want to know the following:

- Who made the change on the CRF.
- When the change was made.
- What was the original entry.
- What is the new information.
- That the change is present on all copies.

These tasks may be accomplished by the following:

- Placing your initials next to the correction.
- Dating the correction.
- Drawing only a single thin line through the previous entry, leaving it legible.
- Inserting the new information above or below the original entry.
- Using a black ballpoint pen to ensure clear copying.

Document Storage

The investigator's copy of the CRFs and the source documents from which they are created must be retained by the investigator in accordance with the regulations. An investigator must retain records required to be maintained under this part of the regulation [21 CFR 312.62(c)] for a period of two years following the date that a marketing application is approved for the drug for the application for which it is being investigated, or, if no application is to be filed or if the application is not approved for such indication, until two years after the investigation is discontinued and the FDA is notified.

The U.S. investigator should be prepared to store his/her medical records (case histories) indefinitely, since no duplicates exist. If case histories are stored by an institution, the investigator should make an effort (by bright-colored labeling and binding) to prevent alteration or premature discarding of these essential research documents.

The original CRFs are stored by the sponsor, and the investigator may eventually dispose of the site copies of these documents after the required retention period has passed. Even then, the cautious investigator will consult the study sponsor before disposing of any documents pertaining to a clinical trial.

In Europe, the topic of case history storage is even more confusing. The European Union Good Clinical Practice Guideline stipulates a retention period of at least 15 years for subject records. However, Directive 91/507/EEC deleted the 15-year rule and left the retention period as "the maximum period of time permitted by the hospital, institution or private practice." In effect, local practice still governs document storage in Europe, and loss of essential documents may well occur.

Electronic CRFs

The increasing sophistication of computers, modems, and software has resulted in the creation of a number of systems permitting electronic data entry directly at the site. The best of these systems permits the CRC to bypass paper copies completely and enter the study data directly into the computer, using predesigned screens and specialized software. The data entry program can be preset to accept only entries that fall within the protocol-specified parameters and will refuse to accept data that are "out of range." For example, if the protocol restricts study eligibility to subjects only over the age of 18, the computer will refuse to accept any birthdate that is less than 17 years and 365 days prior to the current date. Such specialized data entry programs force the CRC to correct obvious errors prior to data entry, a luxury not possible with paper CRFs, and frees the study monitor from basic data checks and permits the performance of higher level tasks.

When the study computer is connected to a telephone line, an additional feature found in some site data entry programs permits the transfer of study data to occur nightly, allowing the sponsor to obtain enrollment data and raw study data within 24 hours of data collection. The data should be printed on the traditional paper format for long-term storage and for auditing purposes. If the study computer is returned to the sponsor and no paper copies exist at the site, the inspector may not be able to "read" the diskette or other electronic storage medium, leaving the inspector with no method of evaluating the accuracy of the study files. Additionally, information stored on diskettes may well deteriorate prior to the expiration of the required document storage date, putting the site in jeopardy in the event of an FDA audit.

A study site computerized data entry system has the advantages of speedy data collection by the sponsor and the performance of basic electronic cross-checks and specification checks at the time of data entry by site personnel. The disadvantages include the considerable cost of equipment and telephone lines, the extra time required to train site personnel and study monitoring personnel, and the vague concern that FDA auditing might be compromised in an electronic data collection system. The cost alone has prevented many sponsors from instituting such a system. Additionally, federal regulation for electronic signatures apply to such systems (refer to Chapter 4).

SITE–DESIGNED CASE REPORT FORMS

Self-Sponsored Studies

For those studies where the investigator is also functioning as the sponsor (the sponsor-investigator), or in single-site studies where the sponsor contracts with the site for additional services, the investigator may have greater control over the appearance and content of the CRFs. The following discussion will aid such an investigator in this task.

As a basis for beginning the design of CRFs, the protocol is thoroughly scrutinized. Every bit of information addressed in the protocol should be captured in the corresponding CRF, including demographic information and study eligibility information. The efficacy data, safety data, and any other parameters mentioned in the protocol are reflected for capture in the CRF.

As a matter of comfort and convenience to the person who is recording the data, the forms should be pleasing to the eye, with easily read type fonts of large size, adequate space for the entry of required information, and a logical progression from the first form to the last, and from the top of each page to the bottom. Consistency in the location of key items among pages will improve the accuracy of completion, as will the insertion of directions and completion advice at every step.

Several basic rules should be observed.

- Do not capture any nonessential information.
- Use innovative design to avoid requiring the user to copy the same information twice.
- Leave abundant "white space" to focus the eye of the user on the essential fields.
- Use precision in language, preferably employing the same terminology in both the protocol and CRF.

COMMON PROBLEMS

The audit assists the investigator to understand what is expected and to detect deficiencies in procedures and in data collection and transcription. Below are some common case history and CRF deficiencies.

Incomplete Case Histories   The source documents, or the patient medical records, should contain ALL the information that is captured on the CRF. After all, the medical record is the source of all the information recorded on the CRF. Too often, the study staff will record information directly onto the CRF and forget to first record the information onto the medical chart. Although the FDA may permit, in some circumstances, data to be recorded directly onto CRFs, the result is an incomplete record.

Study Eligibility Criteria   The inclusion and exclusion criteria as written in the protocol specify the study eligibility criteria. Each item should be present in the source documents, to enable a monitor to verify, from the source documents, that the study subject is indeed eligible for study entry.

Concurrent Medications   The source documents often contain notations concerning concurrent medications that are not transcribed onto the CRF. A well-organized medical record, with a special and distinct location for the recording of all patient medications together with indication for use, will ease the CRC burden of complete concurrent medication reporting.

Adverse Events   These events, if of mild and transient nature, are often not recorded in the patient's medical record. Nevertheless, the CRF must reflect the source documents IN EVERY PARTICULAR, and any concomitant medication or adverse event present on the CRF must be found in the source documents.

Illegible Data   Physicians are notorious for their illegible handwriting. While this may be amusing in the abstract, it becomes frustrating or worse when the records are audited. To avoid any possible errors, each entry in both the source documents and the CRF must be clearly legible. If necessary, dictated office notes may be needed to avoid research errors or censure.

The results of the study will only be as good as the validity of the data. The data collection process must be logical and precisely planned, capturing only the data needed for the study analysis and no more. The data should also be captured in a legible and easily understood format.

REGULATORY REFERENCES

21 CFR 312.62.
21 CFR 11.

Guidelines

Required Recordkeeping in Clinical Investigations (10/95).

Compliance Program Guidance Manual (Clinical Investigators) (8/94).

Guide for Detecting Fraud in Bioresearch Monitoring Inspections (4/93).

See also references for site inspections.

BIBLIOGRAPHY

A Multidisciplinary Approach to Data Standards for Clinical Development. Rebecca Kush, Wayne Kubick, Kaye Fendt, Dave Christiansen, and Judith Sromovsky, *Applied Clinical Trials,* Vol. 9 (6), p. 76, 2000.

Auditing Clinical Data. H. Ruth Pyle, *Applied Clinical Trials*, Vol. 9 (5), p. 65, 2000.

CRF Design and Planned Data Capture. Wendy Bohaychuk, Graham Ball, and Katy Sotirov, *Applied Clinical Trials,* Vol. 8 (6), p. 64, 1999.

Data Collection Forms in Clinical Trials. B. Spilker, Raven Press, New York, 1991.

Designing Case Record Forms for the World Wide Web. Paul Bleicher, Richard Dab, Patricia Giencke, and Jeffrey Klofft, *Applied Clinical Trials,* Vol. 7 (11), p. 36, 1998.

Electronic Exchange of Laboratory Data: Recommendations for Improving Quality and Reducing the Hidden Costs. PMA Task Force, *Drug Information Journal,* Vol. 27 (3), 1993.

Identify Yourself! Computer Security and Authentication. Paul Bleicher, *Applied Clinical Trials,* Vol. 8 (10), p. 40, 1999.

Sign Here Please, Mr. Bond. Paul Bleicher, *Applied Clinical Trials,* Vol. 9 (8), p. 28, 2000.

The Use of Facsimile Technology in a Multicenter Clinical Trial. L. Mondschein, *Drug Information Journal,* Vol. 22 (1), pp. 75–85, 1988.

13

THE STUDY RESULTS

Progress Reports. The investigator shall furnish all reports to the sponsor of the drug who is responsible for collecting and evaluating the results obtained. The sponsor is required under §312.33 to submit annual reports to FDA on the progress of the clinical investigations.

—21 CFR 312.64(b)

Final Report. An investigator shall provide the sponsor with an adequate report shortly after completion of the investigator's participation in the investigation.

—21 CFR 312.64(c)

In this chapter, we discuss the analysis and presentation of study results. We will discuss some basic statistical concepts and how they relate to understanding the results of your trial and also briefly describe the role of the statistician in the analysis of the data obtained in a clinical trial. The contents of a typical study report are presented, together with some tips on distributing or publicizing your results.

ANALYZING THE RESULTS: THE ROLE OF THE STATISTICIAN

For convenience, we divide clinical research into two basic types of clinical studies, *experimental* and *observational*. An *experimental* study is defined as a prospective study in which the investigator has some control over one or more major variables, such as the investigational agent or therapy, and assigns limiting parameters to these variables. An *observational* study is defined as either a prospective or a retrospective study in which the investigator observes and records the relationships among the major variables as they appear in nature.

Clinical studies sponsored by pharmaceutical companies are usually experimental, in that the treatment received is decided by the sponsor according to the protocol. The corporate sponsor will analyze and present the results of its studies without significant contribution from the individual investigator. Since it is difficult for a single individual to design, finance, and perform an experimental study without the support of a pharmaceutical corporate sponsor, a large proportion of clinical studies performed by individual investigators are observational studies. Therefore, the remainder of this chapter will focus on observational study analysis.

In observational studies, the investigator is looking for an association between or among variables. Since such relationships are complex, and involve both the observed variables and others that remain unobserved or even unknown, it is not appropriate to assume that any observed association between variables is necessarily causally related. Every study is, in some way, sampling the real world. That is, during the period of observation, we are obtaining information about a group of people, the study population. This study population reflects the population as a whole and reflects it well or poorly depending on the selection criteria. The statistician is able to measure how closely the sample study population accurately reflects the real world (how closely the results obtained in the observation of the study population would resemble the results of the same study performed on every single person in the world who met the study entrance criteria). Or, to put the same thought in a different perspective, what are the chances that the observed result was not caused by the intervention but merely happened by random chance? The statistician tells us what the odds are that the observed result occurred because of random chance. The statistician can also, in experimental studies, greatly assist in the process of determining the appropriate sample sizes.

PRESENTING THE RESULTS

When you have completed your research on the question that started your efforts, and you have obtained and analyzed the study results, you are eager to share your results with others. That is, after all, the main reason that you began all this effort. You can present your results either orally or in writing. You also can create methodology presentations focusing on your process rather than on your results.

Oral Presentations

Oral presentations require the same discipline and organizational skills that written presentations require, with the added requirement that personal presentation skills are also needed. Like a well-constructed paper, the oral presentation is tailored to the interests of the audience. Therefore, begin by selecting the audience that will hear your presentation and prepare your talk for that listening group. Possible forums for oral presentations include the following:

- Local hospital staffs.
- Local and state professional societies.
- Local and state specific disease support groups.
- National professional societies.
- National specific disease societies.

Your colleagues have invariably been involved at some level in your research, and they will be interested in hearing about your results. They also are, as a rule, more supportive of your efforts and forgiving of your research deficiencies than other audiences.

The following advice is pertinent to each level of audience. Be brief; when the presentation is first scheduled, ask for an estimate of time available to you. Plan to speak for less than all of your allotted time, leaving the remainder for questions and discussion. Be prepared with extra topics for discussion, just in case you have presented your material so well that the audience has no questions.

Ascertain what audiovisual aids are available; transparencies or a writing surface will inexpensively provide the visual material to further convey your thoughts. Remember that more people learn by seeing than by hearing. To be most effective, the visual aid must be clearly legible from the back of the room, attractive, and capable of being completely comprehended in 10 seconds or less.

Organize your thoughts and therefore your presentation. You may well begin by posing the same research question that drove you to perform the study, and the remainder of your talk will supply the answer to that question. In your first five sentences, you should have presented the study question and the methodology by which your study was conducted. That leaves the remainder of the time to focus on the results. Follow by pointing out facets of your research question that remain unanswered, any flaws in your approach, and any limitations in the applicability of your results to a larger population.

Conclude by summarizing the results of your findings. You can then answer questions and discuss your findings in the remainder of the time allotted.

Poster Presentations

There are two main types of written presentations, the *poster* and the *full paper*. The poster presentation at a professional meeting requires less effort initially, and excellent exposure is obtained at minimal cost. You begin by writing and submitting an abstract to the specifications of the meeting organizers. Then, if selected for presentation, you perform the additional effort of creating a visually stimulating poster that summarizes your methods, results, and conclusion.

The poster presentation offers the twin advantages of wide exposure of both you and your work, including the opportunity to travel to the professional meeting at which the poster is to be displayed and the advantage of not having to write a formal paper. The cost of poster materials can be minimal, and your "peer review" takes place informally as observers share comments with you about your work. It offers an excellent venue for the novice researcher to learn the presentation process.

The disadvantages of poster presentations include the necessity to think of your presentation in visual rather than verbal terms. Charts, graphs, "layout," font and color selection, poster background, and the mechanics of attaching your creation to a bulletin board are considerations unique to poster presentations. Your results may be seen only by those who attend the meeting and make the effort to view your poster.

Written Presentations

You will reach a much wider audience by publishing your results as a research paper. Here again, you must tailor the structure of your presentation to the intended audience. A paper that is to be published in your hospital's monthly newsletter to the public should be written in a far less technical fashion than one to be published in a leading specialty journal.

The structure of a written research report usually adheres to the following format:

Abstract   A brief summary of your project, stating the purpose of the study, the methods employed, the main results, and the overall conclusion; the abstract should contain no more than 250 words.

Introduction   A brief discussion of the background and rationale of the study, including the research question that the study was designed to answer.

Methods   A summary of the study design, including the criteria for the selection of subjects, observations or measurements obtained, data collection, and statistical analysis. A description of the study population is usually included as well.

Results   An objective display of the results is presented in this section, avoiding any comments or interpretations at this time. Tables and graphs greatly aid in the presentation of the results.

Discussion   A more subjective discourse, including a statement of the interpretation of the results and conclusions drawn, comparison with previous studies in the same area, and a discussion of the strengths and weaknesses of the research design. Suggestions for further research are also appropriate.

Summary   A shorter version of the abstract, focusing on the results and conclusions.

Bibliography   A sequential listing of all literature cited in your paper.

METHODOLOGY PRESENTATIONS

When you consider presenting your results to other researchers and physicians with an interest in the field of study, remember that they are not only interested in what you learned but also in how you learned it. They are interested in your methodology (the material in the methods section of your written presentations). If the results are of wide or pivotal impact, others will be interested in duplicating the results, to confirm (or fail to confirm, human nature being what it is) your findings. Others will be in the process of designing similar studies and will want to repeat your successes and avoid your mistakes. Often, workers in related fields will find in your methodology an innovation or technique that is transferable to their work. Indeed, many times the investigator finds that the interest in the study methodology will far outweigh the interest in the study results.

Since you will be describing "how you did it," in essence you will be coaching your audience in your technique. This instruction is far better delivered in a visual format than a verbal format because most audiences retain more of what they see than what they hear. Either a poster presentation, a written presentation, or an oral talk with abundant visual aids will be very effective.

The format for methodology presentations is similar to that of the format for the presentation of results:

- Abstract.
- Introduction.
- Methods.
- Results.
- Discussion.
- Summary.
- Bibliography.

However, since your topic is now your methodology, the focus of your presentation will be on the reasons for choosing this study design, and your results will focus on the effectiveness of the study methodology rather than on the results of the study itself. Similarly, your discussion will focus on the strengths and weaknesses of your study methodology rather than on the study results. One study can furnish many more papers on the study methodology than on the study results.

The investigator who pursues a study to its conclusion will find an audience for both the presentation of the results of the study and for a presentation of the methodology of the study. Both topics will be of interest to other researchers and to physicians with an interest in the field of study.

REGULATORY REFERENCE

International Conference on Harmonisation Guideline on Structure and Content of Clinical Study Reports. *Federal Register,* 61 FR 37320, 17 July 1996.

BIBLIOGRAPHY

Journal Submissions Other Than Scientific Papers. B. Bastel. In *Biomedical Communications: Selected AMWA Workshops,* edited by P. Minick, American Medical Writers Association, Bethesda, Md., pp. 139–142, 1994.

Making Effective Presentations. C. Williams. In *Biomedical Communications: Selected AMWA Workshops,* edited by P. Minick, American Medical Writers Association, Bethesda, Md., pp. 84–89, 1994.

Making Effective Slides. E. Stern. In *Biomedical Communications: Selected AMWA Workshops,* edited by P. Minick, American Medical Writers Association, Bethesda, Md., pp. 90–97, 1994.

Organizing the Scientific Journal Paper. R. Iles. In *Biomedical Communications: Selected AMWA Workshops,* edited by P. Minick, American Medical Writers Association, Bethesda, Md., pp. 133–138, 1994.

Reporting on Clinical Study Reports. *Applied Clinical Trials,* Vol. 3 (9), p. 24, 1994.

Survival Analysis: A Practical Approach. *Applied Clinical Trials,* Vol. 5 (5), p. 63, 1996.

Writing Abstracts. J. Eastman and E. Klein. In *Biomedical Communications: Selected AMWA Workshops,* edited by P. Minick. American Medical Writers Association, Bethesda, Md., pp. 143–146, 1994.

14

AUDITS OF CLINICAL RESEARCH SITES

An investigator shall upon request from any properly authorized officer or employee of FDA, at reasonable times, permit such officer or employee to have access to, and copy and verify any records or reports made by the investigator pursuant to §312.62.

—21 CFR 312.68

Clinical research on human subjects is an activity that is regulated by the federal government. Therefore, clinical research records are subject to audit in the same manner and with the same vigorous federal scrutiny as financial records. In addition, by contract and by law, the study sponsor and the Institutional Review Board (IRB) are also authorized to perform a rigorous audit of your research activities. Since an audit is inevitable, the wise investigator will, from the outset, maintain the research documents in accordance with the regulations and contracts.

ORGANIZATIONS AUTHORIZED TO AUDIT CLINICAL RESEARCH FACILITIES

Food and Drug Administration (FDA)

When signing Form FDA 1572, the investigator acknowledges that the FDA may inspect the site's study records. The FDA *Compliance Program Manual* (Part 1, page 2) states that

> the purpose of the compliance program is to (1) assess adherence to FDA regulations; (2) determine the validity of specific studies in support of products pending approval by the FDA, and (3) determine that the rights and safety of subjects used in clinical studies have been properly protected.

The Bioresearch Monitoring Program was established in 1977, and its activities include inspections of clinical investigators, sponsors, biopharmaceutical laboratories, IRBs, and toxicology laboratories. The FDA's routine surveillance inspections occur at randomly selected sites as part of this program and are called study-oriented inspections. Additionally, the FDA routinely audits selected sites that have participated in a pivotal trial (one that is essential to the success of an Investigational New Drug [IND] submission).

The FDA also performs "for cause" inspections (called investigator-oriented inspections). These inspections may be initiated for one or more of the following reasons:

- The site has participated in numerous clinical trials.
- The investigator is conducting a clinical study outside the investigator's particular specialty.
- The investigator had demonstrated safety or efficacy findings that were inconsistent with the findings of other sites.
- The site enrolls too many subjects with a specific disease, given the locale of the investigative site.
- The site reports unusual laboratory findings, either identical values among different subjects or results at divergence with the expected biological differences.
- The sponsor reports to the FDA that the site is not submitting data as required or is in violation of FDA regulations.
- A study subject complains to the FDA about subject right violations.

Study Sponsor

The study sponsor, usually a pharmaceutical company, has an obligation to assure that the investigator and the site research staff are performing the study appropriately, consistently, and in accordance with the protocol and sponsor-designated procedures. Additionally, sponsors will audit sites as part of a quality assurance process or to evaluate the performance of their own monitoring clinical research staff.

Institutional Review Board

As described in the Code of Federal Regulations (CFR), an IRB shall review and have authority to approve, require modifications in, or disapprove all research activities covered by the regulations.

Office for Human Research Protection (OHRP)

OHRP, under the Department of Health and Human Services (DHHS), has the authority to audit institutions conducting human research and to suspend research at the institution where there are violations of human subjects' rights.

Cooperative Groups

Physician investigators may form or join multi-institutional cooperative groups. Such groups are formed to coordinate oncology studies or to coordinate multicenter research with a single institution performing the primary administrative functions. Such groups, by written agreement, assume the right to inspect the study documents of a member site.

FILES ELIGIBLE FOR AUDIT

Records and Storage

The investigator is required to prepare and maintain adequate and accurate case histories designed to record all observations and all other data pertinent to the investigation on each subject treated with the investigational agent or employed as a control in the investigation, under both the CFR and Form FDA 1572. Such case histories should include, at a minimum, adequate information to document that the subject existed, met the entrance criteria as designated in the protocol, underwent the procedures as specified in the protocol, was dispensed the investigational agent in accordance with the provisions of the protocol, and underwent periodic evaluations as prescribed by the protocol, along with unused investigational agent being collected at the end of the study. Any adverse events noted in office or hospital records should be recorded in the Case Report Form (CRF) and treated in accordance with Good Clinical Practice (GCP).

In addition, such case histories and supporting records must be retained by the investigator for a period of two years following the date that a marketing application is approved for the investigational agent or until two years after the investigation is discontinued and the FDA is so notified. The investigator, therefore, should prepare at the outset of each clinical study to create, update, and store these essential documents.

In Europe, the topic of case history storage is even more confusing. The European Union Good Clinical Practice Guideline stipulates a retention period of at least 15 years for subject records. However, Directive 91/507/EEC deleted the 15-year rule and left the retention period as "the maximum period of time permitted by the hospital, institution, or private practice." In effect, local practice still governs document storage in Europe, and loss of essential documents may well occur. The prudent European sponsor or investigator may choose to comply with the 15-year rule.

The investigator is required to maintain adequate records of the disposition of the investigational agent, including dates, quantity, and use by study subjects. At the termination of the clinical study, the investigator is required to return the unused supplies to the sponsor or otherwise dispose of the unused supplies of the drug as directed by the sponsor.

Study Files

Each study generates documents that provide authorization or verification that all pertinent regulations have been followed during the duration of the clinical study. These documents should be kept in a special, separate file, usually called the "study file" or "regulatory file." Since the contents of this file are considered to contain confidential information, access to this file should be permitted only to duly authorized representatives of the FDA, the sponsor (or designated Contract Research Organization [CRO]), the IRB, and your clinical staff. A study file typically includes the following items:

- The monitoring log, which should be signed by every person who visits the study site for the purpose of reviewing any of the clinical study documents. Each visitor should show proper credentials before access to the study files is permitted. The monitoring log therefore will retain a record of every person who, in addition to the study staff, had access to any study records. The monitoring log also serves as documentation that the sponsor's monitor truly did visit the site for the purpose of monitoring study activities.

- A signed copy of each Form FDA 1572 that has been submitted to the sponsor. A new Form FDA 1572 is generated and signed every time any change is made, such as the addition or withdrawal of a subinvestigator or the change of address of any participating clinical laboratory.

- A copy of the current, accurate curriculum vitae (CV) of each investigator listed on Form FDA 1572.

- The study protocol (a copy of every version submitted to the IRB together with all amendments), with each copy signed by the Principal Investigator (PI) as needed.

- The Investigator's Brochure (a copy of each version received by the PI).

- All IND safety reports received by the PI.

- All IRB correspondence, original or a copy.

- A letter to the IRB submitting each version of the protocol, protocol amendments, Investigator's Brochure, Informed Consent Form, and any advertisements to the IRB.

- A letter from the IRB approving each version of the protocol, protocol amendments, Informed Consent Form, and advertisements.

- A letter to the IRB submitting each IND safety report.

- A letter from the IRB reviewing and renewing approval of the study at IRB–specified intervals, minimally, on an annual basis but may be more frequent.

- A letter submitting notification to the IRB of every serious adverse event reported by the site.

- A copy of the IRB membership roster or accreditation number.

- A copy of the Final Study Report.

- Copies of all other general correspondence between the IRB and the PI.

- A copy of each version of the Informed Consent Form that had been approved by the IRB.

- All study correspondence, original or copy, between the site and the sponsor and/or CRO (in a separate section of the study file).

- Laboratory certification documents for each clinical laboratory listed on each version of Form FDA 1572, together with a copy of the laboratory's normal ranges of all laboratory parameters listed in the protocol. If a copy of the CV of the clinical laboratory director can be obtained, it also should be filed.

- Telephone logs or other documentation of each oral conversation with any sponsor or IRB representative.

- The original investigational agent accountability and inventory logs, appropriately signed by the PI. These documents should clearly record who is authorized to dispense or administer the investigational agent. Copies of investigational agent receipts and shipping records should be retained, together with copies of investigational agent accountability documents. Every single tablet, capsule, or drop should be accounted for in these documents. In addition, the source records (office records) should also clearly document the appropriate dispensing of each investigational agent, recording the amount, identification by lot or bottle number of the agent dispensed, the identity of the dispensing person, the instructions given to the subject regarding proper administration of the investigational agent, and an accounting of the amount returned. Copies of drug disposal records must also be retained and available for audit.

- Subject records: A copy of the Informed Consent Form, signed and dated by each subject, must be retained and available for audit. One copy of each version of a blank Case Report Form, together with a copy of each completed Case Report Form for each subject must be retained at the site. All supporting subject clinic notes, hospital records, laboratory results, medical imaging test results, electrocardiogram results, and all other clinical information pertaining to each study subject must be retained by the site and available for audit (although not necessarily in the study files).

PREPARING FOR AN IMPENDING AUDIT

Scheduling

Usually, an inspector will notify the study site in advance of the audit in order to schedule a mutually convenient date and time. During such communication, do your utmost to ascertain the details of the impending audit. Perhaps the audit is only peripheral to a local IRB audit, and the inspector is interested only in your correspondence with the IRB. Perhaps the inspector plans to focus on investigational agent receipt, storage, and disposition. If you can ascertain such information in advance, your preparations will be less extensive than if a full study audit is planned. Also, it is important to ascertain if the audit is routine or "for cause."

In the event of notification of an FDA audit, notify the sponsor or CRO immediately; they will want to schedule visits prior to the audit. Also, schedule appointments with all other appropriate study personnel (subinvestigators, dispensing pharmacists, microbiologists) for the time of the audit. Clear the schedule of the Clinical Research Coordinator (CRC) for the expected duration of the audit and make sure that your own schedule contains ample time to respond to any questions raised by the inspector. Also schedule a secure room in which the inspector can work; make certain that the planned room contains no records, charts, documents, or papers of any sort not specifically requested by the inspector.

Source Documents

If you have been given some indication of the purpose of the audit, then you have time to gather and review source documents for completeness and accuracy. If not, then you can take solace in the fact that you have been maintaining your study file and source documents as the study progressed, and your records are complete and up to date. If you have been negligent, then make certain that the records are complete and accurate prior to the date of the audit.

DURING THE AUDIT

Clinical Research Coordinator

The PI or CRC should, during the initial introductions, inspect the credentials of the inspector. If from the FDA, the inspector will produce a Form FDA 482, "Notice of Inspection."

The investigator and CRC should be courteous and professional at all times, and one should become the constant escort of the inspector during the audit. While every personal courtesy should be extended to the inspector, such as food, drink, and sanitary facilities, the inspector should not be permitted to roam freely about the facility. During the audit, the investigator or CRC will bring into the inspector's room only those documents specifically requested by the inspector. Give the inspector no opportunity to browse or "troll" at liberty through your records.

The investigator or CRC should make written notes of every comment, concern, or question posed by the inspector and attempt to clarify every point and answer every question. The investigator or CRC should not divulge more information or records than those requested or discuss other studies in any way. The investigator or CRC should also object to any request for photographs, signed affidavits, or unreasonable information, such as financial records or home addresses of study subjects.

Principal Investigator

Make yourself available during the audit to answer any questions and to respond to and resist any unreasonable request. Make every attempt to resolve any questions or issues immediately during the audit. Communicate with the sponsor immediately (if a representative is not already present during the audit) in the event of an unreasonable request. In any event, provide a full report to the sponsor after the conclusion of the audit.

AUDIT RESULTS

FDA Audit

If the inspector is from the FDA, Form FDA 483 will be issued prior to the inspector's departure describing any deficiencies or errors noted during the audit. The investigator should immediately respond to any inaccuracies or misconceptions contained in this document. Request that the inspector correct the report on the spot. Follow with a written explanation to the FDA District Office.

The FDA will issue an Establishment Inspection Report (EIR) in approximately six months. Since this document is public information, do your utmost to insure that any inaccuracies are corrected and any errors or problems are corrected or resolved.

One of the following types of letters will be issued by the FDA:

- **NAI** (no action indicated): No findings requiring action.
- **VAI** (voluntary action indicated): Some discrepancies or errors were noted. The investigator should correct these and respond in writing to the FDA District Office documenting these corrections.
- **NAF** (notice of adverse findings): Serious errors or discrepancies were found. The investigator should correct these and respond in writing to the FDA District Office documenting these corrections. Another audit may well be performed to assure correction.
- **OAI** (official action indicated): Serious infractions were found.

Other Audits

The sponsor or the cooperative group may also perform an audit at your site, comparing the source documents with the CRFs and ascertaining the completeness of your study files. Look at the audit as a learning experience. The inspector is neither vengeful nor avaricious but is merely ascertaining that all appropriate regulations and instructions have been followed. Make whatever corrections are appropriate and document those corrections to the satisfaction of the inspector.

Clinical research on human subjects is a strictly regulated activity. The FDA, the study sponsor and/or CRO, and the IRB have the legal right and duty to audit your research activities. Subject case histories, CRFs, documents pertaining to regulatory compliance, and investigational agent accounting records are subject to audit. The audit is not a punishment or a harassment but a legal duty of the auditing agency and should be looked upon as a positive learning experience.

REGULATORY REFERENCES

21 CFR 312.58, 312.68, and 56.109.

FDA Guideline for the Monitoring of Clinical Investigations (1988).

FDA Compliance Program Guidance Manual, Clinical Investigators (8/94).

FDA Compliance Program Guidance Manual, Sponsors, Monitors, and CROs (10/91).

IRB Compliance Program Guidance Manual (6/89).

FDA Inspections of Clinical Investigators (9/98).

Clinical Investigator Regulatory Sanctions (9/98).

Guide for Detecting Fraud in Bioresearch Monitoring Inspections (4/93).

BIBLIOGRAPHY

Detecting Fraud Using Auditing and Biometrical Methods. Jurgen Schmidt, Heiner Gertzen, K. Michael Aschenbrenner, and Steen Ryholt-Jensen, *Applied Clinical Trials,* Vol. 4 (5), p. 50, 1995.

Detection of Gross Negligence, Fraud, and Other Bad Faith Efforts During Field Auditing of Clinical Trial Sites. D. Mackintosh and V. Zepp, *Drug Information Journal,* Vol. 30 (3), pp. 645–653, 1996.

FDA Audit Results as an Investigator Evaluation Tool. Harold Glass, *Applied Clinical Trials,* Vol. 6 (4), p. 42, 1997.

FDA Audits of Clinical Studies. A. Horowitz, *Applied Clinical Trials,* Vol. 1 (5), p. 24, 1992.

FDA Audits of Institutional Review Boards. A. Horowitz, *Applied Clinical Trials,* Vol. 4 (10), pp. 54–59, 1995.

FDA Inspections of Clinical Investigations. *Research Nurse,* Vol. 3 (2), pp. 1–8, 1997.

FDA Inspection of Clinical Research Sponsors and Investigators: Avoiding the Pitfalls. C. S. Lawrence, *Drug Information Journal,* Vol. 22 (2), pp. 207–223, 1988.

FDA's Inspections of Clinical Investigators. M. Bruckheimer, *Drug Information Journal,* Vol. 27 (1), pp. 213–216, 1993.

FDA's Inspections of US and Non-US Clinical Studies. B. Barton, *Drug Information Journal,* Vol. 24 (3), pp. 463–468, 1990.

How to Prepare for a GCP Inspection. Pamela Charnley Nickols, *The Monitor,* Vol. 14 (2), pp. 49–52, 2000.

Investigating Fraud—Again. Frank Wells, *Applied Clinical Trials,* Vol. 9 (2), p. 26, 2000.

Part 6: A Survive and Thrive Approach to Audits and Inspection. Teri Stokes, *Applied Clinical Trials,* Vol. 6 (8), p. 40, 1999.

Site Preparation for an FDA Inspection: A Systematic Approach. *Research Nurse,* Vol. 3 (2), pp. 9–14, 1997.

The Clinical Study Audit Process: 10 Steps to a Successful Audit. Daniel E. Worden, *Applied Clinical Trials,* Vol. 5 (4), p. 50, 1996.

The Role of Quality Assurance in Good Clinical Practice. R. Fischer and K. Schick, *Drug Information Journal,* Vol. 27 (3), pp. 895–901, 1993.

Appendix A
FDA REGULATIONS

CFR, Title 21, Part 50: Protection of Human Subjects

CFR, Title 21, Part 56: Institutional Review Boards

CFR, Title 21, Part 312: Investigational New Drug Applcation
(Subpart D – Responsibilities of Sponsors and Investigators)

CFR, Title 45, Part 46: Protection of Human Subjects
(Subparts B, C, D)

Food and Drug Administration, HHS

50.24 Exception from informed consent requirements for emergency research.
50.25 Elements of informed consent.
50.27 Documentation of informed consent.

AUTHORITY: 21 U.S.C. 321, 346, 346a, 348, 352, 353, 355, 360, 360c–360f, 360h–360j, 371, 379e, 381; 42 U.S.C. 216, 241, 262, 263b–263n.

SOURCE: 45 FR 36390, May 30, 1980, unless otherwise noted.

Subpart A—General Provisions

§ 50.1 Scope.

(a) This part applies to all clinical investigations regulated by the Food and Drug Administration under sections 505(i) and 520(g) of the Federal Food, Drug, and Cosmetic Act, as well as clinical investigations that support applications for research or marketing permits for products regulated by the Food and Drug Administration, including food and color additives, drugs for human use, medical devices for human use, biological products for human use, and electronic products. Additional specific obligations and commitments of, and standards of conduct for, persons who sponsor or monitor clinical investigations involving particular test articles may also be found in other parts (e.g., parts 312 and 812). Compliance with these parts is intended to protect the rights and safety of subjects involved in investigations filed with the Food and Drug Administration pursuant to sections 406, 409, 502, 503, 505, 510, 513–516, 518–520, 721, and 801 of the Federal Food, Drug, and Cosmetic Act and sections 351 and 354–360F of the Public Health Service Act.

(b) References in this part to regulatory sections of the Code of Federal Regulations are to chapter I of title 21, unless otherwise noted.

[45 FR 36390, May 30, 1980; 46 FR 8979, Jan. 27, 1981, as amended at 63 FR 26697, May 13, 1998; 64 FR 399, Jan. 5, 1999]

§ 50.3 Definitions.

As used in this part:

(a) *Act* means the Federal Food, Drug, and Cosmetic Act, as amended (secs. 201—902, 52 Stat. 1040 *et seq.* as amended (21 U.S.C. 321—392)).

(b) *Application for research or marketing permit* includes:

(1) A color additive petition, described in part 71.

(2) A food additive petition, described in parts 171 and 571.

(3) Data and information about a substance submitted as part of the procedures for establishing that the substance is generally recognized as safe for use that results or may reasonably be expected to result, directly or indirectly, in its becoming a component or otherwise affecting the characteristics of any food, described in §§ 170.30 and 570.30.

(4) Data and information about a food additive submitted as part of the procedures for food additives permitted to be used on an interim basis pending additional study, described in § 180.1.

(5) Data and information about a substance submitted as part of the procedures for establishing a tolerance for unavoidable contaminants in food and food-packaging materials, described in section 406 of the act.

(6) An investigational new drug application, described in part 312 of this chapter.

(7) A new drug application, described in part 314.

(8) Data and information about the bioavailability or bioequivalence of drugs for human use submitted as part of the procedures for issuing, amending, or repealing a bioequivalence requirement, described in part 320.

(9) Data and information about an over-the-counter drug for human use submitted as part of the procedures for classifying these drugs as generally recognized as safe and effective and not misbranded, described in part 330.

(10) Data and information about a prescription drug for human use submitted as part of the procedures for classifying these drugs as generally recognized as safe and effective and not misbranded, described in this chapter.

(11) [Reserved]

(12) An application for a biologics license, described in part 601 of this chapter.

(13) Data and information about a biological product submitted as part of the procedures for determining that licensed biological products are safe and effective and not misbranded, described in part 601.

(14) Data and information about an in vitro diagnostic product submitted as part of the procedures for establishing,

amending, or repealing a standard for these products, described in part 809.

(15) An *Application for an Investigational Device Exemption*, described in part 812.

(16) Data and information about a medical device submitted as part of the procedures for classifying these devices, described in section 513.

(17) Data and information about a medical device submitted as part of the procedures for establishing, amending, or repealing a standard for these devices, described in section 514.

(18) An application for premarket approval of a medical device, described in section 515.

(19) A product development protocol for a medical device, described in section 515.

(20) Data and information about an electronic product submitted as part of the procedures for establishing, amending, or repealing a standard for these products, described in section 358 of the Public Health Service Act.

(21) Data and information about an electronic product submitted as part of the procedures for obtaining a variance from any electronic product performance standard, as described in § 1010.4.

(22) Data and information about an electronic product submitted as part of the procedures for granting, amending, or extending an exemption from a radiation safety performance standard, as described in § 1010.5.

(c) *Clinical investigation* means any experiment that involves a test article and one or more human subjects and that either is subject to requirements for prior submission to the Food and Drug Administration under section 505(i) or 520(g) of the act, or is not subject to requirements for prior submission to the Food and Drug Administration under these sections of the act, but the results of which are intended to be submitted later to, or held for inspection by, the Food and Drug Administration as part of an application for a research or marketing permit. The term does not include experiments that are subject to the provisions of part 58 of this chapter, regarding nonclinical laboratory studies.

(d) *Investigator* means an individual who actually conducts a clinical investigation, i.e., under whose immediate direction the test article is administered or dispensed to, or used involving, a subject, or, in the event of an investigation conducted by a team of individuals, is the responsible leader of that team.

(e) *Sponsor* means a person who initiates a clinical investigation, but who does not actually conduct the investigation, i.e., the test article is administered or dispensed to or used involving, a subject under the immediate direction of another individual. A person other than an individual (e.g., corporation or agency) that uses one or more of its own employees to conduct a clinical investigation it has initiated is considered to be a sponsor (not a sponsor-investigator), and the employees are considered to be investigators.

(f) *Sponsor-investigator* means an individual who both initiates and actually conducts, alone or with others, a clinical investigation, i.e., under whose immediate direction the test article is administered or dispensed to, or used involving, a subject. The term does not include any person other than an individual, e.g., corporation or agency.

(g) *Human subject* means an individual who is or becomes a participant in research, either as a recipient of the test article or as a control. A subject may be either a healthy human or a patient.

(h) *Institution* means any public or private entity or agency (including Federal, State, and other agencies). The word *facility* as used in section 520(g) of the act is deemed to be synonymous with the term *institution* for purposes of this part.

(i) *Institutional review board* (IRB) means any board, committee, or other group formally designated by an institution to review biomedical research involving humans as subjects, to approve the initiation of and conduct periodic review of such research. The term has the same meaning as the phrase *institutional review committee* as used in section 520(g) of the act.

(j) *Test article* means any drug (including a biological product for human use), medical device for human use, human food additive, color additive, electronic product, or any other article subject to regulation under the act or under sections 351 and 354–360F of the

Public Health Service Act (42 U.S.C. 262 and 263b–263n).

(k) *Minimal risk* means that the probability and magnitude of harm or discomfort anticipated in the research are not greater in and of themselves than those ordinarily encountered in daily life or during the performance of routine physical or psychological examinations or tests.

(l) *Legally authorized representative* means an individual or judicial or other body authorized under applicable law to consent on behalf of a prospective subject to the subject's particpation in the procedure(s) involved in the research.

(m) *Family member* means any one of the following legally competent persons: Spouse; parents; children (including adopted children); brothers, sisters, and spouses of brothers and sisters; and any individual related by blood or affinity whose close association with the subject is the equivalent of a family relationship.

[45 FR 36390, May 30, 1980, as amended at 46 FR 8950, Jan. 27, 1981; 54 FR 9038, Mar. 3, 1989; 56 FR 28028, June 18, 1991; 61 FR 51528, Oct. 2, 1996; 62 FR 39440, July 23, 1997; 64 FR 399, Jan. 5, 1999; 64 FR 56448, Oct. 20, 1999]

Subpart B—Informed Consent of Human Subjects

SOURCE: 46 FR 8951, Jan. 27, 1981, unless otherwise noted.

§50.20 General requirements for informed consent.

Except as provided in §§50.23 and 50.24, no investigator may involve a human being as a subject in research covered by these regulations unless the investigator has obtained the legally effective informed consent of the subject or the subject's legally authorized representative. An investigator shall seek such consent only under circumstances that provide the prospective subject or the representative sufficient opportunity to consider whether or not to participate and that minimize the possibility of coercion or undue influence. The information that is given to the subject or the representative shall be in language understandable to the subject or the representative. No informed consent, whether oral or writ-

ten, may include any exculpatory language through which the subject or the representative is made to waive or appear to waive any of the subject's legal rights, or releases or appears to release the investigator, the sponsor, the institution, or its agents from liability for negligence.

[46 FR 8951, Jan. 27, 1981, as amended at 64 FR 10942, Mar. 8, 1999]

§50.23 Exception from general requirements.

(a) The obtaining of informed consent shall be deemed feasible unless, before use of the test article (except as provided in paragraph (b) of this section), both the investigator and a physician who is not otherwise participating in the clinical investigation certify in writing all of the following:

(1) The human subject is confronted by a life-threatening situation necessitating the use of the test article.

(2) Informed consent cannot be obtained from the subject because of an inability to communicate with, or obtain legally effective consent from, the subject.

(3) Time is not sufficient to obtain consent from the subject's legal representative.

(4) There is available no alternative method of approved or generally recognized therapy that provides an equal or greater likelihood of saving the life of the subject.

(b) If immediate use of the test article is, in the investigator's opinion, required to preserve the life of the subject, and time is not sufficient to obtain the independent determination required in paragraph (a) of this section in advance of using the test article, the determinations of the clinical investigator shall be made and, within 5 working days after the use of the article, be reviewed and evaluated in writing by a physician who is not participating in the clinical investigation.

(c) The documentation required in paragraph (a) or (b) of this section shall be submitted to the IRB within 5 working days after the use of the test article.

(d)(1) Under 10 U.S.C. 1107(f) the President may waive the prior consent requirement for the administration of an investigational new drug to a member of the armed forces in connection with the member's participation in a particular military operation. The statute specifies that only the President may waive informed consent in this connection and the President may grant such a waiver only if the President determines in writing that obtaining consent: Is not feasible; is contrary to the best interests of the military member; or is not in the interests of national security. The statute further provides that in making a determination to waive prior informed consent on the ground that it is not feasible or the ground that it is contrary to the best interests of the military members involved, the President shall apply the standards and criteria that are set forth in the relevant FDA regulations for a waiver of the prior informed consent requirements of section 505(i)(4) of the Federal Food, Drug, and Cosmetic Act (21 U.S.C. 355(i)(4)). Before such a determination may be made that obtaining informed consent from military personnel prior to the use of an investigational drug (including an antibiotic or biological product) in a specific protocol under an investigational new drug application (IND) sponsored by the Department of Defense (DOD) and limited to specific military personnel involved in a particular military operation is not feasible or is contrary to the best interests of the military members involved the Secretary of Defense must first request such a determination from the President, and certify and document to the President that the following standards and criteria contained in paragraphs (d)(1) through (d)(4) of this section have been met.

(i) The extent and strength of evidence of the safety and effectiveness of the investigational new drug in relation to the medical risk that could be encountered during the military operation supports the drug's administration under an IND.

(ii) The military operation presents a substantial risk that military personnel may be subject to a chemical, biological, nuclear, or other exposure likely to produce death or serious or life-threatening injury or illness.

(iii) There is no available satisfactory alternative therapeutic or preventive treatment in relation to the intended use of the investigational new drug.

(iv) Conditioning use of the investigational new drug on the voluntary participation of each member could significantly risk the safety and health of any individual member who would decline its use, the safety of other military personnel, and the accomplishment of the military mission.

(v) A duly constituted institutional review board (IRB) established and operated in accordance with the requirements of paragraphs (d)(2) and (d)(3) of this section, responsible for review of the study, has reviewed and approved the investigational new drug protocol and the administration of the investigational new drug without informed consent. DOD's request is to include the documentation required by § 56.115(a)(2) of this chapter.

(vi) DOD has explained:

(A) The context in which the investigational drug will be administered, e.g., the setting or whether it will be self-administered or it will be administered by a health professional;

(B) The nature of the disease or condition for which the preventive or therapeutic treatment is intended; and

(C) To the extent there are existing data or information available, information on conditions that could alter the effects of the investigational drug.

(vii) DOD's recordkeeping system is capable of tracking and will be used to track the proposed treatment from supplier to the individual recipient.

(viii) Each member involved in the military operation will be given, prior to the administration of the investigational new drug, a specific written information sheet (including information required by 10 U.S.C. 1107(d)) concerning the investigational new drug, the risks and benefits of its use, potential side effects, and other pertinent information about the appropriate use of the product.

(ix) Medical records of members involved in the military operation will accurately document the receipt by members of the notification required by paragraph (d)(1)(viii) of this section.

(x) Medical records of members involved in the military operation will accurately document the receipt by members of any investigational new drugs in accordance with FDA regulations including part 312 of this chapter.

(xi) DOD will provide adequate followup to assess whether there are beneficial or adverse health consequences that result from the use of the investigational product.

(xii) DOD is pursuing drug development, including a time line, and marketing approval with due diligence.

(xiii) FDA has concluded that the investigational new drug protocol may proceed subject to a decision by the President on the informed consent waiver request.

(xiv) DOD will provide training to the appropriate medical personnel and potential recipients on the specific investigational new drug to be administered prior to its use.

(xv) DOD has stated and justified the time period for which the waiver is needed, not to exceed one year, unless separately renewed under these standards and criteria.

(xvi) DOD shall have a continuing obligation to report to the FDA and to the President any changed circumstances relating to these standards and criteria (including the time period referred to in paragraph (d)(1)(xv) of this section) or that otherwise might affect the determination to use an investigational new drug without informed consent.

(xvii) DOD is to provide public notice as soon as practicable and consistent with classification requirements through notice in the FEDERAL REGISTER describing each waiver of informed consent determination, a summary of the most updated scientific information on the products used, and other pertinent information.

(xviii) Use of the investigational drug without informed consent otherwise conforms with applicable law.

(2) The duly constituted institutional review board, described in paragraph (d)(1)(v) of this section, must include at least 3 nonaffiliated members who shall not be employees or officers of the Federal Government (other than for purposes of membership on the IRB) and shall be required to obtain any necessary security clearances. This IRB shall review the proposed IND protocol at a convened meeting at which a majority of the members are present including at least one member whose primary concerns are in nonscientific areas and, if feasible, including a majority of the nonaffiliated members. The information required by §56.115(a)(2) of this chapter is to be provided to the Secretary of Defense for further review.

(3) The duly constituted institutional review board, described in paragraph (d)(1)(v) of this section, must review and approve:

(i) The required information sheet;

(ii) The adequacy of the plan to disseminate information, including distribution of the information sheet to potential recipients, on the investigational product (e.g., in forms other than written);

(iii) The adequacy of the information and plans for its dissemination to health care providers, including potential side effects, contraindications, potential interactions, and other pertinent considerations; and

(iv) An informed consent form as required by part 50 of this chapter, in those circumstances in which DOD determines that informed consent may be obtained from some or all personnel involved.

(4) DOD is to submit to FDA summaries of institutional review board meetings at which the proposed protocol has been reviewed.

(5) Nothing in these criteria or standards is intended to preempt or limit FDA's and DOD's authority or obligations under applicable statutes and regulations.

[46 FR 8951, Jan. 27, 1981, as amended at 55 FR 52817, Dec. 21, 1990; 64 FR 399, Jan. 5, 1999; 64 FR 54188, Oct. 5, 1999]

§50.24 **Exception from informed consent requirements for emergency research.**

(a) The IRB responsible for the review, approval, and continuing review of the clinical investigation described in this section may approve that investigation without requiring that informed consent of all research subjects be obtained if the IRB (with the concurrence of a licensed physician who is a member of or consultant to the IRB and who is not otherwise participating in the clinical investigation) finds and documents each of the following:

(1) The human subjects are in a life-threatening situation, available treatments are unproven or unsatisfactory, and the collection of valid scientific evidence, which may include evidence obtained through randomized placebo-controlled investigations, is necessary to determine the safety and effectiveness of particular interventions.

(2) Obtaining informed consent is not feasible because:

(i) The subjects will not be able to give their informed consent as a result of their medical condition;

(ii) The intervention under investigation must be administered before consent from the subjects' legally authorized representatives is feasible; and

(iii) There is no reasonable way to identify prospectively the individuals likely to become eligible for participation in the clinical investigation.

(3) Participation in the research holds out the prospect of direct benefit to the subjects because:

(i) Subjects are facing a life-threatening situation that necessitates intervention;

(ii) Appropriate animal and other preclinical studies have been conducted, and the information derived from those studies and related evidence support the potential for the intervention to provide a direct benefit to the individual subjects; and

(iii) Risks associated with the investigation are reasonable in relation to what is known about the medical condition of the potential class of subjects, the risks and benefits of standard therapy, if any, and what is known about the risks and benefits of the proposed intervention or activity.

(4) The clinical investigation could not practicably be carried out without the waiver.

(5) The proposed investigational plan defines the length of the potential therapeutic window based on scientific evidence, and the investigator has committed to attempting to contact a legally authorized representative for each subject within that window of time and, if feasible, to asking the legally authorized representative con-

tacted for consent within that window rather than proceeding without consent. The investigator will summarize efforts made to contact legally authorized representatives and make this information available to the IRB at the time of continuing review.

(6) The IRB has reviewed and approved informed consent procedures and an informed consent document consistent with § 50.25. These procedures and the informed consent document are to be used with subjects or their legally authorized representatives in situations where use of such procedures and documents is feasible. The IRB has reviewed and approved procedures and information to be used when providing an opportunity for a family member to object to a subject's participation in the clinical investigation consistent with paragraph (a)(7)(v) of this section.

(7) Additional protections of the rights and welfare of the subjects will be provided, including, at least:

(i) Consultation (including, where appropriate, consultation carried out by the IRB) with representatives of the communities in which the clinical investigation will be conducted and from which the subjects will be drawn;

(ii) Public disclosure to the communities in which the clinical investigation will be conducted and from which the subjects will be drawn, prior to initiation of the clinical investigation, of plans for the investigation and its risks and expected benefits;

(iii) Public disclosure of sufficient information following completion of the clinical investigation to apprise the community and researchers of the study, including the demographic characteristics of the research population, and its results;

(iv) Establishment of an independent data monitoring committee to exercise oversight of the clinical investigation; and

(v) If obtaining informed consent is not feasible and a legally authorized representative is not reasonably available, the investigator has committed, if feasible, to attempting to contact within the therapeutic window the subject's family member who is not a legally authorized representative, and asking whether he or she objects to the subject's participation in the clinical investigation. The investigator will summarize efforts made to contact family members and make this information available to the IRB at the time of continuing review.

(b) The IRB is responsible for ensuring that procedures are in place to inform, at the earliest feasible opportunity, each subject, or if the subject remains incapacitated, a legally authorized representative of the subject, or if such a representative is not reasonably available, a family member, of the subject's inclusion in the clinical investigation, the details of the investigation and other information contained in the informed consent document. The IRB shall also ensure that there is a procedure to inform the subject, or if the subject remains incapacitated, a legally authorized representative of the subject, or if such a representative is not reasonably available, a family member, that he or she may discontinue the subject's participation at any time without penalty or loss of benefits to which the subject is otherwise entitled. If a legally authorized representative or family member is told about the clinical investigation and the subject's condition improves, the subject is also to be informed as soon as feasible. If a subject is entered into a clinical investigation with waived consent and the subject dies before a legally authorized representative or family member can be contacted, information about the clinical investigation is to be provided to the subject's legally authorized representative or family member, if feasible.

(c) The IRB determinations required by paragraph (a) of this section and the documentation required by paragraph (e) of this section are to be retained by the IRB for at least 3 years after completion of the clinical investigation, and the records shall be accessible for inspection and copying by FDA in accordance with § 56.115(b) of this chapter.

(d) Protocols involving an exception to the informed consent requirement under this section must be performed under a separate investigational new drug application (IND) or investigational device exemption (IDE) that clearly identifies such protocols as protocols that may include subjects who are unable to consent. The submission of those protocols in a separate IND/IDE is required even if an IND for the same drug product or an IDE for the same device already exists. Applications for investigations under this section may not be submitted as amendments under §§ 312.30 or 812.35 of this chapter.

(e) If an IRB determines that it cannot approve a clinical investigation because the investigation does not meet the criteria in the exception provided under paragraph (a) of this section or because of other relevant ethical concerns, the IRB must document its findings and provide these findings promptly in writing to the clinical investigator and to the sponsor of the clinical investigation. The sponsor of the clinical investigation must promptly disclose this information to FDA and to the sponsor's clinical investigators who are participating or are asked to participate in this or a substantially equivalent clinical investigation of the sponsor, and to other IRB's that have been, or are, asked to review this or a substantially equivalent investigation by that sponsor.

[61 FR 51528, Oct. 2, 1996]

§ 50.25 Elements of informed consent.

(a) *Basic elements of informed consent.* In seeking informed consent, the following information shall be provided to each subject:

(1) A statement that the study involves research, an explanation of the purposes of the research and the expected duration of the subject's participation, a description of the procedures to be followed, and identification of any procedures which are experimental.

(2) A description of any reasonably foreseeable risks or discomforts to the subject.

(3) A description of any benefits to the subject or to others which may reasonably be expected from the research.

(4) A disclosure of appropriate alternative procedures or courses of treatment, if any, that might be advantageous to the subject.

(5) A statement describing the extent, if any, to which confidentiality of records identifying the subject will be

maintained and that notes the possibility that the Food and Drug Administration may inspect the records.

(6) For research involving more than minimal risk, an explanation as to whether any compensation and an explanation as to whether any medical treatments are available if injury occurs and, if so, what they consist of, or where further information may be obtained.

(7) An explanation of whom to contact for answers to pertinent questions about the research and research subjects' rights, and whom to contact in the event of a research-related injury to the subject.

(8) A statement that participation is voluntary, that refusal to participate will involve no penalty or loss of benefits to which the subject is otherwise entitled, and that the subject may discontinue participation at any time without penalty or loss of benefits to which the subject is otherwise entitled.

(b) *Additional elements of informed consent.* When appropriate, one or more of the following elements of information shall also be provided to each subject:

(1) A statement that the particular treatment or procedure may involve risks to the subject (or to the embryo or fetus, if the subject is or may become pregnant) which are currently unforeseeable.

(2) Anticipated circumstances under which the subject's participation may be terminated by the investigator without regard to the subject's consent.

(3) Any additional costs to the subject that may result from participation in the research.

(4) The consequences of a subject's decision to withdraw from the research and procedures for orderly termination of participation by the subject.

(5) A statement that significant new findings developed during the course of the research which may relate to the subject's willingness to continue participation will be provided to the subject.

(6) The approximate number of subjects involved in the study.

(c) The informed consent requirements in these regulations are not intended to preempt any applicable Federal, State, or local laws which require additional information to be disclosed for informed consent to be legally effective.

(d) Nothing in these regulations is intended to limit the authority of a physician to provide emergency medical care to the extent the physician is permitted to do so under applicable Federal, State, or local law.

§ 50.27 Documentation of informed consent.

(a) Except as provided in § 56.109(c), informed consent shall be documented by the use of a written consent form approved by the IRB and signed and dated by the subject or the subject's legally authorized representative at the time of consent. A copy shall be given to the person signing the form.

(b) Except as provided in § 56.109(c), the consent form may be either of the following:

(1) A written consent document that embodies the elements of informed consent required by § 50.25. This form may be read to the subject or the subject's legally authorized representative, but, in any event, the investigator shall give either the subject or the representative adequate opportunity to read it before it is signed.

(2) A *short form* written consent document stating that the elements of informed consent required by § 50.25 have been presented orally to the subject or the subject's legally authorized representative. When this method is used, there shall be a witness to the oral presentation. Also, the IRB shall approve a written summary of what is to be said to the subject or the representative. Only the short form itself is to be signed by the subject or the representative. However, the witness shall sign both the short form and a copy of the summary, and the person actually obtaining the consent shall sign a copy of the summary. A copy of the summary shall be given to the subject or the representative in addition to a copy of the short form.

[46 FR 8951, Jan. 27, 1981, as amended at 61 FR 57280, Nov. 5, 1996]

pertaining to the financial interests of clinical investigators who conducted studies on which the application relies and who are not full or part-time employees of the applicant, as follows:

(1) Complete records showing any financial interest or arrangement as described in §54.4(a)(3)(i) paid to such clinical investigators by the sponsor of the covered study.

(2) Complete records showing significant payments of other sorts, as described in §54.4(a)(3)(ii), made by the sponsor of the covered clinical study to the clinical investigator.

(3) Complete records showing any financial interests held by clinical investigators as set forth in §54.4(a)(3)(iii) and (a)(3)(iv).

(b) *Requirements for maintenance of clinical investigators' financial records.*

(1) For any application submitted for a covered product, an applicant shall retain records as described in paragraph (a) of this section for 2 years after the date of approval of the application.

(2) The person maintaining these records shall, upon request from any properly authorized officer or employee of FDA, at reasonable times, permit such officer or employee to have access to and copy and verify these records.

PART 56—INSTITUTIONAL REVIEW BOARDS

Subpart A—General Provisions

AUTHORITY: 21 U.S.C. 321, 346, 346a, 348, 351, 352, 353, 355, 360, 360c–360f, 360h–360j, 371, 379e, 381; 42 U.S.C. 216, 241, 262, 263b–263n.

SOURCE: 46 FR 8975, Jan. 27, 1981, unless otherwise noted.

Subpart A—General Provisions

§56.101 Scope.

(a) This part contains the general standards for the composition, operation, and responsibility of an Institutional Review Board (IRB) that reviews clinical investigations regulated by the Food and Drug Administration under sections 505(i) and 520(g) of the act, as well as clinical investigations that support applications for research or marketing permits for products regulated by the Food and Drug Administration, including food and color additives, drugs for human use, medical devices for human use, biological products for human use, and electronic products. Compliance with this part is intended to protect the rights and welfare of human subjects involved in such investigations.

(b) References in this part to regulatory sections of the Code of Federal Regulations are to chapter I of title 21, unless otherwise noted.

[46 FR 8975, Jan. 27, 1981, as amended at 64 FR 399, Jan. 5, 1999]

§56.102 Definitions.

As used in this part:

(a) *Act* means the Federal Food, Drug, and Cosmetic Act, as amended (secs. 201–902, 52 Stat. 1040 *et seq.*, as amended (21 U.S.C. 321–392)).

(b) *Application for research or marketing permit* includes:

(1) A color additive petition, described in part 71.

(2) Data and information regarding a substance submitted as part of the procedures for establishing that a substance is generally recognized as safe for a use which results or may reasonably be expected to result, directly or indirectly, in its becoming a component or otherwise affecting the characteristics of any food, described in § 170.35.

(3) A food additive petition, described in part 171.

(4) Data and information regarding a food additive submitted as part of the procedures regarding food additives permitted to be used on an interim basis pending additional study, described in § 180.1.

(5) Data and information regarding a substance submitted as part of the procedures for establishing a tolerance for unavoidable contaminants in food and food-packaging materials, described in section 406 of the act.

(6) An investigational new drug application, described in part 312 of this chapter.

(7) A new drug application, described in part 314.

(8) Data and information regarding the bioavailability or bioequivalence of drugs for human use submitted as part of the procedures for issuing, amending, or repealing a bioequivalence requirement, described in part 320.

(9) Data and information regarding an over-the-counter drug for human use submitted as part of the procedures for classifying such drugs as generally recognized as safe and effective and not misbranded, described in part 330.

(10) An application for a biologics license, described in part 601 of this chapter.

(11) Data and information regarding a biological product submitted as part of the procedures for determining that licensed biological products are safe and effective and not misbranded, as described in part 601 of this chapter.

(12) An *Application for an Investigational Device Exemption*, described in parts 812 and 813.

(13) Data and information regarding a medical device for human use sub-

mitted as part of the procedures for classifying such devices, described in part 860.

(14) Data and information regarding a medical device for human use submitted as part of the procedures for establishing, amending, or repealing a standard for such device, described in part 861.

(15) An application for premarket approval of a medical device for human use, described in section 515 of the act.

(16) A product development protocol for a medical device for human use, described in section 515 of the act.

(17) Data and information regarding an electronic product submitted as part of the procedures for establishing, amending, or repealing a standard for such products, described in section 358 of the Public Health Service Act.

(18) Data and information regarding an electronic product submitted as part of the procedures for obtaining a variance from any electronic product performance standard, as described in § 1010.4.

(19) Data and information regarding an electronic product submitted as part of the procedures for granting, amending, or extending an exemption from a radiation safety performance standard, as described in § 1010.5.

(20) Data and information regarding an electronic product submitted as part of the procedures for obtaining an exemption from notification of a radiation safety defect or failure of compliance with a radiation safety performance standard, described in subpart D of part 1003.

(c) *Clinical investigation* means any experiment that involves a test article and one or more human subjects, and that either must meet the requirements for prior submission to the Food and Drug Administration under section 505(i) or 520(g) of the act, or need not meet the requirements for prior submission to the Food and Drug Administration under these sections of the act, but the results of which are intended to be later submitted to, or held for inspection by, the Food and Drug Administration as part of an application for a research or marketing permit. The term does not include experiments that must meet the provisions of part 58, regarding nonclinical laboratory studies.

The terms *research, clinical research, clinical study, study,* and *clinical investigation* are deemed to be synonymous for purposes of this part.

(d) *Emergency use* means the use of a test article on a human subject in a life-threatening situation in which no standard acceptable treatment is available, and in which there is not sufficient time to obtain IRB approval.

(e) *Human subject* means an individual who is or becomes a participant in research, either as a recipient of the test article or as a control. A subject may be either a healthy individual or a patient.

(f) *Institution* means any public or private entity or agency (including Federal, State, and other agencies). The term *facility* as used in section 520(g) of the act is deemed to be synonymous with the term *institution* for purposes of this part.

(g) *Institutional Review Board (IRB)* means any board, committee, or other group formally designated by an institution to review, to approve the initiation of, and to conduct periodic review of, biomedical research involving human subjects. The primary purpose of such review is to assure the protection of the rights and welfare of the human subjects. The term has the same meaning as the phrase *institutional review committee* as used in section 520(g) of the act.

(h) *Investigator* means an individual who actually conducts a clinical investigation (i.e., under whose immediate direction the test article is administered or dispensed to, or used involving, a subject) or, in the event of an investigation conducted by a team of individuals, is the responsible leader of that team.

(i) *Minimal risk* means that the probability and magnitude of harm or discomfort anticipated in the research are not greater in and of themselves than those ordinarily encountered in daily life or during the performance of routine physical or psychological examinations or tests.

(j) *Sponsor* means a person or other entity that initiates a clinical investigation, but that does not actually conduct the investigation, i.e., the test article is administered or dispensed to, or used involving, a subject under the immediate direction of another individual. A person other than an individual (e.g., a corporation or agency) that uses one or more of its own employees to conduct an investigation that it has initiated is considered to be a sponsor (not a sponsor-investigator), and the employees are considered to be investigators.

(k) *Sponsor-investigator* means an individual who both initiates and actually conducts, alone or with others, a clinical investigation, i.e., under whose immediate direction the test article is administered or dispensed to, or used involving, a subject. The term does not include any person other than an individual, e.g., it does not include a corporation or agency. The obligations of a sponsor-investigator under this part include both those of a sponsor and those of an investigator.

(l) *Test article* means any drug for human use, biological product for human use, medical device for human use, human food additive, color additive, electronic product, or any other article subject to regulation under the act or under sections 351 or 354–360F of the Public Health Service Act.

(m) *IRB approval* means the determination of the IRB that the clinical investigation has been reviewed and may be conducted at an institution within the constraints set forth by the IRB and by other institutional and Federal requirements.

[46 FR 8975, Jan. 27, 1981, as amended at 54 FR 9038, Mar. 3, 1989; 56 FR 28028, June 18, 1991; 64 FR 399, Jan. 5, 1999; 64 FR 56448, Oct. 20, 1999; 65 FR 52302, Aug. 29, 2000]

§ 56.103 Circumstances in which IRB review is required.

(a) Except as provided in §§ 56.104 and 56.105, any clinical investigation which must meet the requirements for prior submission (as required in parts 312, 812, and 813) to the Food and Drug Administration shall not be initiated unless that investigation has been reviewed and approved by, and remains subject to continuing review by, an IRB meeting the requirements of this part.

(b) Except as provided in §§ 56.104 and 56.105, the Food and Drug Administration may decide not to consider in support of an application for a research or marketing permit any data or information that has been derived from a clinical investigation that has not been approved by, and that was not subject to initial and continuing review by, an IRB meeting the requirements of this part. The determination that a clinical investigation may not be considered in support of an application for a research or marketing permit does not, however, relieve the applicant for such a permit of any obligation under any other applicable regulations to submit the results of the investigation to the Food and Drug Administration.

(c) Compliance with these regulations will in no way render inapplicable pertinent Federal, State, or local laws or regulations.

[46 FR 8975, Jan. 27, 1981; 46 FR 14340, Feb. 27, 1981]

§ 56.104   Exemptions from IRB requirement.

The following categories of clinical investigations are exempt from the requirements of this part for IRB review:

(a) Any investigation which commenced before July 27, 1981 and was subject to requirements for IRB review under FDA regulations before that date, provided that the investigation remains subject to review of an IRB which meets the FDA requirements in effect before July 27, 1981.

(b) Any investigation commenced before July 27, 1981 and was not otherwise subject to requirements for IRB review under Food and Drug Administration regulations before that date.

(c) Emergency use of a test article, provided that such emergency use is reported to the IRB within 5 working days. Any subsequent use of the test article at the institution is subject to IRB review.

(d) Taste and food quality evaluations and consumer acceptance studies, if wholesome foods without additives are consumed or if a food is consumed that contains a food ingredient at or below the level and for a use found to be safe, or agricultural, chemical, or environmental contaminant at or below the level found to be safe, by the Food and Drug Administration or approved by the Environmental Protection Agency or the Food Safety and In-

spection Service of the U.S. Department of Agriculture.

[46 FR 8975, Jan. 27, 1981, as amended at 56 FR 28028, June 18, 1991]

§ 56.105   Waiver of IRB requirement.

On the application of a sponsor or sponsor-investigator, the Food and Drug Administration may waive any of the requirements contained in these regulations, including the requirements for IRB review, for specific research activities or for classes of research activities, otherwise covered by these regulations.

Subpart B—Organization and Personnel

§ 56.107   IRB membership.

(a) Each IRB shall have at least five members, with varying backgrounds to promote complete and adequate review of research activities commonly conducted by the institution. The IRB shall be sufficiently qualified through the experience and expertise of its members, and the diversity of the members, including consideration of race, gender, cultural backgrounds, and sensitivity to such issues as community attitudes, to promote respect for its advice and counsel in safeguarding the rights and welfare of human subjects. In addition to possessing the professional competence necessary to review the specific research activities, the IRB shall be able to ascertain the acceptability of proposed research in terms of institutional commitments and regulations, applicable law, and standards or professional conduct and practice. The IRB shall therefore include persons knowledgeable in these areas. If an IRB regularly reviews research that involves a vulnerable catgory of subjects, such as children, prisoners, pregnant women, or handicapped or mentally disabled persons, consideration shall be given to the inclusion of one or more individuals who are knowledgeable about and experienced in working with those subjects.

(b) Every nondiscriminatory effort will be made to ensure that no IRB consists entirely of men or entirely of women, including the instituton's consideration of qualified persons of both sexes, so long as no selection is made to the IRB on the basis of gender. No IRB may consist entirely of members of one profession.

(c) Each IRB shall include at least one member whose primary concerns are in the scientific area and at least one member whose primary concerns are in nonscientific areas.

(d) Each IRB shall include at least one member who is not otherwise affiliated with the institution and who is not part of the immediate family of a person who is affiliated with the institution.

(e) No IRB may have a member participate in the IRB's initial or continuing review of any project in which the member has a conflicting interest, except to provide information requested by the IRB.

(f) An IRB may, in its discretion, invite individuals with competence in special areas to assist in the review of complex issues which require expertise beyond or in addition to that available on the IRB. These individuals may not vote with the IRB.

[46 FR 8975, Jan 27, 1981, as amended at 56 FR 28028, June 18, 1991; 56 FR 29756, June 28, 1991]

Subpart C—IRB Functions and Operations

§56.108 IRB functions and operations.

In order to fulfill the requirements of these regulations, each IRB shall:

(a) Follow written procedures: (1) For conducting its initial and continuing review of research and for reporting its findings and actions to the investigator and the institution; (2) for determining which projects require review more often than annually and which projects need verification from sources other than the investigator that no material changes have occurred since previous IRB review; (3) for ensuring prompt reporting to the IRB of changes in research activity; and (4) for ensuring that changes in approved research, during the period for which IRB approval has already been given, may not be initiated without IRB review and approval except where necessary to eliminate apparent immediate hazards to the human subjects.

(b) Follow written procedures for ensuring prompt reporting to the IRB, appropriate institutional officials, and the Food and Drug Administration of: (1) Any unanticipated problems involving risks to human subjects or others; (2) any instance of serious or continuing noncompliance with these regulations or the requirements or determinations of the IRB; or (3) any suspension or termination of IRB approval.

(c) Except when an expedited review procedure is used (see §56.110), review proposed research at convened meetings at which a majority of the members of the IRB are present, including at least one member whose primary concerns are in nonscientific areas. In order for the research to be approved, it shall receive the approval of a majority of those members present at the meeting.

(Information collection requirements in this section were approved by the Office of Management and Budget (OMB) and assigned OMB control number 0910–0130)

[46 FR 8975, Jan. 27, 1981, as amended at 56 FR 28028, June 18, 1991]

§56.109 IRB review of research.

(a) An IRB shall review and have authority to approve, require modifications in (to secure approval), or disapprove all research activities covered by these regulations.

(b) An IRB shall require that information given to subjects as part of informed consent is in accordance with §50.25. The IRB may require that information, in addition to that specifically mentioned in §50.25, be given to the subjects when in the IRB's judgment the information would meaningfully add to the protection of the rights and welfare of subjects.

(c) An IRB shall require documentation of informed consent in accordance with §50.27 of this chapter, except as follows:

(1) The IRB may, for some or all subjects, waive the requirement that the subject, or the subject's legally authorized representative, sign a written consent form if it finds that the research presents no more than minimal risk of

harm to subjects and involves no procedures for which written consent is normally required outside the research context; or

(2) The IRB may, for some or all subjects, find that the requirements in § 50.24 of this chapter for an exception from informed consent for emergency research are met.

(d) In cases where the documentation requirement is waived under paragraph (c)(1) of this section, the IRB may require the investigator to provide subjects with a written statement regarding the research.

(e) An IRB shall notify investigators and the institution in writing of its decision to approve or disapprove the proposed research activity, or of modifications required to secure IRB approval of the research activity. If the IRB decides to disapprove a research activity, it shall include in its written notification a statement of the reasons for its decision and give the investigator an opportunity to respond in person or in writing. For investigations involving an exception to informed consent under § 50.24 of this chapter, an IRB shall promptly notify in writing the investigator and the sponsor of the research when an IRB determines that it cannot approve the research because it does not meet the criteria in the exception provided under § 50.24(a) of this chapter or because of other relevant ethical concerns. The written notification shall include a statement of the reasons for the IRB's determination.

(f) An IRB shall conduct continuing review of research covered by these regulations at intervals appropriate to the degree of risk, but not less than once per year, and shall have authority to observe or have a third party observe the consent process and the research.

(g) An IRB shall provide in writing to the sponsor of research involving an exception to informed consent under § 50.24 of this chapter a copy of information that has been publicly disclosed under § 50.24(a)(7)(ii) and (a)(7)(iii) of this chapter. The IRB shall provide this information to the sponsor promptly so that the sponsor is aware that such disclosure has occurred. Upon receipt, the sponsor shall provide copies of the information disclosed to FDA.

[46 FR 8975, Jan. 27, 1981, as amended at 61 FR 51529, Oct. 2, 1996]

§ 56.110 Expedited review procedures for certain kinds of research involving no more than minimal risk, and for minor changes in approved research.

(a) The Food and Drug Administration has established, and published in the FEDERAL REGISTER, a list of categories of research that may be reviewed by the IRB through an expedited review procedure. The list will be amended, as appropriate, through periodic republication in the FEDERAL REGISTER.

(b) An IRB may use the expedited review procedure to review either or both of the following: (1) Some or all of the research appearing on the list and found by the reviewer(s) to involve no more than minimal risk, (2) minor changes in previously approved research during the period (of 1 year or less) for which approval is authorized. Under an expedited review procedure, the review may be carried out by the IRB chairperson or by one or more experienced reviewers designated by the IRB chairperson from among the members of the IRB. In reviewing the research, the reviewers may exercise all of the authorities of the IRB except that the reviewers may not disapprove the research. A research activity may be disapproved only after review in accordance with the nonexpedited review procedure set forth in § 56.108(c).

(c) Each IRB which uses an expedited review procedure shall adopt a method for keeping all members advised of research proposals which have been approved under the procedure.

(d) The Food and Drug Administration may restrict, suspend, or terminate an institution's or IRB's use of the expedited review procedure when necessary to protect the rights or welfare of subjects.

[46 FR 8975, Jan. 27, 1981, as amended at 56 FR 28029, June 18, 1991]

§ 56.111 Criteria for IRB approval of research.

(a) In order to approve research covered by these regulations the IRB shall

§ 56.112

determine that all of the following requirements are satisfied:

(1) Risks to subjects are minimized: (i) By using procedures which are consistent with sound research design and which do not unnecessarily expose subjects to risk, and (ii) whenever appropriate, by using procedures already being performed on the subjects for diagnostic or treatment purposes.

(2) Risks to subjects are reasonable in relation to anticipated benefits, if any, to subjects, and the importance of the knowledge that may be expected to result. In evaluating risks and benefits, the IRB should consider only those risks and benefits that may result from the research (as distinguished from risks and benefits of therapies that subjects would receive even if not participating in the research). The IRB should not consider possible long-range effects of applying knowledge gained in the research (for example, the possible effects of the research on public policy) as among those research risks that fall within the purview of its responsibility.

(3) Selection of subjects is equitable. In making this assessment the IRB should take into account the purposes of the research and the setting in which the research will be conducted and should be particularly cognizant of the special problems of research involving vulnerable populations, such as children, prisoners, pregnant women, handicapped, or mentally disabled persons, or economically or educationally disadvantaged persons.

(4) Informed consent will be sought from each prospective subject or the subject's legally authorized representative, in accordance with and to the extent required by part 50.

(5) Informed consent will be appropriately documented, in accordance with and to the extent required by § 50.27.

(6) Where appropriate, the research plan makes adequate provision for monitoring the data collected to ensure the safety of subjects.

(7) Where appropriate, there are adequate provisions to protect the privacy of subjects and to maintain the confidentiality of data.

(b) When some or all of the subjects, such as children, prisoners, pregnant women, handicapped, or mentally disabled persons, or economically or educationally disadvantaged persons, are likely to be vulnerable to coercion or undue influence additional safeguards have been included in the study to protect the rights and welfare of these subjects.

[46 FR 8975, Jan. 27, 1981, as amended at 56 FR 28029, June 18, 1991]

§ 56.112 Review by institution.

Research covered by these regulations that has been approved by an IRB may be subject to further appropriate review and approval or disapproval by officials of the institution. However, those officials may not approve the research if it has not been approved by an IRB.

§ 56.113 Suspension or termination of IRB approval of research.

An IRB shall have authority to suspend or terminate approval of research that is not being conducted in accordance with the IRB's requirements or that has been associated with unexpected serious harm to subjects. Any suspension or termination of approval shall include a statement of the reasons for the IRB's action and shall be reported promptly to the investigator, appropriate institutional officials, and the Food and Drug Administration.

§ 56.114 Cooperative research.

In complying with these regulations, institutions involved in multi-institutional studies may use joint review, reliance upon the review of another qualified IRB, or similar arrangements aimed at avoidance of duplication of effort.

Subpart D—Records and Reports

§ 56.115 IRB records.

(a) An institution, or where appropriate an IRB, shall prepare and maintain adequate documentation of IRB activities, including the following:

(1) Copies of all research proposals reviewed, scientific evaluations, if any, that accompany the proposals, approved sample consent documents,

Food and Drug Administration, HHS

§ 56.120

progress reports submitted by investigators, and reports of injuries to subjects.

(2) Minutes of IRB meetings which shall be in sufficient detail to show attendance at the meetings; actions taken by the IRB; the vote on these actions including the number of members voting for, against, and abstaining; the basis for requiring changes in or disapproving research; and a written summary of the discussion of controverted issues and their resolution.

(3) Records of continuing review activities.

(4) Copies of all correspondence between the IRB and the investigators.

(5) A list of IRB members identified by name; earned degrees; representative capacity; indications of experience such as board certifications, licenses, etc., sufficient to describe each member's chief anticipated contributions to IRB deliberations; and any employment or other relationship between each member and the institution; for example: full-time employee, part-time employee, a member of governing panel or board, stockholder, paid or unpaid consultant.

(6) Written procedures for the IRB as required by § 56.108 (a) and (b).

(7) Statements of significant new findings provided to subjects, as required by § 50.25.

(b) The records required by this regulation shall be retained for at least 3 years after completion of the research, and the records shall be accessible for inspection and copying by authorized representatives of the Food and Drug Administration at reasonable times and in a reasonable manner.

(c) The Food and Drug Administration may refuse to consider a clinical investigation in support of an application for a research or marketing permit if the institution or the IRB that reviewed the investigation refuses to allow an inspection under this section.

(Information collection requirements in this section were approved by the Office of Management and Budget (OMB) and assigned OMB control number 0910–0130)

[46 FR 8975, Jan. 27, 1981, as amended at 56 FR 28029, June 18, 1991]

Subpart E—Administrative Actions for Noncompliance

§ 56.120 Lesser administrative actions.

(a) If apparent noncompliance with these regulations in the operation of an IRB is observed by an FDA investigator during an inspection, the inspector will present an oral or written summary of observations to an appropriate representative of the IRB. The Food and Drug Administration may subsequently send a letter describing the noncompliance to the IRB and to the parent institution. The agency will require that the IRB or the parent institution respond to this letter within a time period specified by FDA and describe the corrective actions that will be taken by the IRB, the institution, or both to achieve compliance with these regulations.

(b) On the basis of the IRB's or the institution's response, FDA may schedule a reinspection to confirm the adequacy of corrective actions. In addition, until the IRB or the parent institution takes appropriate corrective action, the agency may:

(1) Withhold approval of new studies subject to the requirements of this part that are conducted at the institution or reviewed by the IRB;

(2) Direct that no new subjects be added to ongoing studies subject to this part;

(3) Terminate ongoing studies subject to this part when doing so would not endanger the subjects; or

(4) When the apparent noncompliance creates a significant threat to the rights and welfare of human subjects, notify relevant State and Federal regulatory agencies and other parties with a direct interest in the agency's action of the deficiencies in the operation of the IRB.

(c) The parent institution is presumed to be responsible for the operation of an IRB, and the Food and Drug Administration will ordinarily direct any administrative action under this subpart against the institution. However, depending on the evidence of responsibility for deficiencies, determined during the investigation, the Food and Drug Administration may restrict its administrative actions to the IRB or to a component of the parent institution determined to be responsible for formal designation of the IRB.

§ 56.121 Disqualification of an IRB or an institution.

(a) Whenever the IRB or the institution has failed to take adequate steps to correct the noncompliance stated in the letter sent by the agency under § 56.120(a), and the Commissioner of Food and Drugs determines that this noncompliance may justify the disqualification of the IRB or of the parent institution, the Commissioner will institute proceedings in accordance with the requirements for a regulatory hearing set forth in part 16.

(b) The Commissioner may disqualify an IRB or the parent institution if the Commissioner determines that:

(1) The IRB has refused or repeatedly failed to comply with any of the regulations set forth in this part, and

(2) The noncompliance adversely affects the rights or welfare of the human subjects in a clinical investigation.

(c) If the Commissioner determines that disqualification is appropriate, the Commissioner will issue an order that explains the basis for the determination and that prescribes any actions to be taken with regard to ongoing clinical research conducted under the review of the IRB. The Food and Drug Administration will send notice of the disqualification to the IRB and the parent institution. Other parties with a direct interest, such as sponsors and clinical investigators, may also be sent a notice of the disqualification. In addition, the agency may elect to publish a notice of its action in the FEDERAL REGISTER.

(d) The Food and Drug Administration will not approve an application for a research permit for a clinical investigation that is to be under the review of a disqualified IRB or that is to be conducted at a disqualified institution, and it may refuse to consider in support of a marketing permit the data from a clinical investigation that was reviewed by a disqualified IRB as conducted at a disqualified institution, unless the IRB or the parent institution is reinstated as provided in § 56.123.

§ 56.122 Public disclosure of information regarding revocation.

A determination that the Food and Drug Administration has disqualified an institution and the administrative record regarding that determination are disclosable to the public under part 20.

§ 56.123 Reinstatement of an IRB or an institution.

An IRB or an institution may be reinstated if the Commissioner determines, upon an evaluation of a written submission from the IRB or institution that explains the corrective action that the institution or IRB plans to take, that the IRB or institution has provided adequate assurance that it will operate in compliance with the standards set forth in this part. Notification of reinstatement shall be provided to all persons notified under § 56.121(c).

§ 56.124 Actions alternative or additional to disqualification.

Disqualification of an IRB or of an institution is independent of, and neither in lieu of nor a precondition to, other proceedings or actions authorized by the act. The Food and Drug Administration may, at any time, through the Department of Justice institute any appropriate judicial proceedings (civil or criminal) and any other appropriate regulatory action, in addition to or in lieu of, and before, at the time of, or after, disqualification. The agency may also refer pertinent matters to another Federal, State, or local government agency for any action that that agency determines to be appropriate.

PART 58—GOOD LABORATORY PRACTICE FOR NONCLINICAL LABORATORY STUDIES

Subpart A—General Provisions

Subpart B—Organization and Personnel

§312.1

AUTHORITY: 21 U.S.C. 321, 331, 351, 352, 353, 355, 371; 42 U.S.C. 262.

SOURCE: 52 FR 8831, Mar. 19, 1987, unless otherwise noted.

Subpart A—General Provisions

§312.1 Scope.

(a) This part contains procedures and requirements governing the use of investigational new drugs, including procedures and requirements for the submission to, and review by, the Food and Drug Administration of investigational new drug applications (IND's). An investigational new drug for which an IND is in effect in accordance with this part is exempt from the premarketing approval requirements that are otherwise applicable and may be shipped lawfully for the purpose of conducting clinical investigations of that drug.

(b) References in this part to regulations in the Code of Federal Regulations are to chapter I of title 21, unless otherwise noted.

§312.2 Applicability.

(a) *Applicability.* Except as provided in this section, this part applies to all clinical investigations of products that are subject to section 505 of the Federal Food, Drug, and Cosmetic Act or to the licensing provisions of the Public Health Service Act (58 Stat. 632, as amended (42 U.S.C. 201 *et seq.*)).

(b) *Exemptions.* (1) The clinical investigation of a drug product that is lawfully marketed in the United States is exempt from the requirements of this part if all the following apply:

(i) The investigation is not intended to be reported to FDA as a well-controlled study in support of a new indication for use nor intended to be used to support any other significant change in the labeling for the drug;

(ii) If the drug that is undergoing investigation is lawfully marketed as a prescription drug product, the investigation is not intended to support a

Food and Drug Administration, HHS   **§312.3**

significant change in the advertising for the product;

(iii) The investigation does not involve a route of administration or dosage level or use in a patient population or other factor that significantly increases the risks (or decreases the acceptability of the risks) associated with the use of the drug product;

(iv) The investigation is conducted in compliance with the requirements for institutional review set forth in part 56 and with the requirements for informed consent set forth in part 50; and

(v) The investigation is conducted in compliance with the requirements of §312.7.

(2)(i) A clinical investigation involving an in vitro diagnostic biological product listed in paragraph (b)(2)(ii) of this section is exempt from the requirements of this part if (a) it is intended to be used in a diagnostic procedure that confirms the diagnosis made by another, medically established, diagnostic product or procedure and (b) it is shipped in compliance with §312.160.

(ii) In accordance with paragraph (b)(2)(i) of this section, the following products are exempt from the requirements of this part: (a) blood grouping serum; (b) reagent red blood cells; and (c) anti-human globulin.

(3) A drug intended solely for tests in vitro or in laboratory research animals is exempt from the requirements of this part if shipped in accordance with §312.160.

(4) FDA will not accept an application for an investigation that is exempt under the provisions of paragraph (b)(1) of this section.

(5) A clinical investigation involving use of a placebo is exempt from the requirements of this part if the investigation does not otherwise require submission of an IND.

(6) A clinical investigation involving an exception from informed consent under §50.24 of this chapter is not exempt from the requirements of this part.

(c) *Bioavailability studies.* The applicability of this part to in vivo bioavailability studies in humans is subject to the provisions of §320.31.

(d) *Unlabeled indication.* This part does not apply to the use in the practice of medicine for an unlabeled indi-

cation of a new drug product approved under part 314 or of a licensed biological product.

(e) *Guidance.* FDA may, on its own initiative, issue guidance on the applicability of this part to particular investigational uses of drugs. On request, FDA will advise on the applicability of this part to a planned clinical investigation.

[52 FR 8831, Mar. 19, 1987, as amended at 61 FR 51529, Oct. 2, 1996; 64 FR 401, Jan. 5, 1999]

§312.3 Definitions and interpretations.

(a) The definitions and interpretations of terms contained in section 201 of the Act apply to those terms when used in this part:

(b) The following definitions of terms also apply to this part:

Act means the Federal Food, Drug, and Cosmetic Act (secs. 201–902, 52 Stat. 1040 *et seq.*, as amended (21 U.S.C. 301–392)).

Clinical investigation means any experiment in which a drug is administered or dispensed to, or used involving, one or more human subjects. For the purposes of this part, an experiment is any use of a drug except for the use of a marketed drug in the course of medical practice.

Contract research organization means a person that assumes, as an independent contractor with the sponsor, one or more of the obligations of a sponsor, e.g., design of a protocol, selection or monitoring of investigations, evaluation of reports, and preparation of materials to be submitted to the Food and Drug Administration.

FDA means the Food and Drug Administration.

IND means an investigational new drug application. For purposes of this part, "IND" is synonymous with "Notice of Claimed Investigational Exemption for a New Drug."

Investigational new drug means a new drug or biological drug that is used in a clinical investigation. The term also includes a biological product that is used in vitro for diagnostic purposes. The terms "investigational drug" and "investigational new drug" are deemed to be synonymous for purposes of this part.

Investigator means an individual who actually conducts a clinical investigation (i.e., under whose immediate direction the drug is administered or dispensed to a subject). In the event an investigation is conducted by a team of individuals, the investigator is the responsible leader of the team. "Subinvestigator" includes any other individual member of that team.

Marketing application means an application for a new drug submitted under section 505(b) of the act or a biologics license application for a biological product submitted under the Public Health Service Act.

Sponsor means a person who takes responsibility for and initiates a clinical investigation. The sponsor may be an individual or pharmaceutical company, governmental agency, academic institution, private organization, or other organization. The sponsor does not actually conduct the investigation unless the sponsor is a sponsor-investigator. A person other than an individual that uses one or more of its own employees to conduct an investigation that it has initiated is a sponsor, not a sponsor-investigator, and the employees are investigators.

Sponsor-Investigator means an individual who both initiates and conducts an investigation, and under whose immediate direction the investigational drug is administered or dispensed. The term does not include any person other than an individual. The requirements applicable to a sponsor-investigator under this part include both those applicable to an investigator and a sponsor.

Subject means a human who participates in an investigation, either as a recipient of the investigational new drug or as a control. A subject may be a healthy human or a patient with a disease.

[52 FR 8831, Mar. 19, 1987, as amended at 64 FR 401, Jan. 5, 1999; 64 FR 56449, Oct. 20, 1999]

§ 312.6 Labeling of an investigational new drug.

(a) The immediate package of an investigational new drug intended for human use shall bear a label with the statement "Caution: New Drug—Limited by Federal (or United States) law to investigational use."

(b) The label or labeling of an investigational new drug shall not bear any statement that is false or misleading in any particular and shall not represent that the investigational new drug is safe or effective for the purposes for which it is being investigated.

§ 312.7 Promotion and charging for investigational drugs.

(a) *Promotion of an investigational new drug.* A sponsor or investigator, or any person acting on behalf of a sponsor or investigator, shall not represent in a promotional context that an investigational new drug is safe or effective for the purposes for which it is under investigation or otherwise promote the drug. This provision is not intended to restrict the full exchange of scientific information concerning the drug, including dissemination of scientific findings in scientific or lay media. Rather, its intent is to restrict promotional claims of safety or effectiveness of the drug for a use for which it is under investigation and to preclude commercialization of the drug before it is approved for commercial distribution.

(b) *Commercial distribution of an investigational new drug.* A sponsor or investigator shall not commercially distribute or test market an investigational new drug.

(c) *Prolonging an investigation.* A sponsor shall not unduly prolong an investigation after finding that the results of the investigation appear to establish sufficient data to support a marketing application.

(d) *Charging for and commercialization of investigational drugs—*(1) *Clinical trials under an IND.* Charging for an investigational drug in a clinical trial under an IND is not permitted without the prior written approval of FDA. In requesting such approval, the sponsor shall provide a full written explanation of why charging is necessary in order for the sponsor to undertake or continue the clinical trial, e.g., why distribution of the drug to test subjects should not be considered part of the normal cost of doing business.

(2) *Treatment protocol or treatment IND.* A sponsor or investigator may charge for an investigational drug for a

Food and Drug Administration, HHS   §312.21

treatment use under a treatment protocol or treatment IND provided: (i) There is adequate enrollment in the ongoing clinical investigations under the authorized IND; (ii) charging does not constitute commercial marketing of a new drug for which a marketing application has not been approved; (iii) the drug is not being commercially promoted or advertised; and (iv) the sponsor of the drug is actively pursuing marketing approval with due diligence. FDA must be notified in writing in advance of commencing any such charges, in an information amendment submitted under §312.31. Authorization for charging goes into effect automatically 30 days after receipt by FDA of the information amendment, unless the sponsor is notified to the contrary.

(3) *Noncommercialization of investigational drug.* Under this section, the sponsor may not commercialize an investigational drug by charging a price larger than that necessary to recover costs of manufacture, research, development, and handling of the investigational drug.

(4) *Withdrawal of authorization.* Authorization to charge for an investigational drug under this section may be withdrawn by FDA if the agency finds that the conditions underlying the authorization are no longer satisfied.

(Collection of information requirements approved by the Office of Management and Budget under control number 0910–0014)

[52 FR 8831, Mar. 19, 1987, as amended at 52 FR 19476, May 22, 1987]

§312.10 Waivers.

(a) A sponsor may request FDA to waive applicable requirement under this part. A waiver request may be submitted either in an IND or in an information amendment to an IND. In an emergency, a request may be made by telephone or other rapid communication means. A waiver request is required to contain at least one of the following:

(1) An explanation why the sponsor's compliance with the requirement is unnecessary or cannot be achieved;

(2) A description of an alternative submission or course of action that satisfies the purpose of the requirement; or

(3) Other information justifying a waiver.

(b) FDA may grant a waiver if it finds that the sponsor's noncompliance would not pose a significant and unreasonable risk to human subjects of the investigation and that one of the following is met:

(1) The sponsor's compliance with the requirement is unnecessary for the agency to evaluate the application, or compliance cannot be achieved;

(2) The sponsor's proposed alternative satisfies the requirement; or

(3) The applicant's submission otherwise justifies a waiver.

(Collection of information requirements approved by the Office of Management and Budget under control number 0910–0014)

[52 FR 8831, Mar. 19, 1987, as amended at 52 FR 23031, June 17, 1987]

Subpart B—Investigational New Drug Application (IND)

§312.20 Requirement for an IND.

(a) A sponsor shall submit an IND to FDA if the sponsor intends to conduct a clinical investigation with an investigational new drug that is subject to §312.2(a).

(b) A sponsor shall not begin a clinical investigation subject to §312.2(a) until the investigation is subject to an IND which is in effect in accordance with §312.40.

(c) A sponsor shall submit a separate IND for any clinical investigation involving an exception from informed consent under §50.24 of this chapter. Such a clinical investigation is not permitted to proceed without the prior written authorization from FDA. FDA shall provide a written determination 30 days after FDA receives the IND or earlier.

[52 FR 8831, Mar. 19, 1987, as amended at 61 FR 51529, Oct. 2, 1996; 62 FR 32479, June 16, 1997]

§312.21 Phases of an investigation.

An IND may be submitted for one or more phases of an investigation. The clinical investigation of a previously untested drug is generally divided into three phases. Although in general the phases are conducted sequentially, they may overlap. These three phases of an investigation are a follows:

(a) *Phase 1*. (1) Phase 1 includes the initial introduction of an investigational new drug into humans. Phase 1 studies are typically closely monitored and may be conducted in patients or normal volunteer subjects. These studies are designed to determine the metabolism and pharmacologic actions of the drug in humans, the side effects associated with increasing doses, and, if possible, to gain early evidence on effectiveness. During Phase 1, sufficient information about the drug's pharmacokinetics and pharmacological effects should be obtained to permit the design of well-controlled, scientifically valid, Phase 2 studies. The total number of subjects and patients included in Phase 1 studies varies with the drug, but is generally in the range of 20 to 80.

(2) Phase 1 studies also include studies of drug metabolism, structure-activity relationships, and mechanism of action in humans, as well as studies in which investigational drugs are used as research tools to explore biological phenomena or disease processes.

(b) *Phase 2*. Phase 2 includes the controlled clinical studies conducted to evaluate the effectiveness of the drug for a particular indication or indications in patients with the disease or condition under study and to determine the common short-term side effects and risks associated with the drug. Phase 2 studies are typically well controlled, closely monitored, and conducted in a relatively small number of patients, usually involving no more than several hundred subjects.

(c) *Phase 3*. Phase 3 studies are expanded controlled and uncontrolled trials. They are performed after preliminary evidence suggesting effectiveness of the drug has been obtained, and are intended to gather the additional information about effectiveness and safety that is needed to evaluate the overall benefit-risk relationship of the drug and to provide an adequate basis for physician labeling. Phase 3 studies usually include from several hundred to several thousand subjects.

§ 312.22 General principles of the IND submission.

(a) FDA's primary objectives in reviewing an IND are, in all phases of the investigation, to assure the safety and rights of subjects, and, in Phase 2 and 3, to help assure that the quality of the scientific evaluation of drugs is adequate to permit an evaluation of the drug's effectiveness and safety. Therefore, although FDA's review of Phase 1 submissions will focus on assessing the safety of Phase 1 investigations, FDA's review of Phases 2 and 3 submissions will also include an assessment of the scientific quality of the clinical investigations and the likelihood that the investigations will yield data capable of meeting statutory standards for marketing approval.

(b) The amount of information on a particular drug that must be submitted in an IND to assure the accomplishment of the objectives described in paragraph (a) of this section depends upon such factors as the novelty of the drug, the extent to which it has been studied previously, the known or suspected risks, and the developmental phase of the drug.

(c) The central focus of the initial IND submission should be on the general investigational plan and the protocols for specific human studies. Subsequent amendments to the IND that contain new or revised protocols should build logically on previous submissions and should be supported by additional information, including the results of animal toxicology studies or other human studies as appropriate. Annual reports to the IND should serve as the focus for reporting the status of studies being conducted under the IND and should update the general investigational plan for the coming year.

(d) The IND format set forth in § 312.23 should be followed routinely by sponsors in the interest of fostering an efficient review of applications. Sponsors are expected to exercise considerable discretion, however, regarding the content of information submitted in each section, depending upon the kind of drug being studied and the nature of the available information. Section 312.23 outlines the information needed

Food and Drug Administration, HHS

for a commercially sponsored IND for a new molecular entity. A sponsor-investigator who uses, as a research tool, an investigational new drug that is already subject to a manufacturer's IND or marketing application should follow the same general format, but ordinarily may, if authorized by the manufacturer, refer to the manufacturer's IND or marketing application in providing the technical information supporting the proposed clinical investigation. A sponsor-investigator who uses an investigational drug not subject to a manufacturer's IND or marketing application is ordinarily required to submit all technical information supporting the IND, unless such information may be referenced from the scientific literature.

§ 312.23   IND content and format.

(a) A sponsor who intends to conduct a clinical investigation subject to this part shall submit an "Investigational New Drug Application" (IND) including, in the following order:

(1) *Cover sheet (Form FDA–1571)*. A cover sheet for the application containing the following:

(i) The name, address, and telephone number of the sponsor, the date of the application, and the name of the investigational new drug.

(ii) Identification of the phase or phases of the clinical investigation to be conducted.

(iii) A commitment not to begin clinical investigations until an IND covering the investigations is in effect.

(iv) A commitment that an Institutional Review Board (IRB) that complies with the requirements set forth in part 56 will be responsible for the initial and continuing review and approval of each of the studies in the proposed clinical investigation and that the investigator will report to the IRB proposed changes in the research activity in accordance with the requirements of part 56.

(v) A commitment to conduct the investigation in accordance with all other applicable regulatory requirements.

(vi) The name and title of the person responsible for monitoring the conduct and progress of the clinical investigations.

(vii) The name(s) and title(s) of the person(s) responsible under § 312.32 for review and evaluation of information relevant to the safety of the drug.

(viii) If a sponsor has transferred any obligations for the conduct of any clinical study to a contract research organization, a statement containing the name and address of the contract research organization, identification of the clinical study, and a listing of the obligations transferred. If all obligations governing the conduct of the study have been transferred, a general statement of this transfer—in lieu of a listing of the specific obligations transferred—may be submitted.

(ix) The signature of the sponsor or the sponsor's authorized representative. If the person signing the application does not reside or have a place of business within the United States, the IND is required to contain the name and address of, and be countersigned by, an attorney, agent, or other authorized official who resides or maintains a place of business within the United States.

(2) *A table of contents*.

(3) *Introductory statement and general investigational plan*. (i) A brief introductory statement giving the name of the drug and all active ingredients, the drug's pharmacological class, the structural formula of the drug (if known), the formulation of the dosage form(s) to be used, the route of administration, and the broad objectives and planned duration of the proposed clinical investigation(s).

(ii) A brief summary of previous human experience with the drug, with reference to other IND's if pertinent, and to investigational or marketing experience in other countries that may be relevant to the safety of the proposed clinical investigation(s).

(iii) If the drug has been withdrawn from investigation or marketing in any country for any reason related to safety or effectiveness, identification of the country(ies) where the drug was withdrawn and the reasons for the withdrawal.

(iv) A brief description of the overall plan for investigating the drug product for the following year. The plan should include the following: (*a*) The rationale for the drug or the research study; (*b*) the indication(s) to be studied; (c) the general approach to be followed in evaluating the drug; (d) the kinds of clinical trials to be conducted in the first year following the submission (if plans are not developed for the entire year, the sponsor should so indicate); (e) the estimated number of patients to be given the drug in those studies; and (f) any risks of particular severity or seriousness anticipated on the basis of the toxicological data in animals or prior studies in humans with the drug or related drugs.

(4) [Reserved]

(5) *Investigator's brochure.* If required under § 312.55, a copy of the investigator's brochure, containing the following information:

(i) A brief description of the drug substance and the formulation, including the structural formula, if known.

(ii) A summary of the pharmacological and toxicological effects of the drug in animals and, to the extent known, in humans.

(iii) A summary of the pharmacokinetics and biological disposition of the drug in animals and, if known, in humans.

(iv) A summary of information relating to safety and effectiveness in humans obtained from prior clinical studies. (Reprints of published articles on such studies may be appended when useful.)

(v) A description of possible risks and side effects to be anticipated on the basis of prior experience with the drug under investigation or with related drugs, and of precautions or special monitoring to be done as part of the investigational use of the drug.

(6) *Protocols.* (i) A protocol for each planned study. (Protocols for studies not submitted initially in the IND should be submitted in accordance with § 312.30(a).) In general, protocols for Phase 1 studies may be less detailed and more flexible than protocols for Phase 2 and 3 studies. Phase 1 protocols should be directed primarily at providing an outline of the investigation—an estimate of the number of patients to be involved, a description of safety exclusions, and a description of the dosing plan including duration, dose, or method to be used in determining dose—and should specify in detail only

those elements of the study that are critical to safety, such as necessary monitoring of vital signs and blood chemistries. Modifications of the experimental design of Phase 1 studies that do not affect critical safety assessments are required to be reported to FDA only in the annual report.

(ii) In Phases 2 and 3, detailed protocols describing all aspects of the study should be submitted. A protocol for a Phase 2 or 3 investigation should be designed in such a way that, if the sponsor anticipates that some deviation from the study design may become necessary as the investigation progresses, alternatives or contingencies to provide for such deviation are built into the protocols at the outset. For example, a protocol for a controlled short-term study might include a plan for an early crossover of nonresponders to an alternative therapy.

(iii) A protocol is required to contain the following, with the specific elements and detail of the protocol reflecting the above distinctions depending on the phase of study:

(a) A statement of the objectives and purpose of the study.

(b) The name and address and a statement of the qualifications (curriculum vitae or other statement of qualifications) of each investigator, and the name of each subinvestigator (e.g., research fellow, resident) working under the supervision of the investigator; the name and address of the research facilities to be used; and the name and address of each reviewing Institutional Review Board.

(c) The criteria for patient selection and for exclusion of patients and an estimate of the number of patients to be studied.

(d) A description of the design of the study, including the kind of control group to be used, if any, and a description of methods to be used to minimize bias on the part of subjects, investigators, and analysts.

(e) The method for determining the dose(s) to be administered, the planned maximum dosage, and the duration of individual patient exposure to the drug.

(f) A description of the observations and measurements to be made to fulfill the objectives of the study.

Food and Drug Administration, HHS   **§ 312.23**

(g) A description of clinical procedures, laboratory tests, or other measures to be taken to monitor the effects of the drug in human subjects and to minimize risk.

(7) *Chemistry, manufacturing, and control information.* (i) As appropriate for the particular investigations covered by the IND, a section describing the composition, manufacture, and control of the drug substance and the drug product. Although in each phase of the investigation sufficient information is required to be submitted to assure the proper identification, quality, purity, and strength of the investigational drug, the amount of information needed to make that assurance will vary with the phase of the investigation, the proposed duration of the investigation, the dosage form, and the amount of information otherwise available. FDA recognizes that modifications to the method of preparation of the new drug substance and dosage form and changes in the dosage form itself are likely as the investigation progresses. Therefore, the emphasis in an initial Phase 1 submission should generally be placed on the identification and control of the raw materials and the new drug substance. Final specifications for the drug substance and drug product are not expected until the end of the investigational process.

(ii) It should be emphasized that the amount of information to be submitted depends upon the scope of the proposed clinical investigation. For example, although stability data are required in all phases of the IND to demonstrate that the new drug substance and drug product are within acceptable chemical and physical limits for the planned duration of the proposed clinical investigation, if very short-term tests are proposed, the supporting stability data can be correspondingly limited.

(iii) As drug development proceeds and as the scale or production is changed from the pilot-scale production appropriate for the limited initial clinical investigations to the larger-scale production needed for expanded clinical trials, the sponsor should submit information amendments to supplement the initial information submitted on the chemistry, manufacturing, and control processes with in-

formation appropriate to the expanded scope of the investigation.

(iv) Reflecting the distinctions described in this paragraph (a)(7), and based on the phase(s) to be studied, the submission is required to contain the following:

(a) *Drug substance.* A description of the drug substance, including its physical, chemical, or biological characteristics; the name and address of its manufacturer; the general method of preparation of the drug substance; the acceptable limits and analytical methods used to assure the identity, strength, quality, and purity of the drug substance; and information sufficient to support stability of the drug substance during the toxicological studies and the planned clinical studies. Reference to the current edition of the United States Pharmacopeia—National Formulary may satisfy relevant requirements in this paragraph.

(b) *Drug product.* A list of all components, which may include reasonable alternatives for inactive compounds, used in the manufacture of the investigational drug product, including both those components intended to appear in the drug product and those which may not appear but which are used in the manufacturing process, and, where applicable, the quantitative composition of the investigational drug product, including any reasonable variations that may be expected during the investigational stage; the name and address of the drug product manufacturer; a brief general description of the manufacturing and packaging procedure as appropriate for the product; the acceptable limits and analytical methods used to assure the identity, strength, quality, and purity of the drug product; and information sufficient to assure the product's stability during the planned clinical studies. Reference to the current edition of the United States Pharmacopeia—National Formulary may satisfy certain requirements in this paragraph.

(c) A brief general description of the composition, manufacture, and control of any placebo used in a controlled clinical trial.

(d) *Labeling.* A copy of all labels and labeling to be provided to each investigator.

(e) *Environmental analysis requirements.* A claim for categorical exclusion under § 25.30 or 25.31 or an environmental assessment under § 25.40.

(8) *Pharmacology and toxicology information.* Adequate information about pharmacological and toxicological studies of the drug involving laboratory animals or in vitro, on the basis of which the sponsor has concluded that it is reasonably safe to conduct the proposed clinical investigations. The kind, duration, and scope of animal and other tests required varies with the duration and nature of the proposed clinical investigations. Guidance documents are available from FDA that describe ways in which these requirements may be met. Such information is required to include the identification and qualifications of the individuals who evaluated the results of such studies and concluded that it is reasonably safe to begin the proposed investigations and a statement of where the investigations were conducted and where the records are available for inspection. As drug development proceeds, the sponsor is required to submit informational amendments, as appropriate, with additional information pertinent to safety.

(i) *Pharmacology and drug disposition.* A section describing the pharmacological effects and mechanism(s) of action of the drug in animals, and information on the absorption, distribution, metabolism, and excretion of the drug, if known.

(ii) *Toxicology.* (a) An integrated summary of the toxicological effects of the drug in animals and in vitro. Depending on the nature of the drug and the phase of the investigation, the description is to include the results of acute, subacute, and chronic toxicity tests; tests of the drug's effects on reproduction and the developing fetus; any special toxicity test related to the drug's particular mode of administration or conditions of use (e.g., inhalation, dermal, or ocular toxicology); and any in vitro studies intended to evaluate drug toxicity.

(b) For each toxicology study that is intended primarily to support the safety of the proposed clinical investigation, a full tabulation of data suitable for detailed review.

(iii) For each nonclinical laboratory study subject to the good laboratory practice regulations under part 58, a statement that the study was conducted in compliance with the good laboratory practice regulations in part 58, or, if the study was not conducted in compliance with those regulations, a brief statement of the reason for the noncompliance.

(9) *Previous human experience with the investigational drug.* A summary of previous human experience known to the applicant, if any, with the investigational drug. The information is required to include the following:

(i) If the investigational drug has been investigated or marketed previously, either in the United States or other countries, detailed information about such experience that is relevant to the safety of the proposed investigation or to the investigation's rationale. If the durg has been the subject of controlled trials, detailed information on such trials that is relevant to an assessment of the drug's effectiveness for the proposed investigational use(s) should also be provided. Any published material that is relevant to the safety of the proposed investigation or to an assessment of the drug's effectiveness for its proposed investigational use should be provided in full. Published material that is less directly relevant may be supplied by a bibliography.

(ii) If the drug is a combination of drugs previously investigated or marketed, the information required under paragraph (a)(9)(i) of this section should be provided for each active drug component. However, if any component in such combination is subject to an approved marketing application or is otherwise lawfully marketed in the United States, the sponsor is not required to submit published material concerning that active drug component unless such material relates directly to the proposed investigational use (including publications relevant to component-component interaction).

(iii) If the drug has been marketed outside the United States, a list of the countries in which the drug has been marketed and a list of the countries in which the drug has been withdrawn from marketing for reasons potentially related to safety or effectiveness.

Food and Drug Administration, HHS   **§ 312.30**

(10) *Additional information.* In certain applications, as described below, information on special topics may be needed. Such information shall be submitted in this section as follows:

(i) *Drug dependence and abuse potential.* If the drug is a psychotropic substance or otherwise has abuse potential, a section describing relevant clinical studies and experience and studies in test animals.

(ii) *Radioactive drugs.* If the drug is a radioactive drug, sufficient data from animal or human studies to allow a reasonable calculation of radiation-absorbed dose to the whole body and critical organs upon administration to a human subject. Phase 1 studies of radioactive drugs must include studies which will obtain sufficient data for dosimetry calculations.

(iii) *Pediatric studies.* Plans for assessing pediatric safety and effectiveness.

(iv) *Other information.* A brief statement of any other information that would aid evaluation of the proposed clinical investigations with respect to their safety or their design and potential as controlled clinical trials to support marketing of the drug.

(11) *Relevant information.* If requested by FDA, any other relevant information needed for review of the application.

(b) *Information previously submitted.* The sponsor ordinarily is not required to resubmit information previously submitted, but may incorporate the information by reference. A reference to information submitted previously must identify the file by name, reference number, volume, and page number where the information can be found. A reference to information submitted to the agency by a person other than the sponsor is required to contain a written statement that authorizes the reference and that is signed by the person who submitted the information.

(c) *Material in a foreign language.* The sponsor shall submit an accurate and complete English translation of each part of the IND that is not in English. The sponsor shall also submit a copy of each original literature publication for which an English translation is submitted.

(d) *Number of copies.* The sponsor shall submit an original and two copies of all submissions to the IND file, including the original submission and all amendments and reports.

(e) *Numbering of IND submissions.* Each submission relating to an IND is required to be numbered serially using a single, three-digit serial number. The initial IND is required to be numbered 000; each subsequent submission (e.g., amendment, report, or correspondence) is required to be numbered chronologically in sequence.

(f) *Identification of exception from informed consent.* If the investigation involves an exception from informed consent under § 50.24 of this chapter, the sponsor shall prominently identify on the cover sheet that the investigation is subject to the requirements in § 50.24 of this chapter.

(Collection of information requirements approved by the Office of Management and Budget under control number 0910–0014)

[52 FR 8831, Mar. 19, 1987, as amended at 52 FR 23031, June 17, 1987; 53 FR 1918, Jan. 25, 1988; 61 FR 51529, Oct. 2, 1996; 62 FR 40599, July 29, 1997; 63 FR 66669, Dec. 2, 1998; 65 FR 56479, Sept. 19, 2000]

§ 312.30 Protocol amendments.

Once an IND is in effect, a sponsor shall amend it as needed to ensure that the clinical investigations are conducted according to protocols included in the application. This section sets forth the provisions under which new protocols may be submitted and changes in previously submitted protocols may be made. Whenever a sponsor intends to conduct a clinical investigation with an exception from informed consent for emergency research as set forth in § 50.24 of this chapter, the sponsor shall submit a separate IND for such investigation.

(a) *New protocol.* Whenever a sponsor intends to conduct a study that is not covered by a protocol already contained in the IND, the sponsor shall submit to FDA a protocol amendment containing the protocol for the study. Such study may begin provided two conditions are met: (1) The sponsor has submitted the protocol to FDA for its review; and (2) the protocol has been approved by the Institutional Review Board (IRB) with responsibility for review and approval of the study in accordance with the requirements of part

56. The sponsor may comply with these two conditions in either order.

(b) *Changes in a protocol.* (1) A sponsor shall submit a protocol amendment describing any change in a Phase 1 protocol that significantly affects the safety of subjects or any change in a Phase 2 or 3 protocol that significantly affects the safety of subjects, the scope of the investigation, or the scientific quality of the study. Examples of changes requiring an amendment under this paragraph include:

(i) Any increase in drug dosage or duration of exposure of individual subjects to the drug beyond that in the current protocol, or any significant increase in the number of subjects under study.

(ii) Any significant change in the design of a protocol (such as the addition or dropping of a control group).

(iii) The addition of a new test or procedure that is intended to improve monitoring for, or reduce the risk of, a side effect or adverse event; or the dropping of a test intended to monitor safety.

(2)(i) A protocol change under paragraph (b)(1) of this section may be made provided two conditions are met:

(*a*) The sponsor has submitted the change to FDA for its review; and

(*b*) The change has been approved by the IRB with responsibility for review and approval of the study. The sponsor may comply with these two conditions in either order.

(ii) Notwithstanding paragraph (b)(2)(i) of this section, a protocol change intended to eliminate an apparent immediate hazard to subjects may be implemented immediately provided FDA is subsequently notified by protocol amendment and the reviewing IRB is notified in accordance with § 56.104(c).

(c) *New investigator.* A sponsor shall submit a protocol amendment when a new investigator is added to carry out a previously submitted protocol, except that a protocol amendment is not required when a licensed practitioner is added in the case of a treatment protocol under § 312.34. Once the investigator is added to the study, the investigational drug may be shipped to the

investigator and the investigator may begin participating in the study. The sponsor shall notify FDA of the new investigator within 30 days of the investigator being added.

(d) *Content and format.* A protocol amendment is required to be prominently identified as such (i.e., "Protocol Amendment: New Protocol", "Protocol Amendment: Change in Protocol", or "Protocol Amendment: New Investigator"), and to contain the following:

(1)(i) In the case of a new protocol, a copy of the new protocol and a brief description of the most clinically significant differences between it and previous protocols.

(ii) In the case of a change in protocol, a brief description of the change and reference (date and number) to the submission that contained the protocol.

(iii) In the case of a new investigator, the investigator's name, the qualifications to conduct the investigation, reference to the previously submitted protocol, and all additional information about the investigator's study as is required under § 312.23(a)(6)(iii)(*b*).

(2) Reference, if necessary, to specific technical information in the IND or in a concurrently submitted information amendment to the IND that the sponsor relies on to support any clinically significant change in the new or amended protocol. If the reference is made to supporting information already in the IND, the sponsor shall identify by name, reference number, volume, and page number the location of the information.

(3) If the sponsor desires FDA to comment on the submission, a request for such comment and the specific questions FDA's response should address.

(e) *When submitted.* A sponsor shall submit a protocol amendment for a new protocol or a change in protocol before its implementation. Protocol amendments to add a new investigator or to provide additional information about investigators may be grouped and submitted at 30-day intervals.

Food and Drug Administration, HHS

When several submissions of new protocols or protocol changes are anticipated during a short period, the sponsor is encouraged, to the extent feasible, to include these all in a single submission.

(Collection of information requirements approved by the Office of Management and Budget under control number 0910–0014)

[52 FR 8831, Mar. 19, 1987, as amended at 52 FR 23031, June 17, 1987; 53 FR 1918, Jan. 25, 1988; 61 FR 51530, Oct. 2, 1996]

§312.31   Information amendments.

(a) *Requirement for information amendment.* A sponsor shall report in an information amendment essential information on the IND that is not within the scope of a protocol amendment, IND safety reports, or annual report. Examples of information requiring an information amendment include:

(1) New toxicology, chemistry, or other technical information; or

(2) A report regarding the discontinuance of a clinical investigation.

(b) *Content and format of an information amendment.* An information amendment is required to bear prominent identification of its contents (e.g., "Information Amendment: Chemistry, Manufacturing, and Control", "Information Amendment: Pharmacology-Toxicology", "Information Amendment: Clinical"), and to contain the following:

(1) A statement of the nature and purpose of the amendment.

(2) An organized submission of the data in a format appropriate for scientific review.

(3) If the sponsor desires FDA to comment on an information amendment, a request for such comment.

(c) *When submitted.* Information amendments to the IND should be submitted as necessary but, to the extent feasible, not more than every 30 days.

(Collection of information requirements approved by the Office of Management and Budget under control number 0910–0014)

[52 FR 8831, Mar. 19, 1987, as amended at 52 FR 23031, June 17, 1987; 53 FR 1918, Jan. 25, 1988]

§312.32   IND safety reports.

(a) *Definitions.* The following definitions of terms apply to this section:-

Associated with the use of the drug. There is a reasonable possibility that the experience may have been caused by the drug.

Disability. A substantial disruption of a person's ability to conduct normal life functions.

Life-threatening adverse drug experience. Any adverse drug experience that places the patient or subject, in the view of the investigator, at immediate risk of death from the reaction as it occurred, i.e., it does not include a reaction that, had it occurred in a more severe form, might have caused death.

Serious adverse drug experience: Any adverse drug experience occurring at any dose that results in any of the following outcomes: Death, a life-threatening adverse drug experience, inpatient hospitalization or prolongation of existing hospitalization, a persistent or significant disability/incapacity, or a congenital anomaly/birth defect. Important medical events that may not result in death, be life-threatening, or require hospitalization may be considered a serious adverse drug experience when, based upon appropriate medical judgment, they may jeopardize the patient or subject and may require medical or surgical intervention to prevent one of the outcomes listed in this definition. Examples of such medical events include allergic bronchospasm requiring intensive treatment in an emergency room or at home, blood dyscrasias or convulsions that do not result in inpatient hospitalization, or the development of drug dependency or drug abuse.

Unexpected adverse drug experience: Any adverse drug experience, the specificity or severity of which is not consistent with the current investigator brochure; or, if an investigator brochure is not required or available, the specificity or severity of which is not consistent with the risk information described in the general investigational plan or elsewhere in the current application, as amended. For example, under this definition, hepatic necrosis would be unexpected (by virtue of greater severity) if the investigator brochure only referred to elevated hepatic enzymes or hepatitis. Similarly, cerebral thromboembolism and cerebral vasculitis would be unexpected (by virtue of greater specificity) if the investigator brochure only listed cerebral vascular accidents. "Unexpected," as used in this definition, refers to an adverse drug experience that has not been previously observed (e.g., included in the investigator brochure) rather than from the perspective of such experience not being anticipated from the pharmacological properties of the pharmaceutical product.

(b) *Review of safety information.* The sponsor shall promptly review all information relevant to the safety of the drug obtained or otherwise received by the sponsor from any source, foreign or domestic, including information derived from any clinical or epidemiological investigations, animal investigations, commercial marketing experience, reports in the scientific literature, and unpublished scientific papers, as well as reports from foreign regulatory authorities that have not already been previously reported to the agency by the sponsor.

(c) *IND safety reports.* (1) *Written reports*—(i) The sponsor shall notify FDA and all participating investigators in a written IND safety report of:

(A) Any adverse experience associated with the use of the drug that is both serious and unexpected; or

(B) Any finding from tests in laboratory animals that suggests a significant risk for human subjects including reports of mutagenicity, teratogenicity, or carcinogenicity. Each notification shall be made as soon as possible and in no event later than 15 calendar days after the sponsor's initial receipt of the information. Each written notification may be submitted on FDA Form 3500A or in a narrative format (foreign events may be submitted either on an FDA Form 3500A or, if preferred, on a CIOMS I form; reports from animal or epidemiological studies shall be submitted in a narrative format) and shall bear prominent identification of its contents, i.e., "IND Safety Report." Each written notification to FDA shall be transmitted to the FDA new drug review division in the Center for Drug Evaluation and Research or the product review division in the Center for Biologics Evaluation and Research that has responsibility for review of the IND. If FDA deter-

mines that additional data are needed, the agency may require further data to be submitted.

(ii) In each written IND safety report, the sponsor shall identify all safety reports previously filed with the IND concerning a similar adverse experience, and shall analyze the significance of the adverse experience in light of the previouos, similar reports.

(2) *Telephone and facsimile transmission safety reports.* The sponsor shall also notify FDA by telephone or by facsimile transmission of any unexpected fatal or life-threatening experience associated with the use of the drug as soon as possible but in no event later than 7 calendar days after the sponsor's initial receipt of the information. Each telephone call or facsimile transmission to FDA shall be transmitted to the FDA new drug review division in the Center for Drug Evaluation and Research or the product review division in the Center for Biologics Evaluation and Research that has responsibility for review of the IND.

(3) *Reporting format or frequency.* FDA may request a sponsor to submit IND safety reports in a format or at a frequency different than that required under this paragraph. The sponsor may also propose and adopt a different reporting format or frequency if the change is agreed to in advance by the director of the new drug review division in the Center for Drug Evaluation and Research or the director of the products review division in the Center for Biologics Evaluation and Research which is responsible for review of the IND.

(4) A sponsor of a clinical study of a marketed drug is not required to make a safety report for any adverse experience associated with use of the drug that is not from the clinical study itself.

(d) *Followup.* (1) The sponsor shall promptly investigate all safety information received by it.

(2) Followup information to a safety report shall be submitted as soon as the relevant information is available.

(3) If the results of a sponsor's investigation show that an adverse drug experience not initially determined to be reportable under paragraph (c) of this section is so reportable, the sponsor

shall report such experience in a written safety report as soon as possible, but in no event later than 15 calendar days after the determination is made.

(4) Results of a sponsor's investigation of other safety information shall be submitted, as appropriate, in an information amendment or annual report.

(e) *Disclaimer.* A safety report or other information submitted by a sponsor under this part (and any release by FDA of that report or information) does not necessarily reflect a conclusion by the sponsor or FDA that the report or information constitutes an admission that the drug caused or contributed to an adverse experience. A sponsor need not admit, and may deny, that the report or information submitted by the sponsor constitutes an admission that the drug caused or contributed to an adverse experience.

(Collection of information requirements approved by the Office of Management and Budget under control number 0910–0014)

[52 FR 8831, Mar. 19, 1987, as amended at 52 FR 23031, June 17, 1987; 55 FR 11579, Mar. 29, 1990; 62 FR 52250, Oct. 7, 1997]

§312.33 Annual reports.

A sponsor shall within 60 days of the anniversary date that the IND went into effect, submit a brief report of the progress of the investigation that includes:

(a) *Individual study information.* A brief summary of the status of each study in progress and each study completed during the previous year. The summary is required to include the following information for each study:

(1) The title of the study (with any appropriate study identifiers such as protocol number), its purpose, a brief statement identifying the patient population, and a statement as to whether the study is completed.

(2) The total number of subjects initially planned for inclusion in the study; the number entered into the study to date, tabulated by age group, gender, and race; the number whose participation in the study was completed as planned; and the number who dropped out of the study for any reason.

(3) If the study has been completed, or if interim results are known, a brief description of any available study results.

(b) *Summary information.* Information obtained during the previous year's clinical and nonclinical investigations, including:

(1) A narrative or tabular summary showing the most frequent and most serious adverse experiences by body system.

(2) A summary of all IND safety reports submitted during the past year.

(3) A list of subjects who died during participation in the investigation, with the cause of death for each subject.

(4) A list of subjects who dropped out during the course of the investigation in association with any adverse experience, whether or not thought to be drug related.

(5) A brief description of what, if anything, was obtained that is pertinent to an understanding of the drug's actions, including, for example, information about dose response, information from controlled trails, and information about bioavailability.

(6) A list of the preclinical studies (including animal studies) completed or in progress during the past year and a summary of the major preclinical findings.

(7) A summary of any significant manufacturing or microbiological changes made during the past year.

(c) A description of the general investigational plan for the coming year to replace that submitted 1 year earlier. The general investigational plan shall contain the information required under §312.23(a)(3)(iv).

(d) If the investigator brochure has been revised, a description of the revision and a copy of the new brochure.

(e) A description of any significant Phase 1 protocol modifications made during the previous year and not previously reported to the IND in a protocol amendment.

(f) A brief summary of significant foreign marketing developments with the drug during the past year, such as approval of marketing in any country or withdrawal or suspension from marketing in any country.

(g) If desired by the sponsor, a log of any outstanding business with respect to the IND for which the sponsor requests or expects a reply, comment, or meeting.

(Collection of information requirements approved by the Office of Management and Budget under control number 0910-0014)

[52 FR 8831, Mar. 19, 1987, as amended at 52 FR 23031, June 17, 1987; 63 FR 6862, Feb. 11, 1998]

§ 312.34 Treatment use of an investigational new drug.

(a) *General.* A drug that is not approved for marketing may be under clinical investigation for a serious or immediately life-threatening disease condition in patients for whom no comparable or satisfactory alternative drug or other therapy is available. During the clinical investigation of the drug, it may be appropriate to use the drug in the treatment of patients not in the clinical trials, in accordance with a treatment protocol or treatment IND. The purpose of this section is to facilitate the availability of promising new drugs to desperately ill patients as early in the drug development process as possible, before general marketing begins, and to obtain additional data on the drug's safety and effectiveness. In the case of a serious disease, a drug ordinarily may be made available for treatment use under this section during Phase 3 investigations or after all clinical trials have been completed; however, in appropriate circumstances, a drug may be made available for treatment use during Phase 2. In the case of an immediately life-threatening disease, a drug may be made available for treatment use under this section earlier than Phase 3, but ordinarily not earlier than Phase 2. For purposes of this section, the "treatment use" of a drug includes the use of a drug for diagnostic purposes. If a protocol for an investigational drug meets the criteria of this section, the protocol is to be submitted as a treatment protocol under the provisions of this section.

(b) *Criteria.* (1) FDA shall permit an investigational drug to be used for a treatment use under a treatment protocol or treatment IND if:

(i) The drug is intended to treat a serious or immediately life-threatening disease;

(ii) There is no comparable or satisfactory alternative drug or other therapy available to treat that stage of the disease in the intended patient population;

(iii) The drug is under investigation in a controlled clinical trial under an IND in effect for the trial, or all clinical trials have been completed; and

(iv) The sponsor of the controlled clinical trial is actively pursuing marketing approval of the investigational drug with due diligence.

(2) *Serious disease.* For a drug intended to treat a serious disease, the Commissioner may deny a request for treatment use under a treatment protocol or treatment IND if there is insufficient evidence of safety and effectiveness to support such use.

(3) *Immediately life-threatening disease.* (i) For a drug intended to treat an immediately life-threatening disease, the Commissioner may deny a request for treatment use of an investigational drug under a treatment protocol or treatment IND if the available scientific evidence, taken as a whole, fails to provide a reasonable basis for concluding that the drug:

(A) May be effective for its intended use in its intended patient population; or

(B) Would not expose the patients to whom the drug is to be administered to an unreasonable and significant additional risk of illness or injury.

(ii) For the purpose of this section, an "immediately life-threatening" disease means a stage of a disease in which there is a reasonable likelihood that death will occur within a matter of months or in which premature death is likely without early treatment.

(c) *Safeguards.* Treatment use of an investigational drug is conditioned on the sponsor and investigators complying with the safeguards of the IND process, including the regulations governing informed consent (21 CFR part 50) and institutional review boards (21 CFR part 56) and the applicable provisions of part 312, including distribution of the drug through qualified experts, maintenance of adequate manufacturing facilities, and submission of IND safety reports.

(d) *Clinical hold.* FDA may place on clinical hold a proposed or ongoing

Food and Drug Administration, HHS §312.35

treatment protocol or treatment IND in accordance with §312.42.

[52 FR 19476, May 22, 1987, as amended at 57 FR 13248, Apr. 15, 1992]

§312.35 Submissions for treatment use.

(a) *Treatment protocol submitted by IND sponsor.* Any sponsor of a clinical investigation of a drug who intends to sponsor a treatment use for the drug shall submit to FDA a treatment protocol under §312.34 if the sponsor believes the criteria of §312.34 are satisfied. If a protocol is not submitted under §312.34, but FDA believes that the protocol should have been submitted under this section, FDA may deem the protocol to be submitted under §312.34. A treatment use under a treatment protocol may begin 30 days after FDA receives the protocol or on earlier notification by FDA that the treatment use described in the protocol may begin.

(1) A treatment protocol is required to contain the following:

(i) The intended use of the drug.

(ii) An explanation of the rationale for use of the drug, including, as appropriate, either a list of what available regimens ordinarily should be tried before using the investigational drug or an explanation of why the use of the investigational drug is preferable to the use of available marketed treatments.

(iii) A brief description of the criteria for patient selection.

(iv) The method of administration of the drug and the dosages.

(v) A description of clinical procedures, laboratory tests, or other measures to monitor the effects of the drug and to minimize risk.

(2) A treatment protocol is to be supported by the following:

(i) Informational brochure for supplying to each treating physician.

(ii) The technical information that is relevant to safety and effectiveness of the drug for the intended treatment purpose. Information contained in the sponsor's IND may be incorporated by reference.

(iii) A commitment by the sponsor to assure compliance of all participating investigators with the informed consent requirements of 21 CFR part 50.

(3) A licensed practioner who receives an investigational drug for treatment use under a treatment protocol is an "investigator" under the protocol and is responsible for meeting all applicable investigator responsibilities under this part and 21 CFR parts 50 and 56.

(b) *Treatment IND submitted by licensed practitioner.* (1) If a licensed medical practitioner wants to obtain an investigational drug subject to a controlled clinical trial for a treatment use, the practitioner should first attempt to obtain the drug from the sponsor of the controlled trial under a treatment protocol. If the sponsor of the controlled clinical investigation of the drug will not establish a treatment protocol for the drug under paragraph (a) of this section, the licensed medical practitioner may seek to obtain the drug from the sponsor and submit a treatment IND to FDA requesting authorization to use the investigational drug for treatment use. A treatment use under a treatment IND may begin 30 days after FDA receives the IND or on earlier notification by FDA that the treatment use under the IND may begin. A treatment IND is required to contain the following:

(i) A cover sheet (Form FDA 1571) meeting §312.23(g)(1).

(ii) Information (when not provided by the sponsor) on the drug's chemistry, manufacturing, and controls, and prior clinical and nonclinical experience with the drug submitted in accordance with §312.23. A sponsor of a clinical investigation subject to an IND who supplies an investigational drug to a licensed medical practitioner for purposes of a separate treatment clinical investigation shall be deemed to authorize the incorporation-by-reference of the technical information contained in the sponsor's IND into the medical practitioner's treatment IND.

(iii) A statement of the steps taken by the practitioner to obtain the drug under a treatment protocol from the drug sponsor.

(iv) A treatment protocol containing the same information listed in paragraph (a)(1) of this section.

(v) A statement of the practitioner's qualifications to use the investigational drug for the intended treatment use.

§ 312.36

(vi) The practitioner's statement of familiarity with information on the drug's safety and effectiveness derived from previous clinical and nonclinical experience with the drug.

(vii) Agreement to report to FDA safety information in accordance with § 312.32.

(2) A licensed practitioner who submits a treatment IND under this section is the sponsor-investigator for such IND and is responsible for meeting all applicable sponsor and investigator responsibilities under this part and 21 CFR parts 50 and 56.

(Collection of information requirements approved by the Office of Management and Budget under control number 0910–0014)

[52 FR 19477, May 22, 1987, as amended at 57 FR 13249, Apr. 15, 1992]

§ 312.36 Emergency use of an investigational new drug.

Need for an investigational drug may arise in an emergency situation that does not allow time for submission of an IND in accordance with § 312.23 or § 312.34. In such a case, FDA may authorize shipment of the drug for a specified use in advance of submission of an IND. A request for such authorization may be transmitted to FDA by telephone or other rapid communication means. For investigational biological drugs, the request should be directed to the Division of Biological Investigational New Drugs (HFB–230), Center for Biologics Evaluation and Research, 8800 Rockville Pike, Bethesda, MD 20892, 301–443–4864. For all other investigational drugs, the request for authorization should be directed to the Document Management and Reporting Branch (HFD–53), Center for Drug Evaluation and Research, 5600 Fishers Lane, Rockville, MD 20857, 301–443–4320. After normal working hours, eastern standard time, the request should be directed to the FDA Division of Emergency and Epidemiological Operations, 202–857–8400. Except in extraordinary circumstances, such authorization will be conditioned on the sponsor making an appropriate IND submission as soon

as practicable after receiving the authorization.

(Collection of information requirements approved by the Office of Management and Budget under control number 0910–0014)

[52 FR 8831, Mar. 19, 1987, as amended at 52 FR 23031, June 17, 1987; 55 FR 11579, Mar. 29, 1990]

§ 312.38 Withdrawal of an IND.

(a) At any time a sponsor may withdraw an effective IND without prejudice.

(b) If an IND is withdrawn, FDA shall be so notified, all clinical investigations conducted under the IND shall be ended, all current investigators notified, and all stocks of the drug returned to the sponsor or otherwise disposed of at the request of the sponsor in accordance with § 312.59.

(c) If an IND is withdrawn because of a safety reason, the sponsor shall promptly so inform FDA, all participating investigators, and all reviewing Institutional Review Boards, together with the reasons for such withdrawal.

(Collection of information requirements approved by the Office of Management and Budget under control number 0910–0014)

[52 FR 8831, Mar. 19, 1987, as amended at 52 FR 23031, June 17, 1987]

Subpart C—Administrative Actions

§ 312.40 General requirements for use of an investigational new drug in a clinical investigation.

(a) An investigational new drug may be used in a clinical investigation if the following conditions are met:

(1) The sponsor of the investigation submits an IND for the drug to FDA; the IND is in effect under paragraph (b) of this section; and the sponsor complies with all applicable requirements in this part and parts 50 and 56 with respect to the conduct of the clinical investigations; and

(2) Each participating investigator conducts his or her investigation in compliance with the requirements of this part and parts 50 and 56.

(b) An IND goes into effect:

(1) Thirty days after FDA receives the IND, unless FDA notifies the sponsor that the investigations described in the IND are subject to a clinical hold under §312.42; or

(2) On earlier notification by FDA that the clinical investigations in the IND may begin. FDA will notify the sponsor in writing of the date it receives the IND.

(c) A sponsor may ship an investigational new drug to investigators named in the IND:

(1) Thirty days after FDA receives the IND; or

(2) On earlier FDA authorization to ship the drug.

(d) An investigator may not administer an investigational new drug to human subjects until the IND goes into effect under paragraph (b) of this section.

§312.41 Comment and advice on an IND.

(a) FDA may at any time during the course of the investigation communicate with the sponsor orally or in writing about deficiencies in the IND or about FDA's need for more data or information.

(b) On the sponsor's request, FDA will provide advice on specific matters relating to an IND. Examples of such advice may include advice on the adequacy of technical data to support an investigational plan, on the design of a clinical trial, and on whether proposed investigations are likely to produce the data and information that is needed to meet requirements for a marketing application.

(c) Unless the communication is accompanied by a clinical hold order under §312.42, FDA communications with a sponsor under this section are solely advisory and do not require any modification in the planned or ongoing clinical investigations or response to the agency.

(Collection of information requirements approved by the Office of Management and Budget under control number 0910–0014)

[52 FR 8831, Mar. 19, 1987, as amended at 52 FR 23031, June 17, 1987]

§312.42 Clinical holds and requests for modification.

(a) *General.* A clinical hold is an order issued by FDA to the sponsor to delay a proposed clinical investigation or to suspend an ongoing investigation. The clinical hold order may apply to one or more of the investigations covered by an IND. When a proposed study is placed on clinical hold, subjects may not be given the investigational drug. When an ongoing study is placed on clinical hold, no new subjects may be recruited to the study and placed on the investigational drug; patients already in the study should be taken off therapy involving the investigational drug unless specifically permitted by FDA in the interest of patient safety.

(b) *Grounds for imposition of clinical hold*—(1) *Clinical hold of a Phase 1 study under an IND.* FDA may place a proposed or ongoing Phase 1 investigation on clinical hold if it finds that:

(i) Human subjects are or would be exposed to an unreasonable and significant risk of illness or injury;

(ii) The clinical investigators named in the IND are not qualified by reason of their scientific training and experience to conduct the investigation described in the IND;

(iii) The investigator brochure is misleading, erroneous, or materially incomplete; or

(iv) The IND does not contain sufficient information required under §312.23 to assess the risks to subjects of the proposed studies.

(v) The IND is for the study of an investigational drug intended to treat a life-threatening disease or condition that affects both genders, and men or women with reproductive potential who have the disease or condition being studied are excluded from eligibility because of a risk or potential risk from use of the investigational drug of reproductive toxicity (i.e., affecting reproductive organs) or developmental toxicity (i.e., affecting potential offspring). The phrase "women with reproductive potential" does not include pregnant women. For purposes of this paragraph, "life-threatening illnesses or diseases" are defined as "diseases or conditions where the likelihood of death is high unless the course of the disease is interrupted." The clinical hold would not apply under this paragraph to clinical studies conducted:

(A) Under special circumstances, such as studies pertinent only to one gender (e.g., studies evaluating the excretion of a drug in semen or the effects on menstrual function);

(B) Only in men or women, as long as a study that does not exclude members of the other gender with reproductive potential is being conducted concurrently, has been conducted, or will take place within a reasonable time agreed upon by the agency; or

(C) Only in subjects who do not suffer from the disease or condition for which the drug is being studied.

(2) *Clinical hold of a Phase 2 or 3 study under an IND.* FDA may place a proposed or ongoing Phase 2 or 3 investigation on clinical hold if it finds that:

(i) Any of the conditions in paragraphs (b)(1)(i) through (b)(1)(v) of this section apply; or

(ii) The plan or protocol for the investigation is clearly deficient in design to meet its stated objectives.

(3) Clinical hold of a treatment IND or treatment protocol.

(i) *Proposed use.* FDA may place a proposed treatment IND or treatment protocol on clinical hold if it is determined that:

(A) The pertinent criteria in § 312.34(b) for permitting the treatment use to begin are not satisfied; or

(B) The treatment protocol or treatment IND does not contain the information required under § 312.35 (a) or (b) to make the specified determination under § 312.34(b).

(ii) *Ongoing use.* FDA may place an ongoing treatment protocol or treatment IND on clinical hold if it is determined that:

(A) There becomes available a comparable or satisfactory alternative drug or other therapy to treat that stage of the disease in the intended patient population for which the investigational drug is being used;

(B) The investigational drug is not under investigation in a controlled clinical trial under an IND in effect for the trial and not all controlled clinical trials necessary to support a marketing application have been completed, or a clinical study under the IND has been placed on clinical hold:

(C) The sponsor of the controlled clinical trial is not pursuing marketing approval with due diligence;

(D) If the treatment IND or treatment protocol is intended for a serious disease, there is insufficient evidence of safety and effectiveness to support such use; or

(E) If the treatment protocol or treatment IND was based on an immediately life-threatening disease, the available scientific evidence, taken as a whole, fails to provide a reasonable basis for concluding that the drug:

(*1*) May be effective for its intended use in its intended population; or

(*2*) Would not expose the patients to whom the drug is to be administered to an unreasonable and significant additional risk of illness or injury.

(iii) FDA may place a proposed or ongoing treatment IND or treatment protocol on clinical hold if it finds that any of the conditions in paragraph (b)(4)(i) through (b)(4)(viii) of this section apply.

(4) *Clinical hold of any study that is not designed to be adequate and well-controlled.* FDA may place a proposed or ongoing investigation that is not designed to be adequate and well-controlled on clinical hold if it finds that:

(i) Any of the conditions in paragraph (b)(1) or (b)(2) of this section apply; or

(ii) There is reasonable evidence the investigation that is not designed to be adequate and well-controlled is impeding enrollment in, or otherwise interfering with the conduct or completion of, a study that is designed to be an adequate and well-controlled investigation of the same or another investigational drug; or

(iii) Insufficient quantities of the investigational drug exist to adequately conduct both the investigation that is not designed to be adequate and well-controlled and the investigations that are designed to be adequate and well-controlled; or

(iv) The drug has been studied in one or more adequate and well-controlled investigations that strongly suggest lack of effectiveness; or

(v) Another drug under investigation or approved for the same indication and available to the same patient population has demonstrated a better potential benefit/risk balance; or

(vi) The drug has received marketing approval for the same indication in the same patient population; or

(vii) The sponsor of the study that is designed to be an adequate and well-controlled investigation is not actively pursuing marketing approval of the investigational drug with due diligence; or

(viii) The Commissioner determines that it would not be in the public interest for the study to be conducted or continued. FDA ordinarily intends that clinical holds under paragraphs (b)(4)(ii), (b)(4)(iii) and (b)(4)(v) of this section would only apply to additional enrollment in nonconcurrently controlled trials rather than eliminating continued access to individuals already receiving the investigational drug.

(5) *Clinical hold of any investigation involving an exception from informed consent under § 50.24 of this chapter.* FDA may place a proposed or ongoing investigation involving an exception from informed consent under § 50.24 of this chapter on clinical hold if it is determined that:

(i) Any of the conditions in paragraphs (b)(1) or (b)(2) of this section apply; or

(ii) The pertinent criteria in § 50.24 of this chapter for such an investigation to begin or continue are not submitted or not satisfied.

(6) Clinical hold of any investigation involving an exception from informed consent under § 50.23(d) of this chapter. FDA may place a proposed or ongoing investigation involving an exception from informed consent under § 50.23(d) of this chapter on clinical hold if it is determined that:

(i) Any of the conditions in paragraphs (b)(1) or (b)(2) of this section apply; or

(ii) A determination by the President to waive the prior consent requirement for the administration of an investigational new drug has not been made.

(c) *Discussion of deficiency.* Whenever FDA concludes that a deficiency exists in a clinical investigation that may be grounds for the imposition of clinical hold FDA will, unless patients are exposed to immediate and serious risk, attempt to discuss and satisfactorily resolve the matter with the sponsor before issuing the clinical hold order.

(d) *Imposition of clinical hold.* The clinical hold order may be made by telephone or other means of rapid communication or in writing. The clinical hold order will identify the studies under the IND to which the hold applies, and will briefly explain the basis for the action. The clinical hold order will be made by or on behalf of the Division Director with responsibility for review of the IND. As soon as possible, and no more than 30 days after imposition of the clinical hold, the Division Director will provide the sponsor a written explanation of the basis for the hold.

(e) *Resumption of clinical investigations.* An investigation may only resume after FDA (usually the Division Director, or the Director's designee, with responsibility for review of the IND) has notified the sponsor that the investigation may proceed. Resumption of the affected investigation(s) will be authorized when the sponsor corrects the deficiency(ies) previously cited or otherwise satisfies the agency that the investigation(s) can proceed. FDA may notify a sponsor of its determination regarding the clinical hold by telephone or other means of rapid communication. If a sponsor of an IND that has been placed on clinical hold requests in writing that the clinical hold be removed and submits a complete response to the issue(s) identified in the clinical hold order, FDA shall respond in writing to the sponsor within 30-calendar days of receipt of the request and the complete response. FDA's response will either remove or maintain the clinical hold, and will state the reasons for such determination. Notwithstanding the 30-calendar day response time, a sponsor may not proceed with a clinical trial on which a clinical hold has been imposed until the sponsor has been notified by FDA that the hold has been lifted.

(f) *Appeal.* If the sponsor disagrees with the reasons cited for the clinical hold, the sponsor may request reconsideration of the decision in accordance with § 312.48.

(g) *Conversion of IND on clinical hold to inactive status.* If all investigations covered by an IND remain on clinical hold for 1 year or more, the IND may be placed on inactive status by FDA under §312.45.

[52 FR 8831, Mar. 19, 1987, as amended at 52 FR 19477, May 22, 1987; 57 FR 13249, Apr. 15, 1992; 61 FR 51530, Oct. 2, 1996; 63 FR 68678, Dec. 14, 1998; 64 FR 54189, Oct. 5, 1999; 65 FR 34971, June 1, 2000]

§312.44 Termination.

(a) *General.* This section describes the procedures under which FDA may terminate an IND. If an IND is terminated, the sponsor shall end all clinical investigations conducted under the IND and recall or otherwise provide for the disposition of all unused supplies of the drug. A termination action may be based on deficiencies in the IND or in the conduct of an investigation under an IND. Except as provided in paragraph (d) of this section, a termination shall be preceded by a proposal to terminate by FDA and an opportunity for the sponsor to respond. FDA will, in general, only initiate an action under this section after first attempting to resolve differences informally or, when appropriate, through the clinical hold procedures described in §312.42.

(b) *Grounds for termination*—(1) *Phase 1.* FDA may propose to terminate an IND during Phase 1 if it finds that:

(i) Human subjects would be exposed to an unreasonable and significant risk of illness or injury.

(ii) The IND does not contain sufficient information required under §312.23 to assess the safety to subjects of the clinical investigations.

(iii) The methods, facilities, and controls used for the manufacturing, processing, and packing of the investigational drug are inadequate to establish and maintain appropriate standards of identity, strength, quality, and purity as needed for subject safety.

(iv) The clinical investigations are being conducted in a manner substantially different than that described in the protocols submitted in the IND.

(v) The drug is being promoted or distributed for commercial purposes not justified by the requirements of the investigation or permitted by §312.7.

(vi) The IND, or any amendment or report to the IND, contains an untrue statement of a material fact or omits material information required by this part.

(vii) The sponsor fails promptly to investigate and inform the Food and Drug Administration and all investigators of serious and unexpected adverse experiences in accordance with §312.32 or fails to make any other report required under this part.

(viii) The sponsor fails to submit an accurate annual report of the investigations in accordance with §312.33.

(ix) The sponsor fails to comply with any other applicable requirement of this part, part 50, or part 56.

(x) The IND has remained on inactive status for 5 years or more.

(xi) The sponsor fails to delay a proposed investigation under the IND or to suspend an ongoing investigation that has been placed on clinical hold under §312.42(b)(4).

(2) *Phase 2 or 3.* FDA may propose to terminate an IND during Phase 2 or Phase 3 if FDA finds that:

(i) Any of the conditions in paragraphs (b)(1)(i) through (b)(1)(xi) of this section apply; or

(ii) The investigational plan or protocol(s) is not reasonable as a bona fide scientific plan to determine whether or not the drug is safe and effective for use; or

(iii) There is convincing evidence that the drug is not effective for the purpose for which it is being investigated.

(3) FDA may propose to terminate a treatment IND if it finds that:

(i) Any of the conditions in paragraphs (b)(1)(i) through (x) of this section apply; or

(ii) Any of the conditions in §312.42(b)(3) apply.

(c) *Opportunity for sponsor response.* (1) If FDA proposes to terminate an IND, FDA will notify the sponsor in writing, and invite correction or explanation within a period of 30 days.

(2) On such notification, the sponsor may provide a written explanation or correction or may request a conference with FDA to provide the requested explanation or correction. If the sponsor does not respond to the notification

---

within the allocated time, the IND shall be terminated.

(3) If the sponsor responds but FDA does not accept the explanation or correction submitted, FDA shall inform the sponsor in writing of the reason for the nonacceptance and provide the sponsor with an opportunity for a regulatory hearing before FDA under part 16 on the question of whether the IND should be terminated. The sponsor's request for a regulatory hearing must be made within 10 days of the sponsor's receipt of FDA's notification of nonacceptance.

(d) *Immediate termination of IND.* Notwithstanding paragraphs (a) through (c) of this section, if at any time FDA concludes that continuation of the investigation presents an immediate and substantial danger to the health of individuals, the agency shall immediately, by written notice to the sponsor from the Director of the Center for Drug Evaluation and Research or the Director of the Center for Biologics Evaluation and Research, terminate the IND. An IND so terminated is subject to reinstatement by the Director on the basis of additional submissions that eliminate such danger. If an IND is terminated under this paragraph, the agency will afford the sponsor an opportunity for a regulatory hearing under part 16 on the question of whether the IND should be reinstated.

(Collection of information requirements approved by the Office of Management and Budget under control number 0910–0014)

[52 FR 8831, Mar. 19, 1987, as amended at 52 FR 23031, June 17, 1987; 55 FR 11579, Mar. 29, 1990; 57 FR 13249, Apr. 15, 1992]

§312.45 Inactive status.

(a) If no subjects are entered into clinical studies for a period of 2 years or more under an IND, or if all investigations under an IND remain on clinical hold for 1 year or more, the IND may be placed by FDA on inactive status. This action may be taken by FDA either on request of the sponsor or on FDA's own initiative. If FDA seeks to act on its own initiative under this section, it shall first notify the sponsor in writing of the proposed inactive status. Upon receipt of such notification, the sponsor shall have 30 days to respond

as to why the IND should continue to remain active.

(b) If an IND is placed on inactive status, all investigators shall be so notified and all stocks of the drug shall be returned or otherwise disposed of in accordance with §312.59.

(c) A sponsor is not required to submit annual reports to an IND on inactive status. An inactive IND is, however, still in effect for purposes of the public disclosure of data and information under §312.130.

(d) A sponsor who intends to resume clinical investigation under an IND placed on inactive status shall submit a protocol amendment under §312.30 containing the proposed general investigational plan for the coming year and appropriate protocols. If the protocol amendment relies on information previously submitted, the plan shall reference such information. Additional information supporting the proposed investigation, if any, shall be submitted in an information amendment. Notwithstanding the provisions of §312.30, clinical investigations under an IND on inactive status may only resume (1) 30 days after FDA receives the protocol amendment, unless FDA notifies the sponsor that the investigations described in the amendment are subject to a clinical hold under §312.42, or (2) on earlier notification by FDA that the clinical investigations described in the protocol amendment may begin.

(e) An IND that remains on inactive status for 5 years or more may be terminated under §312.44.

(Collection of information requirements approved by the Office of Management and Budget under control number 0910–0014)

[52 FR 8831, Mar. 19, 1987, as amended at 52 FR 23031, June 17, 1987]

§312.47 Meetings.

(a) *General.* Meetings between a sponsor and the agency are frequently useful in resolving questions and issues raised during the course of a clinical investigation. FDA encourages such meetings to the extent that they aid in the evaluation of the drug and in the solution of scientific problems concerning the drug, to the extent that FDA's resources permit. The general principle underlying the conduct of such meetings is that there should be free, full, and open communication about any scientific or medical question that may arise during the clinical investigation. These meetings shall be conducted and documented in accordance with part 10.

(b) *"End-of-Phase 2" meetings and meetings held before submission of a marketing application.* At specific times during the drug investigation process, meetings between FDA and a sponsor can be especially helpful in minimizing wasteful expenditures of time and money and thus in speeding the drug development and evaluation process. In particular, FDA has found that meetings at the end of Phase 2 of an investigation (end-of-Phase 2 meetings) are of considerable assistance in planning later studies and that meetings held near completion of Phase 3 and before submission of a marketing application ("pre-NDA" meetings) are helpful in developing methods of presentation and submission of data in the marketing application that facilitate review and allow timely FDA response.

(1) *End-of-Phase 2 meetings—*(i) *Purpose.* The purpose of an end-of-phase 2 meeting is to determine the safety of proceeding to Phase 3, to evaluate the Phase 3 plan and protocols and the adequacy of current studies and plans to assess pediatric safety and effectiveness, and to identify any additional information necessary to support a marketing application for the uses under investigation.

(ii) *Eligibility for meeting.* While the end-of-Phase 2 meeting is designed primarily for IND's involving new molecular entities or major new uses of marketed drugs, a sponsor of any IND may request and obtain an end-of-Phase 2 meeting.

(iii) *Timing.* To be most useful to the sponsor, end-of-Phase 2 meetings should be held before major commitments of effort and resources to specific Phase 3 tests are made. The scheduling of an end-of-Phase 2 meeting is not, however, intended to delay the transition of an investigation from Phase 2 to Phase 3.

(iv) *Advance information.* At least 1 month in advance of an end-of-Phase 2 meeting, the sponsor should submit background information on the sponsor's plan for Phase 3, including sum-

maries of the Phase 1 and 2 investigations, the specific protocols for Phase 3 clinical studies, plans for any additional nonclinical studies, plans for pediatric studies, including a time line for protocol finalization, enrollment, completion, and data analysis, or information to support any planned request for waiver or deferral of pediatric studies, and, if available, tentative labeling for the drug. The recommended contents of such a submission are described more fully in FDA Staff Manual Guide 4850.7 that is publicly available under FDA's public information regulations in part 20.

(v) *Conduct of meeting.* Arrangements for an end-of-Phase 2 meeting are to be made with the division in FDA's Center for Drug Evaluation and Research or the Center for Biologics Evaluation and Research which is responsible for review of the IND. The meeting will be scheduled by FDA at a time convenient to both FDA and the sponsor. Both the sponsor and FDA may bring consultants to the meeting. The meeting should be directed primarily at establishing agreement between FDA and the sponsor of the overall plan for Phase 3 and the objectives and design of particular studies. The adequacy of the technical information to support Phase 3 studies and/or a marketing application may also be discussed. FDA will also provide its best judgment, at that time, of the pediatric studies that will be required for the drug product and whether their submission will be deferred until after approval. Agreements reached at the meeting on these matters will be recorded in minutes of the conference that will be taken by FDA in accordance with § 10.65 and provided to the sponsor. The minutes along with any other written material provided to the sponsor will serve as a permanent record of any agreements reached. Barring a significant scientific development that requires otherwise, studies conducted in accordance with the agreement shall be presumed to be sufficient in objective and design for the purpose of obtaining marketing approval for the drug.

(2) *"Pre-NDA" and "pre-BLA" meetings.* FDA has found that delays associated with the initial review of a marketing application may be reduced by

Food and Drug Administration, HHS   **§312.48**

exchanges of information about a proposed marketing application. The primary purpose of this kind of exchange is to uncover any major unresolved problems, to identify those studies that the sponsor is relying on as adequate and well-controlled to establish the drug's effectiveness, to identify the status of ongoing or needed studies adequate to assess pediatric safety and effectiveness, to acquaint FDA reviewers with the general information to be submitted in the marketing application (including technical information), to discuss appropriate methods for statistical analysis of the data, and to discuss the best approach to the presentation and formatting of data in the marketing application. Arrangements for such a meeting are to be initiated by the sponsor with the division responsible for review of the IND. To permit FDA to provide the sponsor with the most useful advice on preparing a marketing application, the sponsor should submit to FDA's reviewing division at least 1 month in advance of the meeting the following information:

(i) A brief summary of the clinical studies to be submitted in the application.

(ii) A proposed format for organizing the submission, including methods for presenting the data.

(iii) Information on the status of needed or ongoing pediatric studies.

(iv) Any other information for discussion at the meeting.

(Collection of information requirements approved by the Office of Management and Budget under control number 0910–0014)

[52 FR 8831, Mar. 19, 1987, as amended at 52 FR 23031, June 17, 1987; 55 FR 11580, Mar. 29, 1990; 63 FR 66669, Dec. 2, 1998]

§312.48 Dispute resolution.

(a) *General.* The Food and Drug Administration is committed to resolving differences between sponsors and FDA reviewing divisions with respect to requirements for IND's as quickly and amicably as possible through the cooperative exchange of information and views.

(b) *Administrative and procedural issues.* When administrative or procedural disputes arise, the sponsor should first attempt to resolve the matter

with the division in FDA's Center for Drug Evaluation and Research or Center for Biologics Evaluation and Research which is responsible for review of the IND, beginning with the consumer safety officer assigned to the application. If the dispute is not resolved, the sponsor may raise the matter with the person designated as ombudsman, whose function shall be to investigate what has happened and to facilitate a timely and equitable resolution. Appropriate issues to raise with the ombudsman include resolving difficulties in scheduling meetings and obtaining timely replies to inquiries. Further details on this procedure are contained in FDA Staff Manual Guide 4820.7 that is publicly available under FDA's public information regulations in part 20.

(c) *Scientific and medical disputes.* (1) When scientific or medical disputes arise during the drug investigation process, sponsors should discuss the matter directly with the responsible reviewing officials. If necessary, sponsors may request a meeting with the appropriate reviewing officials and management representatives in order to seek a resolution. Requests for such meetings shall be directed to the director of the division in FDA's Center for Drug Evaluation and Research or Center for Biologics Evaluation and Research which is responsible for review of the IND. FDA will make every attempt to grant requests for meetings that involve important issues and that can be scheduled at mutually convenient times.

(2) The "end-of-Phase 2" and "pre-NDA" meetings described in §312.47(b) will also provide a timely forum for discussing and resolving scientific and medical issues on which the sponsor disagrees with the agency.

(3) In requesting a meeting designed to resolve a scientific or medical dispute, applicants may suggest that FDA seek the advice of outside experts, in which case FDA may, in its discretion, invite to the meeting one or more of its advisory committee members or other consultants, as designated by the agency. Applicants may rely on, and may

bring to any meeting, their own consultants. For major scientific and medical policy issues not resolved by informal meetings, FDA may refer the matter to one of its standing advisory committees for its consideration and recommendations.

[52 FR 8831, Mar. 19, 1987, as amended at 55 FR 11580, Mar. 29, 1990]

Subpart D—Responsibilities of Sponsors and Investigators

§ 312.50 General responsibilities of sponsors.

Sponsors are responsibile for selecting qualified investigators, providing them with the information they need to conduct an investigation properly, ensuring proper monitoring of the investigation(s), ensuring that the investigation(s) is conducted in accordance with the general investigational plan and protocols contained in the IND, maintaining an effective IND with respect to the investigations, and ensuring that FDA and all participating investigators are promptly informed of significant new adverse effects or risks with respect to the drug. Additional specific responsibilities of sponsors are described elsewhere in this part.

§ 312.52 Transfer of obligations to a contract research organization.

(a) A sponsor may transfer responsibility for any or all of the obligations set forth in this part to a contract research organization. Any such transfer shall be described in writing. If not all obligations are transferred, the writing is required to describe each of the obligations being assumed by the contract research organization. If all obligations are transferred, a general statement that all obligations have been transferred is acceptable. Any obligation not covered by the written description shall be deemed not to have been transferred.

(b) A contract research organization that assumes any obligation of a sponsor shall comply with the specific regulations in this chapter applicable to this obligation and shall be subject to the same regulatory action as a sponsor for failure to comply with any obligation assumed under these regulations. Thus, all references to "sponsor"

in this part apply to a contract research organization to the extent that it assumes one or more obligations of the sponsor.

§ 312.53 Selecting investigators and monitors.

(a) *Selecting investigators.* A sponsor shall select only investigators qualified by training and experience as appropriate experts to investigate the drug.

(b) *Control of drug.* A sponsor shall ship investigational new drugs only to investigators participating in the investigation.

(c) *Obtaining information from the investigator.* Before permitting an investigator to begin participation in an investigation, the sponsor shall obtain the following:

(1) A signed investigator statement (Form FDA–1572) containing:

(i) The name and address of the investigator;

(ii) The name and code number, if any, of the protocol(s) in the IND identifying the study(ies) to be conducted by the investigator;

(iii) The name and address of any medical school, hospital, or other research facility where the clinical investigation(s) will be conducted;

(iv) The name and address of any clinical laboratory facilities to be used in the study;

(v) The name and address of the IRB that is responsible for review and approval of the study(ies);

(vi) A commitment by the investigator that he or she:

(*a*) Will conduct the study(ies) in accordance with the relevant, current protocol(s) and will only make changes in a protocol after notifying the sponsor, except when necessary to protect the safety, the rights, or welfare of subjects;

(*b*) Will comply with all requirements regarding the obligations of clinical investigators and all other pertinent requirements in this part;

(*c*) Will personally conduct or supervise the described investigation(s);

(*d*) Will inform any potential subjects that the drugs are being used for investigational purposes and will ensure that the requirements relating to obtaining informed consent (21 CFR part

50) and institutional review board review and approval (21 CFR part 56) are met;

(*e*) Will report to the sponsor adverse experiences that occur in the course of the investigation(s) in accordance with § 312.64;

(*f*) Has read and understands the information in the investigator's brochure, including the potential risks and side effects of the drug; and

(*g*) Will ensure that all associates, colleagues, and employees assisting in the conduct of the study(ies) are informed about their obligations in meeting the above commitments.

(vii) A commitment by the investigator that, for an investigation subject to an institutional review requirement under part 56, an IRB that complies with the requirements of that part will be responsible for the initial and continuing review and approval of the clinical investigation and that the investigator will promptly report to the IRB all changes in the research activity and all unanticipated problems involving risks to human subjects or others, and will not make any changes in the research without IRB approval, except where necessary to eliminate apparent immediate hazards to the human subjects.

(viii) A list of the names of the subinvestigators (e.g., research fellows, residents) who will be assisting the investigator in the conduct of the investigation(s).

(2) *Curriculum vitae.* A curriculum vitae or other statement of qualifications of the investigator showing the education, training, and experience that qualifies the investigator as an expert in the clinical investigation of the drug for the use under investigation.

(3) *Clinical protocol.* (i) For Phase 1 investigations, a general outline of the planned investigation including the estimated duration of the study and the maximum number of subjects that will be involved.

(ii) For Phase 2 or 3 investigations, an outline of the study protocol including an approximation of the number of subjects to be treated with the drug and the number to be employed as controls, if any; the clinical uses to be investigated; characteristics of subjects by age, sex, and condition; the kind of clinical observations and laboratory tests to be conducted; the estimated duration of the study; and copies or a description of case report forms to be used.

(4) *Financial disclosure information.* Sufficient accurate financial information to allow the sponsor to submit complete and accurate certification or disclosure statements required under part 54 of this chapter. The sponsor shall obtain a commitment from the clinical investigator to promptly update this information if any relevant changes occur during the course of the investigation and for 1 year following the completion of the study.

(d) *Selecting monitors.* A sponsor shall select a monitor qualified by training and experience to monitor the progress of the investigation.

(Collection of information requirements approved by the Office of Management and Budget under control number 0910–0014)

[52 FR 8831, Mar. 19, 1987, as amended at 52 FR 23031, June 17, 1987; 61 FR 57280, Nov. 5, 1996; 63 FR 5252, Feb. 2, 1998]

§ 312.54   Emergency research under § 50.24 of this chapter.

(a) The sponsor shall monitor the progress of all investigations involving an exception from informed consent under § 50.24 of this chapter. When the sponsor receives from the IRB information concerning the public disclosures required by § 50.24(a)(7)(ii) and (a)(7)(iii) of this chapter, the sponsor promptly shall submit to the IND file and to Docket Number 95S–0158 in the Dockets Management Branch (HFA–305), Food and Drug Administration, 12420 Parklawn Dr., rm. 1–23, Rockville, MD 20857, copies of the information that was disclosed, identified by the IND number.

(b) The sponsor also shall monitor such investigations to identify when an IRB determines that it cannot approve the research because it does not meet the criteria in the exception in § 50.24(a) of this chapter or because of other relevant ethical concerns. The sponsor promptly shall provide this information in writing to FDA, investigators who are asked to participate in this or a substantially equivalent clinical investigation, and other IRB's that are asked to review this or a substantially equivalent investigation.

[61 FR 51530, Oct. 2, 1996]

§ 312.55 Informing investigators.

(a) Before the investigation begins, a sponsor (other than a sponsor-investigator) shall give each participating clinical investigator an investigator brochure containing the information described in § 312.23(a)(5).

(b) The sponsor shall, as the overall investigation proceeds, keep each participating investigator informed of new observations discovered by or reported to the sponsor on the drug, particularly with respect to adverse effects and safe use. Such information may be distributed to investigators by means of periodically revised investigator brochures, reprints or published studies, reports or letters to clinical investigators, or other appropriate means. Important safety information is required to be relayed to investigators in accordance with § 312.32.

(Collection of information requirements approved by the Office of Management and Budget under control number 0910–0014)

[52 FR 8831, Mar. 19, 1987, as amended at 52 FR 23031, June 17, 1987]

§ 312.56 Review of ongoing investigations.

(a) The sponsor shall monitor the progress of all clinical investigations being conducted under its IND.

(b) A sponsor who discovers that an investigator is not complying with the signed agreement (Form FDA–1572), the general investigational plan, or the requirements of this part or other applicable parts shall promptly either secure compliance or discontinue shipments of the investigational new drug to the investigator and end the investigator's participation in the investigation. If the investigator's participation in the investigation is ended, the sponsor shall require that the investigator dispose of or return the investigational drug in accordance with the requirements of § 312.59 and shall notify FDA.

(c) The sponsor shall review and evaluate the evidence relating to the safety and effectiveness of the drug as it is obtained from the investigator. The sponsors shall make such reports to FDA regarding information relevant to the safety of the drug as are required under § 312.32. The sponsor shall make annual reports on the progress of the investigation in accordance with § 312.33.

(d) A sponsor who determines that its investigational drug presents an unreasonable and significant risk to subjects shall discontinue those investigations that present the risk, notify FDA, all institutional review boards, and all investigators who have at any time participated in the investigation of the discontinuance, assure the disposition of all stocks of the drug outstanding as required by § 312.59, and furnish FDA with a full report of the sponsor's actions. The sponsor shall discontinue the investigation as soon as possible, and in no event later than 5 working days after making the determination that the investigation should be discontinued. Upon request, FDA will confer with a sponsor on the need to discontinue an investigation.

(Collection of information requirements approved by the Office of Management and Budget under control number 0910–0014)

[52 FR 8831, Mar. 19, 1987, as amended at 52 FR 23031, June 17, 1987]

§ 312.57 Recordkeeping and record retention.

(a) A sponsor shall maintain adequate records showing the receipt, shipment, or other disposition of the investigational drug. These records are required to include, as appropriate, the name of the investigator to whom the drug is shipped, and the date, quantity, and batch or code mark of each such shipment.

(b) A sponsor shall maintain complete and accurate records showing any financial interest in § 54.4(a)(3)(i), (a)(3)(ii), (a)(3)(iii), and (a)(3)(iv) of this chapter paid to clinical investigators by the sponsor of the covered study. A sponsor shall also maintain complete and accurate records concerning all other financial interests of investigators subject to part 54 of this chapter.

(c) A sponsor shall retain the records and reports required by this part for 2 years after a marketing application is approved for the drug; or, if an application is not approved for the drug, until 2 years after shipment and delivery of

Food and Drug Administration, HHS

the drug for investigational use is discontinued and FDA has been so notified.

(d) A sponsor shall retain reserve samples of any test article and reference standard identified in, and used in any of the bioequivalence or bioavailability studies described in, § 320.38 or § 320.63 of this chapter, and release the reserve samples to FDA upon request, in accordance with, and for the period specified in § 320.38.

(Collection of information requirements approved by the Office of Management and Budget under control number 0910–0014)

[52 FR 8831, Mar. 19, 1987, as amended at 52 FR 23031, June 17, 1987; 58 FR 25926, Apr. 28, 1993; 63 FR 5252, Feb. 2, 1998]

§ 312.58 Inspection of sponsor's records and reports.

(a) *FDA inspection.* A sponsor shall upon request from any properly authorized officer or employee of the Food and Drug Administration, at reasonable times, permit such officer or employee to have access to and copy and verify any records and reports relating to a clinical investigation conducted under this part. Upon written request by FDA, the sponsor shall submit the records or reports (or copies of them) to FDA. The sponsor shall discontinue shipments of the drug to any investigator who has failed to maintain or make available records or reports of the investigation as required by this part.

(b) *Controlled substances.* If an investigational new drug is a substance listed in any schedule of the Controlled Substances Act (21 U.S.C. 801; 21 CFR part 1308), records concerning shipment, delivery, receipt, and disposition of the drug, which are required to be kept under this part or other applicable parts of this chapter shall, upon the request of a properly authorized employee of the Drug Enforcement Administration of the U.S. Department of Justice, be made available by the investigator or sponsor to whom the request is made, for inspection and copying. In addition, the sponsor shall assure that adequate precautions are taken, including storage of the investigational drug in a securely locked, substantially constructed cabinet, or

other securely locked, substantially constructed enclosure, access to which is limited, to prevent theft or diversion of the substance into illegal channels of distribution.

§ 312.59 Disposition of unused supply of investigational drug.

The sponsor shall assure the return of all unused supplies of the investigational drug from each individual investigator whose participation in the investigation is discontinued or terminated. The sponsor may authorize alternative disposition of unused supplies of the investigational drug provided this alternative disposition does not expose humans to risks from the drug. The sponsor shall maintain written records of any disposition of the drug in accordance with § 312.57.

(Collection of information requirements approved by the Office of Management and Budget under control number 0910–0014)

[52 FR 8831, Mar. 19, 1987, as amended at 52 FR 23031, June 17, 1987]

§ 312.60 General responsibilities of investigators.

An investigator is responsible for ensuring that an investigation is conducted according to the signed investigator statement, the investigational plan, and applicable regulations; for protecting the rights, safety, and welfare of subjects under the investigator's care; and for the control of drugs under investigation. An investigator shall, in accordance with the provisions of part 50 of this chapter, obtain the informed consent of each human subject to whom the drug is administered, except as provided in §§ 50.23 or 50.24 of this chapter. Additional specific responsibilities of clinical investigators are set forth in this part and in parts 50 and 56 of this chapter.

[52 FR 8831, Mar. 19, 1987, as amended at 61 FR 51530, Oct. 2, 1996]

§ 312.61 Control of the investigational drug.

An investigator shall administer the drug only to subjects under the investigator's personal supervision or under the supervision of a subinvestigator responsible to the investigator. The investigator shall not supply the investigational drug to any person not authorized under this part to receive it.

§ 312.62 Investigator recordkeeping and record retention.

(a) *Disposition of drug.* An investigator is required to maintain adequate records of the disposition of the drug, including dates, quantity, and use by subjects. If the investigation is terminated, suspended, discontinued, or completed, the investigator shall return the unused supplies of the drug to the sponsor, or otherwise provide for disposition of the unused supplies of the drug under § 312.59.

(b) *Case histories.* An investigator is required to prepare and maintain adequate and accurate case histories that record all observations and other data pertinent to the investigation on each individual administered the investigational drug or employed as a control in the investigation. Case histories include the case report forms and supporting data including, for example, signed and dated consent forms and medical records including, for example, progress notes of the physician, the individual's hospital chart(s), and the nurses' notes. The case history for each individual shall document that informed consent was obtained prior to participation in the study.

(c) *Record retention.* An investigator shall retain records required to be maintained under this part for a period of 2 years following the date a marketing application is approved for the drug for the indication for which it is being investigated; or, if no application is to be filed or if the application is not approved for such indication, until 2 years after the investigation is discontinued and FDA is notified.

(Collection of information requirements approved by the Office of Management and Budget under control number 0910–0014)

[52 FR 8831, Mar. 19, 1987, as amended at 52 FR 23031, June 17, 1987; 61 FR 57280, Nov. 5, 1996]

§ 312.64 Investigator reports.

(a) *Progress reports.* The investigator shall furnish all reports to the sponsor of the drug who is responsible for collecting and evaluating the results obtained. The sponsor is required under § 312.33 to submit annual reports to FDA on the progress of the clinical investigations.

(b) *Safety reports.* An investigator shall promptly report to the sponsor any adverse effect that may reasonably be regarded as caused by, or probably caused by, the drug. If the adverse effect is alarming, the investigator shall report the adverse effect immediately.

(c) *Final report.* An investigator shall provide the sponsor with an adequate report shortly after completion of the investigator's participation in the investigation.

(d) *Financial disclosure reports.* The clinical investigator shall provide the sponsor with sufficient accurate financial information to allow an applicant to submit complete and accurate certification or disclosure statements as required under part 54 of this chapter. The clinical investigator shall promptly update this information if any relevant changes occur during the course of the investigation and for 1 year following the completion of the study.

(Collection of information requirements approved by the Office of Management and Budget under control number 0910–0014)

[52 FR 8831, Mar. 19, 1987, as amended at 52 FR 23031, June 17, 1987; 63 FR 5252, Feb. 2, 1998]

§ 312.66 Assurance of IRB review.

An investigator shall assure that an IRB that complies with the requirements set forth in part 56 will be responsible for the initial and continuing review and approval of the proposed clinical study. The investigator shall also assure that he or she will promptly report to the IRB all changes in the research activity and all unanticipated problems involving risk to human subjects or others, and that he or she will not make any changes in the research without IRB approval, except where necessary to eliminate apparent immediate hazards to human subjects.

(Collection of information requirements approved by the Office of Management and Budget under control number 0910–0014)

[52 FR 8831, Mar. 19, 1987, as amended at 52 FR 23031, June 17, 1987]

§312.68 Inspection of investigator's records and reports.

An investigator shall upon request from any properly authorized officer or employee of FDA, at reasonable times, permit such officer or employee to have access to, and copy and verify any records or reports made by the investigator pursuant to §312.62. The investigator is not required to divulge subject names unless the records of particular individuals require a more detailed study of the cases, or unless there is reason to believe that the records do not represent actual case studies, or do not represent actual results obtained.

§312.69 Handling of controlled substances.

If the investigational drug is subject to the Controlled Substances Act, the investigator shall take adequate precautions, including storage of the investigational drug in a securely locked, substantially constructed cabinet, or other securely locked, substantially constructed enclosure, access to which is limited, to prevent theft or diversion of the substance into illegal channels of distribution.

§312.70 Disqualification of a clinical investigator.

(a) If FDA has information indicating that an investigator (including a sponsor-investigator) has repeatedly or deliberately failed to comply with the requirements of this part, part 50, or part 56 of this chapter, or has submitted to FDA or to the sponsor false information in any required report, the Center for Drug Evaluation and Research or the Center for Biologics Evaluation and Research will furnish the investigator written notice of the matter complained of and offer the investigator an opportunity to explain the matter in writing, or, at the option of the investigator, in an informal conference. If an explanation is offered but not accepted by the Center for Drug Evaluation and Research or the Center for Biologics Evaluation and Research, the investigator will be given an opportunity for a regulatory hearing under part 16 on the question of whether the investigator is entitled to receive investigational new drugs.

(b) After evaluating all available information, including any explanation presented by the investigator, if the Commissioner determines that the investigator has repeatedly or deliberately failed to comply with the requirements of this part, part 50, or part 56 of this chapter, or has deliberately or repeatedly submitted false information to FDA or to the sponsor in any required report, the Commissioner will notify the investigator and the sponsor of any investigation in which the investigator has been named as a participant that the investigator is not entitled to receive investigational drugs. The notification will provide a statement of basis for such determination.

(c) Each IND and each approved application submitted under part 314 containing data reported by an investigator who has been determined to be ineligible to receive investigational drugs will be examined to determine whether the investigator has submitted unreliable data that are essential to the continuation of the investigation or essential to the approval of any marketing application.

(d) If the Commissioner determines, after the unreliable data submitted by the investigator are eliminated from consideration, that the data remaining are inadequate to support a conclusion that it is reasonably safe to continue the investigation, the Commissioner will notify the sponsor who shall have an opportunity for a regulatory hearing under part 16. If a danger to the public health exists, however, the Commissioner shall terminate the IND immediately and notify the sponsor of the determination. In such case, the sponsor shall have an opportunity for a regulatory hearing before FDA under part 16 on the question of whether the IND should be reinstated.

(e) If the Commissioner determines, after the unreliable data submitted by the investigator are eliminated from consideration, that the continued approval of the drug product for which the data were submitted cannot be justified, the Commissioner will proceed to withdraw approval of the drug product in accordance with the applicable provisions of the act.

§ 312.80

(f) An investigator who has been determined to be ineligible to receive investigational drugs may be reinstated as eligible when the Commissioner determines that the investigator has presented adequate assurances that the investigator will employ investigatioal drugs solely in compliance with the provisions of this part and of parts 50 and 56.

(Collection of information requirements approved by the Office of Management and Budget under control number 0910-0014)

[52 FR 8831, Mar. 19, 1987, as amended at 52 FR 23031, June 17, 1987; 55 FR 11580, Mar. 29, 1990; 62 FR 46876, Sept. 5, 1997]

Subpart E—Drugs Intended to Treat Life-threatening and Severely-debilitating Illnesses

AUTHORITY: 21 U.S.C. 351, 352, 353, 355, 371; 42 U.S.C. 262.

SOURCE: 53 FR 41523, Oct. 21, 1988, unless otherwise noted.

§ 312.80 Purpose.

The purpose of this section is to establish procedures designed to expedite the development, evaluation, and marketing of new therapies intended to treat persons with life-threatening and severely-debilitating illnesses, especially where no satisfactory alternative therapy exists. As stated § 314.105(c) of this chapter, while the statutory standards of safety and effectiveness apply to all drugs, the many kinds of drugs that are subject to them, and the wide range of uses for those drugs, demand flexibility in applying the standards. The Food and Drug Administration (FDA) has determined that it is appropriate to exercise the broadest flexibility in applying the statutory standards, while preserving appropriate guarantees for safety and effectiveness. These procedures reflect the recognition that physicians and patients are generally willing to accept greater risks or side effects from products that treat life-threatening and severely-debilitating illnesses, than they would accept from products that treat less serious illnesses. These procedures also reflect the recognition that the benefits of the drug need to be evaluated in light of the severity of the dis-

ease being treated. The procedure outlined in this section should be interpreted consistent with that purpose.

§ 312.81 Scope.

This section applies to new drug and biological products that are being studied for their safety and effectiveness in treating life-threatening or severely-debilitating diseases.

(a) For purposes of this section, the term "life-threatening" means:

(1) Diseases or conditions where the likelihood of death is high unless the course of the disease is interrupted; and

(2) Diseases or conditions with potentially fatal outcomes, where the end point of clinical trial analysis is survival.

(b) For purposes of this section, the term "severely debilitating" means diseases or conditions that cause major irreversible morbidity.

(c) Sponsors are encouraged to consult with FDA on the applicability of these procedures to specific products.

[53 FR 41523, Oct. 21, 1988, as amended at 64 FR 401, Jan. 5, 1999]

§ 312.82 Early consultation.

For products intended to treat life-threatening or severely-debilitating illnesses, sponsors may request to meet with FDA-reviewing officials early in the drug development process to review and reach agreement on the design of necessary preclinical and clinical studies. Where appropriate, FDA will invite to such meetings one or more outside expert scientific consultants or advisory committee members. To the extent FDA resources permit, agency reviewing officials will honor requests for such meetings

(a) *Pre-investigational new drug (IND) meetings.* Prior to the submission of the initial IND, the sponsor may request a meeting with FDA-reviewing officials. The primary purpose of this meeting is to review and reach agreement on the design of animal studies needed to initiate human testing. The meeting may also provide an opportunity for discussing the scope and design of phase 1 testing, plans for studying the drug product in pediatric populations, and the best approach for presentation and formatting of data in the IND.

Food and Drug Administration, HHS   **§ 312.85**

(b) *End-of-phase 1 meetings.* When data from phase 1 clinical testing are available, the sponsor may again request a meeting with FDA-reviewing officials. The primary purpose of this meeting is to review and reach agreement on the design of phase 2 controlled clinical trials, with the goal that such testing will be adequate to provide sufficient data on the drug's safety and effectiveness to support a decision on its approvability for marketing, and to discuss the need for, as well as the design and timing of, studies of the drug in pediatric patients. For drugs for life-threatening diseases, FDA will provide its best judgment, at that time, whether pediatric studies will be required and whether their submission will be deferred until after approval. The procedures outlined in § 312.47(b)(1) with respect to end-of-phase 2 conferences, including documentation of agreements reached, would also be used for end-of-phase 1 meetings.

[53 FR 41523, Oct. 21, 1988, as amended at 63 FR 66669, Dec. 2, 1998]

§ 312.83 Treatment protocols.

If the preliminary analysis of phase 2 test results appears promising, FDA may ask the sponsor to submit a treatment protocol to be reviewed under the procedures and criteria listed in §§ 312.34 and 312.35. Such a treatment protocol, if requested and granted, would normally remain in effect while the complete data necessary for a marketing application are being assembled by the sponsor and reviewed by FDA (unless grounds exist for clinical hold of ongoing protocols, as provided in § 312.42(b)(3)(ii)).

§ 312.84 Risk-benefit analysis in review of marketing applications for drugs to treat life-threatening and severely-debilitating illnesses.

(a) FDA's application of the statutory standards for marketing approval shall recognize the need for a medical risk-benefit judgment in making the final decision on approvability. As part of this evaluation, consistent with the statement of purpose in § 312.80, FDA will consider whether the benefits of the drug outweigh the known and potential risks of the drug and the need to answer remaining questions about risks and benefits of the drug, taking into consideration the severity of the disease and the absence of satisfactory alternative therapy.

(b) In making decisions on whether to grant marketing approval for products that have been the subject of an end-of-phase 1 meeting under § 312.82, FDA will usually seek the advice of outside expert scientific consultants or advisory committees. Upon the filing of such a marketing application under § 314.101 or part 601 of this chapter, FDA will notify the members of the relevant standing advisory committee of the application's filing and its availability for review.

(c) If FDA concludes that the data presented are not sufficient for marketing approval, FDA will issue (for a drug) a not approvable letter pursuant to § 314.120 of this chapter, or (for a biologic) a deficiencies letter consistent with the biological product licensing procedures. Such letter, in describing the deficiencies in the application, will address why the results of the research design agreed to under § 312.82, or in subsequent meetings, have not provided sufficient evidence for marketing approval. Such letter will also describe any recommendations made by the advisory committee regarding the application.

(d) Marketing applications submitted under the procedures contained in this section will be subject to the requirements and procedures contained in part 314 or part 600 of this chapter, as well as those in this subpart.

§ 312.85 Phase 4 studies.

Concurrent with marketing approval, FDA may seek agreement from the sponsor to conduct certain post-marketing (phase 4) studies to delineate additional information about the drug's risks, benefits, and optimal use. These studies could include, but would not be limited to, studying different doses or schedules of administration than were used in phase 2 studies, use of the drug in other patient populations or other stages of the disease, or use of the drug over a longer period of time.

§312.86 Focused FDA regulatory research.

At the discretion of the agency, FDA may undertake focused regulatory research on critical rate-limiting aspects of the preclinical, chemical/manufacturing, and clinical phases of drug development and evaluation. When initiated, FDA will undertake such research efforts as a means for meeting a public health need in facilitating the development of therapies to treat life-threatening or severely debilitating illnesses.

§312.87 Active monitoring of conduct and evaluation of clinical trials.

For drugs covered under this section, the Commissioner and other agency officials will monitor the progress of the conduct and evaluation of clinical trials and be involved in facilitating their appropriate progress.

§312.88 Safeguards for patient safety.

All of the safeguards incorporated within parts 50, 56, 312, 314, and 600 of this chapter designed to ensure the safety of clinical testing and the safety of products following marketing approval apply to drugs covered by this section. This includes the requirements for informed consent (part 50 of this chapter) and institutional review boards (part 56 of this chapter). These safeguards further include the review of animal studies prior to initial human testing (§312.23), and the monitoring of adverse drug experiences through the requirements of IND safety reports (§312.32), safety update reports during agency review of a marketing application (§314.50 of this chapter), and postmarketing adverse reaction reporting (§314.80 of this chapter).

Subpart F—Miscellaneous

§312.110 Import and export requirements.

(a) *Imports*. An investigational new drug offered for import into the United States complies with the requirements of this part if it is subject to an IND that is in effect for it under §312.40 and: (1) The consignee in the United States is the sponsor of the IND; (2) the consignee is a qualified investigator named in the IND; or (3) the consignee

is the domestic agent of a foreign sponsor, is responsible for the control and distribution of the investigational drug, and the IND identifies the consignee and describes what, if any, actions the consignee will take with respect to the investigational drug.

(b) *Exports*. An investigational new drug intended for export from the United States complies with the requirements of this part as follows:

(1) If an IND is in effect for the drug under §312.40 and each person who receives the drug is an investigator named in the application; or

(2) If FDA authorizes shipment of the drug for use in a clinical investigation. Authorization may be obtained as follows:

(i) Through submission to the International Affairs Staff (HFY–50), Associate Commissioner for Health Affairs, Food and Drug Administration, 5600 Fishers Lane, Rockville, MD 20857, of a written request from the person that seeks to export the drug. A request must provide adequate information about the drug to satisfy FDA that the drug is appropriate for the proposed investigational use in humans, that the drug will be used for investigational purposes only, and that the drug may be legally used by that consignee in the importing country for the proposed investigational use. The request shall specify the quantity of the drug to be shipped per shipment and the frequency of expected shipments. If FDA authorizes exportation under this paragraph, the agency shall concurrently notify the government of the importing country of such authorization.

(ii) Through submission to the International Affairs Staff (HFY–50), Associate Commissioner for Health Affairs, Food and Drug Administration, 5600 Fishers Lane, Rockville, MD 20857, of a formal request from an authorized official of the government of the country to which the drug is proposed to be shipped. A request must specify that the foreign government has adequate information about the drug and the proposed investigational use, that the drug will be used for investigational purposes only, and that the foreign government is satisfied that the drug may legally be used by the intended consignee in that country. Such a request shall specify the quantity of drug to be shipped per shipment and the frequency of expected shipments.

(iii) Authorization to export an investigational drug under paragraph (b)(2)(i) or (ii) of this section may be revoked by FDA if the agency finds that the conditions underlying its authorization are not longer met.

(3) This paragraph applies only where the drug is to be used for the purpose of clinical investigation.

(4) This paragraph does not apply to the export of new drugs (including biological products, antibiotic drugs, and insulin) approved or authorized for export under section 802 of the act (21 U.S.C. 382) or section 351(h)(1)(A) of the Public Health Service Act (42 U.S.C. 262(h)(1)(A)).

(Collection of information requirements approved by the Office of Management and Budget under control number 0910-0014)

[52 FR 8831, Mar. 19, 1987, as amended at 52 FR 23031, June 17, 1987; 64 FR 401, Jan. 5, 1999]

§ 312.120 Foreign clinical studies not conducted under an IND.

(a) *Introduction.* This section describes the criteria for acceptance by FDA of foreign clinical studies not conducted under an IND. In general, FDA accepts such studies provided they are well designed, well conducted, performed by qualified investigators, and conducted in accordance with ethical principles acceptable to the world community. Studies meeting these criteria may be utilized to support clinical investigations in the United States and/ or marketing approval. Marketing approval of a new drug based solely on foreign clinical data is governed by § 314.106.

(b) *Data submissions.* A sponsor who wishes to rely on a foreign clinical study to support an IND or to support an application for marketing approval shall submit to FDA the following information:

(1) A description of the investigator's qualifications;

(2) A description of the research facilities;

(3) A detailed summary of the protocol and results of the study, and,

should FDA request, case records maintained by the investigator or additional background data such as hospital or other institutional records;

(4) A description of the drug substance and drug product used in the study, including a description of components, formulation, specifications, and bioavailability of the specific drug product used in the clinical study, if available; and

(5) If the study is intended to support the effectiveness of a drug product, information showing that the study is adequate and well controlled under § 314.126.

(c) *Conformance with ethical principles.* (1) Foreign clinical research is required to have been conducted in accordance with the ethical principles stated in the "Declaration of Helsinki" (see paragraph (c)(4) of this section) or the laws and regulations of the country in which the research was conducted, whichever represents the greater protection of the individual.

(2) For each foreign clinical study submitted under this section, the sponsor shall explain how the research conformed to the ethical principles contained in the "Declaration of Helsinki" or the foreign country's standards, whichever were used. If the foreign country's standards were used, the sponsor shall explain in detail how those standards differ from the "Declaration of Helsinki" and how they offer greater protection.

(3) When the research has been approved by an independent review committee, the sponsor shall submit to FDA documentation of such review and approval, including the names and qualifications of the members of the committee. In this regard, a "review committee" means a committee composed of scientists and, where practicable, individuals who are otherwise qualified (e.g., other health professionals or laymen). The investigator may not vote on any aspect of the review of his or her protocol by a review committee.

(4) The "Declaration of Helsinki" states as follows:

91

RECOMMENDATIONS GUIDING PHYSICIANS IN BIOMEDICAL RESEARCH INVOLVING HUMAN SUBJECTS

Introduction

It is the mission of the physician to safeguard the health of the people. His or her knowledge and conscience are dedicated to the fulfillment of this mission.

The Declaration of Geneva of the World Medical Association binds the physician with the words, "The health of my patient will be my first consideration," and the International Code of Medical Ethics declares that, "A physician shall act only in the patient's interest when providing medical care which might have the effect of weakening the physical and mental condition of the patient."

The purpose of biomedical research involving human subjects must be to improve diagnostic, therapeutic and prophylactic procedures and the understanding of the aetiology and pathogenesis of disease.

In current medical practice most diagnostic, therapeutic or prophylactic procedures involve hazards. This applies especially to biomedical research.

Medical progress is based on research which ultimately must rest in part on experimentation involving human subjects.

In the field of biomedical research a fundamental distinction must be recognized between medical research in which the aim is essentially diagnostic or therapeutic for a patient, and medical research, the essential object of which is purely scientific and without implying direct diagnostic or therapeutic value to the person subjected to the research.

Special caution must be exercised in the conduct of research which may affect the environment, and the welfare of animals used for research must be respected.

Because it is essential that the results of laboratory experiments be applied to human beings to further scientific knowledge and to help suffering humanity, the World Medical Association has prepared the following recommendations as a guide to every physician in biomedical research involving human subjects. They should be kept under review in the future. It must be stressed that the standards as drafted are only a guide to physicians all over the world. Physicians are not relieved from criminal, civil and ethical responsibilities under the laws of their own countries.

I. Basic Principles

1. Biomedical research involving human subjects must conform to generally accepted scientific principles and should be based on adequately performed laboratory and animal experimentation and on a thorough knowledge of the scientific literature.

2. The design and performance of each experimental procedure involving human subjects should be clearly formulated in an experimental protocol which should be transmitted for consideration, comment and guidance to a specially appointed committee independent of the investigator and the sponsor provided that this independent committee is in conformity with the laws and regulations of the country in which the research experiment is performed.

3. Biomedical research involving human subjects should be conducted only by scientifically qualified persons and under the supervision of a clinically competent medical person. The responsibility for the human subject must always rest with a medically qualified person and never rest on the subject of the research, even though the subject has given his or her consent.

4. Biomedical research involving human subjects cannot legitimately be carried out unless the importance of the objective is in proportion to the inherent risk to the subject.

5. Every biomedical research project involving human subjects should be preceded by careful assessment of predictable risks in comparison with foreseeable benefits to the subject or to others. Concern for the interests of the subject must always prevail over the interests of science and society.

6. The right of the research subject to safeguard his or her integrity must always be respected. Every precaution should be taken to respect the privacy of the subject and to minimize the impact of the study on the subject's physical and mental integrity and on the personality of the subject.

7. Physicians should abstain from engaging in research projects involving human subjects unless they are satisfied that the hazards involved are believed to be predictable. Physicians should cease any investigation if the hazards are found to outweigh the potential benefits.

8. In publication of the results of his or her research, the physician is obliged to preserve the accuracy of the results. Reports of experimentation not in accordance with the principles laid down in this Declaration should not be accepted for publication.

9. In any research on human beings, each potential subject must be adequately informed of the aims, methods, anticipated benefits and potential hazards of the study and the discomfort it may entail. He or she should be informed that he or she is at liberty to abstain from participation in the study and that he or she is free to withdraw his or her consent to participation at any time. The physician should then obtain the subject's freely-given informed consent, preferably in writing.

10. When obtaining informed consent for the research project the physician should be particularly cautious if the subject is in a

Food and Drug Administration, HHS   **§ 312.130**

dependent relationship to him or her or may consent under duress. In that case the informed consent should be obtained by a physician who is not engaged in the investigation and who is completely independent of this official relationship.

11. In case of legal incompetence, informed consent should be obtained from the legal guardian in accordance with national legislation. Where physical or mental incapacity makes it impossible to obtain informed consent, or when the subject is a minor, permission from the responsible relative replaces that of the subject in accordance with national legislation.

Whenever the minor child is in fact able to give a consent, the minor's consent must be obtained in addition to the consent of the minor's legal guardian.

12. The research protocol should always contain a statement of the ethical considerations involved and should indicate that the principles enunciated in the present Declaration are complied with.

II. Medical Research Combined with Professional Care (Clinical Research)

1. In the treatment of the sick person, the physician must be free to use a new diagnostic and therapeutic measure, if in his or her judgment it offers hope of saving life, reestablishing health or alleviating suffering.

2. The potential benefits, hazards and discomfort of a new method should be weighed against the advantages of the best current diagnostic and therapeutic methods.

3. In any medical study, every patient—including those of a control group, if any—should be assured of the best proven diagnostic and therapeutic method.

4. The refusal of the patient to participate in a study must never interfere with the physician-patient relationship.

5. If the physician considers it essential not to obtain informed consent, the specific reasons for this proposal should be stated in the experimental protocol for transmission to the independent committee (I, 2).

6. The physician can combine medical research with professional care, the objective being the acquisition of new medical knowledge, only to the extent that medical research is justified by its potential diagnostic or therapeutic value for the patient.

III. Non-Therapeutic Biomedical Research Involving Human Subjects (Non-Clinical Biomedical Research)

1. In the purely scientific application of medical research carried out on a human being, it is the duty of the physician to remain the protector of the life and health of that person on whom biomedical research is being carried out.

2. The subjects should be volunteers—either healthy persons or patients for whom

the experimental design is not related to the patient's illness.

3. The investigator or the investigating team should discontinue the research if in his/her or their judgment it may, if continued, be harmful to the individual.

4. In research on man, the interest of science and society should never take precedence over considerations related to the well-being of the subject.

(Collection of information requirements approved by the Office of Management and Budget under control number 0910-0014)

[52 FR 8831, Mar. 19, 1987, as amended at 52 FR 23031, June 17, 1987; 56 FR 22113, May 14, 1991; 64 FR 401, Jan. 5, 1999]

§ 312.130 Availability for public disclosure of data and information in an IND.

(a) The existence of an investigational new drug application will not be disclosed by FDA unless it has previously been publicly disclosed or acknowledged.

(b) The availability for public disclosure of all data and information in an investigational new drug application for a new drug will be handled in accordance with the provisions established in § 314.430 for the confidentiality of data and information in applications submitted in part 314. The availability for public disclosure of all data and information in an investigational new drug application for a biological product will be governed by the provisions of §§ 601.50 and 601.51.

(c) Notwithstanding the provisions of § 314.430, FDA shall disclose upon request to an individual to whom an investigational new drug has been given a copy of any IND safety report relating to the use in the individual.

(d) The availability of information required to be publicly disclosed for investigations involving an exception from informed consent under § 50.24 of this chapter will be handled as follows: Persons wishing to request the publicly disclosable information in the IND that was required to be filed in Docket Number 95S-0158 in the Dockets Management Branch (HFA-305), Food and Drug Administration, 12420 Parklawn Dr., rm. 1-23, Rockville, MD 20857, shall

§312.140

submit a request under the Freedom of Information Act.

[52 FR 8831, Mar. 19, 1987. Redesignated at 53 FR 41523, Oct. 21, 1988, as amended at 61 FR 51530, Oct. 2, 1996; 64 FR 401, Jan. 5, 1999]

§312.140 Address for correspondence.

(a) Except as provided in paragraph (b) of this section, a sponsor shall send an initial IND submission to the Central Document Room, Center for Drug Evaluation and Research, Food and Drug Administration, Park Bldg., Rm. 214, 12420 Parklawn Dr., Rockville, MD 20852. On receiving the IND, FDA will inform the sponsor which one of the divisions in the Center for Drug Evaluation and Research or the Center for Biologics Evaluation and Research is responsible for the IND. Amendments, reports, and other correspondence relating to matters covered by the IND should be directed to the appropriate division. The outside wrapper of each submission shall state what is contained in the submission, for example, "IND Application", "Protocol Amendment", etc.

(b) Applications for the products listed below should be submitted to the Division of Biological Investigational New Drugs (HFB-230), Center for Biologics Evaluation and Research, Food and Drug Administration, 8800 Rockville Pike, Bethesda, MD 20892. (1) Products subject to the licensing provisions of the Public Health Service Act of July 1, 1944 (58 Stat. 682, as amended (42 U.S.C. 201 *et seq.*)) or subject to part 600; (2) ingredients packaged together with containers intended for the collection, processing, or storage of blood or blood components; (3) urokinase products; (4) plasma volume expanders and hydroxyethyl starch for leukapheresis; and (5) coupled antibodies, i.e., products that consist of an antibody component coupled with a drug or radionuclide component in which both components provide a pharmacological effect but the biological component determines the site of action.

(c) All correspondence relating to biological products for human use which are also radioactive drugs shall be submitted to the Division of Oncology and Radiopharmaceutical Drug Products (HFD–150), Center for Drug Evaluation and Research, Food and Drug Administration, 5600 Fishers Lane, Rockville, MD 20857, except that applications for coupled antibodies shall be submitted in accordance with paragraph (b) of this section.

(d) All correspondence relating to export of an investigational drug under §312.110(b)(2) shall be submitted to the International Affairs Staff (HFY–50), Office of Health Affairs, Food and Drug Administration, 5600 Fishers Lane, Rockville, MD 20857.

(Collection of information requirements approved by the Office of Management and Budget under control number 0910–0014)

[52 FR 8831, Mar. 19, 1987, as amended at 52 FR 23031, June 17, 1987; 55 FR 11580, Mar. 29, 1990]

§312.145 Guidance documents.

(a) FDA has made available guidance documents under §10.115 of this chapter to help you to comply with certain requirements of this part.

(b) The Center for Drug Evaluation and Research (CDER) and the Center for Biologics Evaluation and Research (CBER) maintain lists of guidance documents that apply to the centers' regulations. The lists are maintained on the Internet and are published annually in the FEDERAL REGISTER. A request for a copy of the CDER list should be directed to the Office of Training and Communications, Division of Communications Management, Drug Information Branch (HFD–210), Center for Drug Evaluation and Research, Food and Drug Administration, 5600 Fishers Lane, Rockville, MD 20857. A request for a copy of the CBER list should be directed to the Office of Communication, Training, and Manufacturers Assistance (HFM–40), Center for Biologics Evaluation and Research, Food and Drug Administration, 1401 Rockville Pike, Rockville, MD 20852–1448.

[65 FR 56479, Sept. 19, 2000]

Food and Drug Administration, HHS

Subpart G—Drugs for Investigational Use in Laboratory Research Animals or In Vitro Tests

§ 312.160 Drugs for investigational use in laboratory research animals or in vitro tests.

(a) *Authorization to ship.* (1)(i) A person may ship a drug intended solely for tests in vitro or in animals used only for laboratory research purposes if it is labeled as follows:

CAUTION: Contains a new drug for investigational use only in laboratory research animals, or for tests in vitro. Not for use in humans.

(ii) A person may ship a biological product for investigational in vitro diagnostic use that is listed in § 312.2(b)(2)(ii) if it is labeled as follows:

CAUTION: Contains a biological product for investigational in vitro diagnostic tests only.

(2) A person shipping a drug under paragraph (a) of this section shall use due diligence to assure that the consignee is regularly engaged in conducting such tests and that the shipment of the new drug will actually be used for tests in vitro or in animals used only for laboratory research.

(3) A person who ships a drug under paragraph (a) of this section shall maintain adequate records showing the name and post office address of the expert to whom the drug is shipped and the date, quantity, and batch or code mark of each shipment and delivery. Records of shipments under paragraph (a)(1)(i) of this section are to be maintained for a period of 2 years after the shipment. Records and reports of data and shipments under paragraph (a)(1)(ii) of this section are to be maintained in accordance with § 312.57(b). The person who ships the drug shall upon request from any properly authorized officer or employee of the Food and Drug Administration, at reasonable times, permit such officer or employee to have access to and copy and verify records required to be maintained under this section.

(b) *Termination of authorization to ship.* FDA may terminate authorization to ship a drug under this section if it finds that:

Pt. 314

(1) The sponsor of the investigation has failed to comply with any of the conditions for shipment established under this section; or

(2) The continuance of the investigation is unsafe or otherwise contrary to the public interest or the drug is used for purposes other than bona fide scientific investigation. FDA will notify the person shipping the drug of its finding and invite immediate correction. If correction is not immediately made, the person shall have an opportunity for a regulatory hearing before FDA pursuant to part 16.

(c) *Disposition of unused drug.* The person who ships the drug under paragraph (a) of this section shall assure the return of all unused supplies of the drug from individual investigators whenever the investigation discontinues or the investigation is terminated. The person who ships the drug may authorize in writing alternative disposition of unused supplies of the drug provided this alternative disposition does not expose humans to risks from the drug, either directly or indirectly (e.g., through food-producing animals). The shipper shall maintain records of any alternative disposition.

(Collection of information requirements approved by the Office of Management and Budget under control number 0910–0014)

[52 FR 8831, Mar. 19, 1987, as amended at 52 FR 23031, June 17, 1987. Redesignated at 53 FR 41523, Oct. 21, 1988]

PART 314—APPLICATIONS FOR FDA APPROVAL TO MARKET A NEW DRUG

Subpart A—General Provisions

Subpart B—Applications

Department of Health and Human Services   **§46.101**

46.404 Research not involving greater than minimal risk.

46.405 Research involving greater than minimal risk but presenting the prospect of direct benefit to the individual subjects.

46.406 Research involving greater than minimal risk and no prospect of direct benefit to individual subjects, but likely to yield generalizable knowledge about the subject's disorder or condition.

46.407 Research not otherwise approvable which presents an opportunity to understand, prevent, or alleviate a serious problem affecting the health or welfare of children.

46.408 Requirements for permission by parents or guardians and for assent by children.

46.409 Wards.

AUTHORITY: 5 U.S.C. 301; 42 U.S.C. 289.

EDITORIAL NOTE: The Department of Health and Human Services issued a notice of waiver regarding the requirements set forth in part 46, relating to protection of human subjects, as they pertain to demonstration projects, approved under section 1115 of the Social Security Act, which test the use of cost—sharing, such as deductibles, copayment and coinsurance, in the Medicaid program. For further information see 47 FR 9208, Mar. 4, 1982.

Subpart A—Basic HHS Policy for Protection of Human Research Subjects

AUTHORITY: 5 U.S.C. 301; 42 U.S.C. 289, 42 U.S.C. 300v-1(b).

SOURCE: 56 FR 28012, 28022, June 18, 1991, unless otherwise noted.

§46.101 To what does this policy apply?

(a) Except as provided in paragraph (b) of this section, this policy applies to all research involving human subjects conducted, supported or otherwise subject to regulation by any federal department or agency which takes appropriate administrative action to make the policy applicable to such research. This includes research conducted by federal civilian employees or military personnel, except that each department or agency head may adopt such procedural modifications as may be appropriate from an administrative standpoint. It also includes research conducted, supported, or otherwise subject to regulation by the federal government outside the United States.

(1) Research that is conducted or supported by a federal department or agency, whether or not it is regulated as defined in §46.102(e), must comply with all sections of this policy.

(2) Research that is neither conducted nor supported by a federal department or agency but is subject to regulation as defined in §46.102(e) must be reviewed and approved, in compliance with §46.101, §46.102, and §46.107 through §46.117 of this policy, by an institutional review board (IRB) that operates in accordance with the pertinent requirements of this policy.

(b) Unless otherwise required by department or agency heads, research activities in which the only involvement of human subjects will be in one or more of the following categories are exempt from this policy:

(1) Research conducted in established or commonly accepted educational settings, involving normal educational practices, such as (i) research on regular and special education instructional strategies, or (ii) research on the effectiveness of or the comparison among instructional techniques, curricula, or classroom management methods.

(2) Research involving the use of educational tests (cognitive, diagnostic, aptitude, achievement), survey procedures, interview procedures or observation of public behavior, unless:

(i) Information obtained is recorded in such a manner that human subjects can be identified, directly or through identifiers linked to the subjects; and (ii) any disclosure of the human subjects' responses outside the research could reasonably place the subjects at risk of criminal or civil liability or be damaging to the subjects' financial standing, employability, or reputation.

(3) Research involving the use of educational tests (cognitive, diagnostic, aptitude, achievement), survey procedures, interview procedures, or observation of public behavior that is not exempt under paragraph (b)(2) of this section, if:

(i) The human subjects are elected or appointed public officials or candidates for public office; or (ii) federal statute(s) require(s) without exception that the confidentiality of the personally identifiable information will be maintained throughout the research and thereafter.

(4) Research, involving the collection or study of existing data, documents, records, pathological specimens, or diagnostic specimens, if these sources are publicly available or if the information is recorded by the investigator in such a manner that subjects cannot be identified, directly or through identifiers linked to the subjects.

(5) Research and demonstration projects which are conducted by or subject to the approval of department or agency heads, and which are designed to study, evaluate, or otherwise examine:

(i) Public benefit or service programs; (ii) procedures for obtaining benefits or services under those programs; (iii) possible changes in or alternatives to those programs or procedures; or (iv) possible changes in methods or levels of payment for benefits or services under those programs.

(6) Taste and food quality evaluation and consumer acceptance studies, (i) if wholesome foods without additives are consumed or (ii) if a food is consumed that contains a food ingredient at or below the level and for a use found to be safe, or agricultural chemical or environmental contaminant at or below the level found to be safe, by the Food and Drug Administration or approved by the Environmental Protection Agency or the Food Safety and Inspection Service of the U.S. Department of Agriculture.

(c) Department or agency heads retain final judgment as to whether a particular activity is covered by this policy.

(d) Department or agency heads may require that specific research activities or classes of research activities conducted, supported, or otherwise subject to regulation by the department or agency but not otherwise covered by this policy, comply with some or all of the requirements of this policy.

(e) Compliance with this policy requires compliance with pertinent federal laws or regulations which provide additional protections for human subjects.

(f) This policy does not affect any state or local laws or regulations which may otherwise be applicable and which provide additional protections for human subjects.

(g) This policy does not affect any foreign laws or regulations which may otherwise be applicable and which provide additional protections to human subjects of research.

(h) When research covered by this policy takes place in foreign countries, procedures normally followed in the foreign countries to protect human subjects may differ from those set forth in this policy. [An example is a foreign institution which complies with guidelines consistent with the World Medical Assembly Declaration (Declaration of Helsinki amended 1989) issued either by sovereign states or by an organization whose function for the protection of human research subjects is internationally recognized.] In these circumstances, if a department or agency head determines that the procedures prescribed by the institution afford protections that are at least equivalent to those provided in this policy, the department or agency head may approve the substitution of the foreign procedures in lieu of the procedural requirements provided in this policy. Except when otherwise required by statute, Executive Order, or the department or agency head, notices of these actions as they occur will be published in the FEDERAL REGISTER or will be otherwise published as provided in department or agency procedures.

(i) Unless otherwise required by law, department or agency heads may waive the applicability of some or all of the provisions of this policy to specific research activities or classes of research activities otherwise covered by this policy. Except when otherwise required by statute or Executive Order, the department or agency head shall forward advance notices of these actions to the Office for Protection from Research Risks, Department of Health and Human Services (HHS), and shall also publish them in the FEDERAL REGISTER

Department of Health and Human Services   **§ 46.102**

or in such other manner as provided in department or agency procedures.[1]

[56 FR 28012, 28022, June 18, 1991; 56 FR 29756, June 28, 1991]

§ 46.102   Definitions.

(a) *Department or agency head* means the head of any federal department or agency and any other officer or employee of any department or agency to whom authority has been delegated.

(b) *Institution* means any public or private entity or agency (including federal, state, and other agencies).

(c) *Legally authorized representative* means an individual or judicial or other body authorized under applicable law to consent on behalf of a prospective subject to the subject's participation in the procedure(s) involved in the research.

(d) *Research* means a systematic investigation, including research development, testing and evaluation, designed to develop or contribute to generalizable knowledge. Activities which meet this definition constitute research for purposes of this policy, whether or not they are conducted or supported under a program which is considered research for other purposes. For example, some demonstration and service programs may include research activities.

(e) *Research subject to regulation*, and similar terms are intended to encompass those research activities for which a federal department or agency has specific responsibility for regulating as a research activity, (for example, Investigational New Drug requirements administered by the Food and Drug Ad-

---

[1]Institutions with HHS-approved assurances on file will abide by provisions of title 45 CFR part 46 subparts A–D. Some of the other Departments and Agencies have incorporated all provisions of title 45 CFR part 46 into their policies and procedures as well. However, the exemptions at 45 CFR 46.101(b) do not apply to research involving prisoners, fetuses, pregnant women, or human in vitro fertilization, subparts B and C. The exemption at 45 CFR 46.101(b)(2), for research involving survey or interview procedures or observation of public behavior, does not apply to research with children, subpart D, except for research involving observations of public behavior when the investigator(s) do not participate in the activities being observed.

ministration). It does not include research activities which are incidentally regulated by a federal department or agency solely as part of the department's or agency's broader responsibility to regulate certain types of activities whether research or non-research in nature (for example, Wage and Hour requirements administered by the Department of Labor).

(f) *Human subject* means a living individual about whom an investigator (whether professional or student) conducting research obtains

(1) Data through intervention or interaction with the individual, or

(2) Identifiable private information.

Intervention includes both physical procedures by which data are gathered (for example, venipuncture) and manipulations of the subject or the subject's environment that are performed for research purposes. Interaction includes communication or interpersonal contact between investigator and subject. *Private information* includes information about behavior that occurs in a context in which an individual can reasonably expect that no observation or recording is taking place, and information which has been provided for specific purposes by an individual and which the individual can reasonably expect will not be made public (for example, a medical record). Private information must be individually identifiable (i.e., the identity of the subject is or may readily be ascertained by the investigator or associated with the information) in order for obtaining the information to constitute research involving human subjects.

(g) *IRB* means an institutional review board established in accord with and for the purposes expressed in this policy.

(h) *IRB approval* means the determination of the IRB that the research has been reviewed and may be conducted at an institution within the constraints set forth by the IRB and by other institutional and federal requirements.

(i) *Minimal risk* means that the probability and magnitude of harm or discomfort anticipated in the research are not greater in and of themselves than those ordinarily encountered in daily

§ 46.103

life or during the performance of routine physical or psychological examinations or tests.

(j) *Certification* means the official notification by the institution to the supporting department or agency, in accordance with the requirements of this policy, that a research project or activity involving human subjects has been reviewed and approved by an IRB in accordance with an approved assurance.

§ 46.103 **Assuring compliance with this policy—research conducted or supported by any Federal Department or Agency.**

(a) Each institution engaged in research which is covered by this policy and which is conducted or supported by a federal department or agency shall provide written assurance satisfactory to the department or agency head that it will comply with the requirements set forth in this policy. In lieu of requiring submission of an assurance, individual department or agency heads shall accept the existence of a current assurance, appropriate for the research in question, on file with the Office for Protection from Research Risks, HHS, and approved for federalwide use by that office. When the existence of an HHS-approved assurance is accepted in lieu of requiring submission of an assurance, reports (except certification) required by this policy to be made to department and agency heads shall also be made to the Office for Protection from Research Risks, HHS.

(b) Departments and agencies will conduct or support research covered by this policy only if the institution has an assurance approved as provided in this section, and only if the institution has certified to the department or agency head that the research has been reviewed and approved by an IRB provided for in the assurance, and will be subject to continuing review by the IRB. Assurances applicable to federally supported or conducted research shall at a minimum include:

(1) A statement of principles governing the institution in the discharge of its responsibilities for protecting the rights and welfare of human subjects of research conducted at or sponsored by the institution, regardless of whether the research is subject to federal regu-

lation. This may include an appropriate existing code, declaration, or statement of ethical principles, or a statement formulated by the institution itself. This requirement does not preempt provisions of this policy applicable to department- or agency-supported or regulated research and need not be applicable to any research exempted or waived under § 46.101 (b) or (i).

(2) Designation of one or more IRBs established in accordance with the requirements of this policy, and for which provisions are made for meeting space and sufficient staff to support the IRB's review and recordkeeping duties.

(3) A list of IRB members identified by name; earned degrees; representative capacity; indications of experience such as board certifications, licenses, etc., sufficient to describe each member's chief anticipated contributions to IRB deliberations; and any employment or other relationship between each member and the institution; for example: full-time employee, part-time employee, member of governing panel or board, stockholder, paid or unpaid consultant. Changes in IRB membership shall be reported to the department or agency head, unless in accord with § 46.103(a) of this policy, the existence of an HHS-approved assurance is accepted. In this case, change in IRB membership shall be reported to the Office for Protection from Research Risks, HHS.

(4) Written procedures which the IRB will follow (i) for conducting its initial and continuing review of research and for reporting its findings and actions to the investigator and the institution; (ii) for determining which projects require review more often than annually and which projects need verification from sources other than the investigators that no material changes have occurred since previous IRB review; and (iii) for ensuring prompt reporting to the IRB of proposed changes in a research activity, and for ensuring that such changes in approved research, during the period for which IRB approval has already been given, may not be initiated without IRB review and approval except when necessary to

Department of Health and Human Services   **§46.107**

eliminate apparent immediate hazards to the subject.

(5) Written procedures for ensuring prompt reporting to the IRB, appropriate institutional officials, and the department or agency head of (i) any unanticipated problems involving risks to subjects or others or any serious or continuing noncompliance with this policy or the requirements or determinations of the IRB and (ii) any suspension or termination of IRB approval.

(c) The assurance shall be executed by an individual authorized to act for the institution and to assume on behalf of the institution the obligations imposed by this policy and shall be filed in such form and manner as the department or agency head prescribes.

(d) The department or agency head will evaluate all assurances submitted in accordance with this policy through such officers and employees of the department or agency and such experts or consultants engaged for this purpose as the department or agency head determines to be appropriate. The department or agency head's evaluation will take into consideration the adequacy of the proposed IRB in light of the anticipated scope of the institution's research activities and the types of subject populations likely to be involved, the appropriateness of the proposed initial and continuing review procedures in light of the probable risks, and the size and complexity of the institution.

(e) On the basis of this evaluation, the department or agency head may approve or disapprove the assurance, or enter into negotiations to develop an approvable one. The department or agency head may limit the period during which any particular approved assurance or class of approved assurances shall remain effective or otherwise condition or restrict approval.

(f) Certification is required when the research is supported by a federal department or agency and not otherwise exempted or waived under §46.101 (b) or (i). An institution with an approved assurance shall certify that each application or proposal for research covered by the assurance and by §46.103 of this Policy has been reviewed and approved by the IRB. Such certification must be submitted with the application or pro-

posal or by such later date as may be prescribed by the department or agency to which the application or proposal is submitted. Under no condition shall research covered by §46.103 of the Policy be supported prior to receipt of the certification that the research has been reviewed and approved by the IRB. Institutions without an approved assurance covering the research shall certify within 30 days after receipt of a request for such a certification from the department or agency, that the application or proposal has been approved by the IRB. If the certification is not submitted within these time limits, the application or proposal may be returned to the institution.

(Approved by the Office of Management and Budget under control number 9999–0020)

[56 FR 28012, 28022, June 18, 1991; 56 FR 29756, June 28, 1991]

§§46.104—46.106   [Reserved]

§46.107   IRB membership.

(a) Each IRB shall have at least five members, with varying backgrounds to promote complete and adequate review of research activities commonly conducted by the institution. The IRB shall be sufficiently qualified through the experience and expertise of its members, and the diversity of the members, including consideration of race, gender, and cultural backgrounds and sensitivity to such issues as community attitudes, to promote respect for its advice and counsel in safeguarding the rights and welfare of human subjects. In addition to possessing the professional competence necessary to review specific research activities, the IRB shall be able to ascertain the acceptability of proposed research in terms of institutional commitments and regulations, applicable law, and standards of professional conduct and practice. The IRB shall therefore include persons knowledgeable in these areas. If an IRB regularly reviews research that involves a vulnerable category of subjects, such as children, prisoners, pregnant women, or handicapped or mentally disabled persons, consideration shall be given to the inclusion of one or more individuals who are knowledgeable about and experienced in working with these subjects.

§ 46.108

(b) Every nondiscriminatory effort will be made to ensure that no IRB consists entirely of men or entirely of women, including the institution's consideration of qualified persons of both sexes, so long as no selection is made to the IRB on the basis of gender. No IRB may consist entirely of members of one profession.

(c) Each IRB shall include at least one member whose primary concerns are in scientific areas and at least one member whose primary concerns are in nonscientific areas.

(d) Each IRB shall include at least one member who is not otherwise affiliated with the institution and who is not part of the immediate family of a person who is affiliated with the institution.

(e) No IRB may have a member participate in the IRB's initial or continuing review of any project in which the member has a conflicting interest, except to provide information requested by the IRB.

(f) An IRB may, in its discretion, invite individuals with competence in special areas to assist in the review of issues which require expertise beyond or in addition to that available on the IRB. These individuals may not vote with the IRB.

§ 46.108 IRB functions and operations.

In order to fulfill the requirements of this policy each IRB shall:

(a) Follow written procedures in the same detail as described in § 46.103(b)(4) and, to the extent required by, § 46.103(b)(5).

(b) Except when an expedited review procedure is used (see § 46.110), review proposed research at convened meetings at which a majority of the members of the IRB are present, including at least one member whose primary concerns are in nonscientific areas. In order for the research to be approved, it shall receive the approval of a majority of those members present at the meeting.

§ 46.109 IRB review of research.

(a) An IRB shall review and have authority to approve, require modifications in (to secure approval), or disapprove all research activities covered by this policy.

(b) An IRB shall require that information given to subjects as part of informed consent is in accordance with § 46.116. The IRB may require that information, in addition to that specifically mentioned in § 46.116, be given to the subjects when in the IRB's judgment the information would meaningfully add to the protection of the rights and welfare of subjects.

(c) An IRB shall require documentation of informed consent or may waive documentation in accordance with § 46.117.

(d) An IRB shall notify investigators and the institution in writing of its decision to approve or disapprove the proposed research activity, or of modifications required to secure IRB approval of the research activity. If the IRB decides to disapprove a research activity, it shall include in its written notification a statement of the reasons for its decision and give the investigator an opportunity to respond in person or in writing.

(e) An IRB shall conduct continuing review of research covered by this policy at intervals appropriate to the degree of risk, but not less than once per year, and shall have authority to observe or have a third party observe the consent process and the research.

(Approved by the Office of Management and Budget under control number 9999–0020)

§ 46.110 Expedited review procedures for certain kinds of research involving no more than minimal risk, and for minor changes in approved research.

(a) The Secretary, HHS, has established, and published as a Notice in the FEDERAL REGISTER, a list of categories of research that may be reviewed by the IRB through an expedited review procedure. The list will be amended, as appropriate after consultation with other departments and agencies, through periodic republication by the Secretary, HHS, in the FEDERAL REGISTER. A copy of the list is available from the Office for Protection from Research Risks, National Institutes of Health, HHS, Bethesda, Maryland 20892.

(b) An IRB may use the expedited review procedure to review either or both of the following:

(1) Some or all of the research appearing on the list and found by the reviewer(s) to involve no more than minimal risk,

(2) Minor changes in previously approved research during the period (of one year or less) for which approval is authorized.

Under an expedited review procedure, the review may be carried out by the IRB chairperson or by one or more experienced reviewers designated by the chairperson from among members of the IRB. In reviewing the research, the reviewers may exercise all of the authorities of the IRB except that the reviewers may not disapprove the research. A research activity may be disapproved only after review in accordance with the non-expedited procedure set forth in §46.108(b).

(c) Each IRB which uses an expedited review procedure shall adopt a method for keeping all members advised of research proposals which have been approved under the procedure.

(d) The department or agency head may restrict, suspend, terminate, or choose not to authorize an institution's or IRB's use of the expedited review procedure.

§46.111 Criteria for IRB approval of research.

(a) In order to approve research covered by this policy the IRB shall determine that all of the following requirements are satisfied:

(1) Risks to subjects are minimized: (i) By using procedures which are consistent with sound research design and which do not unnecessarily expose subjects to risk, and (ii) whenever appropriate, by using procedures already being performed on the subjects for diagnostic or treatment purposes.

(2) Risks to subjects are reasonable in relation to anticipated benefits, if any, to subjects, and the importance of the knowledge that may reasonably be expected to result. In evaluating risks and benefits, the IRB should consider only those risks and benefits that may result from the research (as distinguished from risks and benefits of therapies subjects would receive even if not participating in the research). The IRB should not consider possible long-range effects of applying knowledge gained in the research (for example, the possible effects of the research on public policy) as among those research risks that fall within the purview of its responsibility.

(3) Selection of subjects is equitable. In making this assessment the IRB should take into account the purposes of the research and the setting in which the research will be conducted and should be particularly cognizant of the special problems of research involving vulnerable populations, such as children, prisoners, pregnant women, mentally disabled persons, or economically or educationally disadvantaged persons.

(4) Informed consent will be sought from each prospective subject or the subject's legally authorized representative, in accordance with, and to the extent required by §46.116.

(5) Informed consent will be appropriately documented, in accordance with, and to the extent required by §46.117.

(6) When appropriate, the research plan makes adequate provision for monitoring the data collected to ensure the safety of subjects.

(7) When appropriate, there are adequate provisions to protect the privacy of subjects and to maintain the confidentiality of data.

(b) When some or all of the subjects are likely to be vulnerable to coercion or undue influence, such as children, prisoners, pregnant women, mentally disabled persons, or economically or educationally disadvantaged persons, additional safeguards have been included in the study to protect the rights and welfare of these subjects.

§46.112 Review by institution.

Research covered by this policy that has been approved by an IRB may be subject to further appropriate review and approval or disapproval by officials of the institution. However, those officials may not approve the research if it has not been approved by an IRB.

§46.113 Suspension or termination of IRB approval of research.

An IRB shall have authority to suspend or terminate approval of research that is not being conducted in accordance with the IRB's requirements or

§ 46.114

that has been associated with unexpected serious harm to subjects. Any suspension or termination of approval shall include a statement of the reasons for the IRB's action and shall be reported promptly to the investigator, appropriate institutional officials, and the department or agency head.

(Approved by the Office of Management and Budget under control number 9999–0020)

§ 46.114 Cooperative research.

Cooperative research projects are those projects covered by this policy which involve more than one institution. In the conduct of cooperative research projects, each institution is responsible for safeguarding the rights and welfare of human subjects and for complying with this policy. With the approval of the department or agency head, an institution participating in a cooperative project may enter into a joint review arrangement, rely upon the review of another qualified IRB, or make similar arrangements for avoiding duplication of effort.

§ 46.115 IRB records.

(a) An institution, or when appropriate an IRB, shall prepare and maintain adequate documentation of IRB activities, including the following:

(1) Copies of all research proposals reviewed, scientific evaluations, if any, that accompany the proposals, approved sample consent documents, progress reports submitted by investigators, and reports of injuries to subjects.

(2) Minutes of IRB meetings which shall be in sufficient detail to show attendance at the meetings; actions taken by the IRB; the vote on these actions including the number of members voting for, against, and abstaining; the basis for requiring changes in or disapproving research; and a written summary of the discussion of controverted issues and their resolution.

(3) Records of continuing review activities.

(4) Copies of all correspondence between the IRB and the investigators.

(5) A list of IRB members in the same detail as described is § 46.103(b)(3).

(6) Written procedures for the IRB in the same detail as described in § 46.103(b)(4) and § 46.103(b)(5).

(7) Statements of significant new findings provided to subjects, as required by § 46.116(b)(5).

(b) The records required by this policy shall be retained for at least 3 years, and records relating to research which is conducted shall be retained for at least 3 years after completion of the research. All records shall be accessible for inspection and copying by authorized representatives of the department or agency at reasonable times and in a reasonable manner.

(Approved by the Office of Management and Budget under control number 9999–0020)

§ 46.116 General requirements for informed consent.

Except as provided elsewhere in this policy, no investigator may involve a human being as a subject in research covered by this policy unless the investigator has obtained the legally effective informed consent of the subject or the subject's legally authorized representative. An investigator shall seek such consent only under circumstances that provide the prospective subject or the representative sufficient opportunity to consider whether or not to participate and that minimize the possibility of coercion or undue influence. The information that is given to the subject or the representative shall be in language understandable to the subject or the representative. No informed consent, whether oral or written, may include any exculpatory language through which the subject or the representative is made to waive or appear to waive any of the subject's legal rights, or releases or appears to release the investigator, the sponsor, the institution or its agents from liability for negligence.

(a) Basic elements of informed consent. Except as provided in paragraph (c) or (d) of this section, in seeking informed consent the following information shall be provided to each subject:

(1) A statement that the study involves research, an explanation of the purposes of the research and the expected duration of the subject's participation, a description of the procedures to be followed, and identification of any procedures which are experimental;

(2) A description of any reasonably foreseeable risks or discomforts to the subject;

(3) A description of any benefits to the subject or to others which may reasonably be expected from the research;

(4) A disclosure of appropriate alternative procedures or courses of treatment, if any, that might be advantageous to the subject;

(5) A statement describing the extent, if any, to which confidentiality of records identifying the subject will be maintained;

(6) For research involving more than minimal risk, an explanation as to whether any compensation and an explanation as to whether any medical treatments are available if injury occurs and, if so, what they consist of, or where further information may be obtained;

(7) An explanation of whom to contact for answers to pertinent questions about the research and research subjects' rights, and whom to contact in the event of a research-related injury to the subject; and

(8) A statement that participation is voluntary, refusal to participate will involve no penalty or loss of benefits to which the subject is otherwise entitled, and the subject may discontinue participation at any time without penalty or loss of benefits to which the subject is otherwise entitled.

(b) Additional elements of informed consent. When appropriate, one or more of the following elements of information shall also be provided to each subject:

(1) A statement that the particular treatment or procedure may involve risks to the subject (or to the embryo or fetus, if the subject is or may become pregnant) which are currently unforeseeable;

(2) Anticipated circumstances under which the subject's participation may be terminated by the investigator without regard to the subject's consent;

(3) Any additional costs to the subject that may result from participation in the research;

(4) The consequences of a subject's decision to withdraw from the research and procedures for orderly termination of participation by the subject;

(5) A statement that significant new findings developed during the course of the research which may relate to the subject's willingness to continue participation will be provided to the subject; and

(6) The approximate number of subjects involved in the study.

(c) An IRB may approve a consent procedure which does not include, or which alters, some or all of the elements of informed consent set forth above, or waive the requirement to obtain informed consent provided the IRB finds and documents that:

(1) The research or demonstration project is to be conducted by or subject to the approval of state or local government officials and is designed to study, evaluate, or otherwise examine: (i) Public benefit of service programs; (ii) procedures for obtaining benefits or services under those programs; (iii) possible changes in or alternatives to those programs or procedures; or (iv) possible changes in methods or levels of payment for benefits or services under those programs; and

(2) The research could not practicably be carried out without the waiver or alteration.

(d) An IRB may approve a consent procedure which does not include, or which alters, some or all of the elements of informed consent set forth in this section, or waive the requirements to obtain informed consent provided the IRB finds and documents that:

(1) The research involves no more than minimal risk to the subjects;

(2) The waiver or alteration will not adversely affect the rights and welfare of the subjects;

(3) The research could not practicably be carried out without the waiver or alteration; and

(4) Whenever appropriate, the subjects will be provided with additional pertinent information after participation.

(e) The informed consent requirements in this policy are not intended to preempt any applicable federal, state, or local laws which require additional information to be disclosed in order for informed consent to be legally effective.

(f) Nothing in this policy is intended to limit the authority of a physician to

§46.117

provide emergency medical care, to the extent the physician is permitted to do so under applicable federal, state, or local law.

(Approved by the Office of Management and Budget under control number 9999-0020)

§46.117 Documentation of informed consent.

(a) Except as provided in paragraph (c) of this section, informed consent shall be documented by the use of a written consent form approved by the IRB and signed by the subject or the subject's legally authorized representative. A copy shall be given to the person signing the form.

(b) Except as provided in paragraph (c) of this section, the consent form may be either of the following:

(1) A written consent document that embodies the elements of informed consent required by §46.116. This form may be read to the subject or the subject's legally authorized representative, but in any event, the investigator shall give either the subject or the representative adequate opportunity to read it before it is signed; or

(2) A short form written consent document stating that the elements of informed consent required by §46.116 have been presented orally to the subject or the subject's legally authorized representative. When this method is used, there shall be a witness to the oral presentation. Also, the IRB shall approve a written summary of what is to be said to the subject or the representative. Only the short form itself is to be signed by the subject or the representative. However, the witness shall sign both the short form and a copy of the summary, and the person actually obtaining consent shall sign a copy of the summary. A copy of the summary shall be given to the subject or the representative, in addition to a copy of the short form.

(c) An IRB may waive the requirement for the investigator to obtain a signed consent form for some or all subjects if it finds either:

(1) That the only record linking the subject and the research would be the consent document and the principal risk would be potential harm resulting from a breach of confidentiality. Each

subject will be asked whether the subject wants documentation linking the subject with the research, and the subject's wishes will govern; or

(2) That the research presents no more than minimal risk of harm to subjects and involves no procedures for which written consent is normally required outside of the research context.

In cases in which the documentation requirement is waived, the IRB may require the investigator to provide subjects with a written statement regarding the research.

(Approved by the Office of Management and Budget under control number 9999-0020)

§46.118 Applications and proposals lacking definite plans for involvement of human subjects.

Certain types of applications for grants, cooperative agreements, or contracts are submitted to departments or agencies with the knowledge that subjects may be involved within the period of support, but definite plans would not normally be set forth in the application or proposal. These include activities such as institutional type grants when selection of specific projects is the institution's responsibility; research training grants in which the activities involving subjects remain to be selected; and projects in which human subjects' involvement will depend upon completion of instruments, prior animal studies, or purification of compounds. These applications need not be reviewed by an IRB before an award may be made. However, except for research exempted or waived under §46.101 (b) or (i), no human subjects may be involved in any project supported by these awards until the project has been reviewed and approved by the IRB, as provided in this policy, and certification submitted, by the institution, to the department or agency.

§46.119 Research undertaken without the intention of involving human subjects.

In the event research is undertaken without the intention of involving human subjects, but it is later proposed to involve human subjects in the research, the research shall first be reviewed and approved by an IRB, as provided in this policy, a certification submitted, by the institution, to the department or agency, and final approval given to the proposed change by the department or agency.

§46.120 Evaluation and disposition of applications and proposals for research to be conducted or supported by a Federal Department or Agency.

(a) The department or agency head will evaluate all applications and proposals involving human subjects submitted to the department or agency through such officers and employees of the department or agency and such experts and consultants as the department or agency head determines to be appropriate. This evaluation will take into consideration the risks to the subjects, the adequacy of protection against these risks, the potential benefits of the research to the subjects and others, and the importance of the knowledge gained or to be gained.

(b) On the basis of this evaluation, the department or agency head may approve or disapprove the application or proposal, or enter into negotiations to develop an approvable one.

§46.121 [Reserved]

§46.122 Use of Federal funds.

Federal funds administered by a department or agency may not be expended for research involving human subjects unless the requirements of this policy have been satisfied.

§46.123 Early termination of research support: Evaluation of applications and proposals.

(a) The department or agency head may require that department or agency support for any project be terminated or suspended in the manner prescribed in applicable program requirements, when the department or agency head finds an institution has materially failed to comply with the terms of this policy.

(b) In making decisions about supporting or approving applications or proposals covered by this policy the department or agency head may take into account, in addition to all other eligibility requirements and program criteria, factors such as whether the applicant has been subject to a termination or suspension under paragarph (a) of this section and whether the aplicant or the person or persons who would direct or has have directed the scientific and technical aspects of an activity has have, in the judgment of the department or agency head, materially failed to discharge responsibility for the protection of the rights and welfare of human subjects (whether or not the research was subject to federal regulation).

§46.124 Conditions.

With respect to any research project or any class of research projects the department or agency head may impose additional conditions prior to or at the time of approval when in the judgment of the department or agency head additional conditions are necessary for the protection of human subjects.

Subpart B—Additional Protections Pertaining to Research, Development, and Related Activities Involving Fetuses, Pregnant Women, and Human In Vitro Fertilization

SOURCE: 40 FR 33528, Aug. 8, 1975, unless otherwise noted.

§46.201 Applicability.

(a) The regulations in this subpart are applicable to all Department of Health and Human Services grants and contracts supporting research, development, and related activities involving: (1) The fetus, (2) pregnant women, and (3) human in vitro fertilization.

(b) Nothing in this subpart shall be construed as indicating that compliance with the procedures set forth herein will in any way render inapplicable pertinent State or local laws bearing upon activities covered by this subpart.

(c) The requirements of this subpart are in addition to those imposed under the other subparts of this part.

§ 46.202

§ 46.202 **Purpose.**

It is the purpose of this subpart to provide additional safeguards in reviewing activities to which this subpart is applicable to assure that they conform to appropriate ethical standards and relate to important societal needs.

§ 46.203 **Definitions.**

As used in this subpart:

(a) *Secretary* means the Secretary of Health and Human Services and any other officer or employee of the Department of Health and Human Services to whom authority has been delegated.

(b) *Pregnancy* encompasses the period of time from confirmation of implantation (through any of the presumptive signs of pregnancy, such as missed menses, or by a medically acceptable pregnancy test), until expulsion or extraction of the fetus.

(c) *Fetus* means the product of conception from the time of implantation (as evidenced by any of the presumptive signs of pregnancy, such as missed menses, or a medically acceptable pregnancy test), until a determination is made, following expulsion or extraction of the fetus, that it is viable.

(d) *Viable* as it pertains to the fetus means being able, after either spontaneous or induced delivery, to survive (given the benefit of available medical therapy) to the point of independently maintaining heart beat and respiration. The Secretary may from time to time, taking into account medical advances, publish in the FEDERAL REGISTER guidelines to assist in determining whether a fetus is viable for purposes of this subpart. If a fetus is viable after delivery, it is a premature infant.

(e) *Nonviable fetus* means a fetus *ex utero* which, although living, is not viable.

(f) *Dead fetus* means a fetus *ex utero* which exhibits neither heartbeat, spontaneous respiratory activity, spontaneous movement of voluntary muscles, nor pulsation of the umbilical cord (if still attached).

(g) *In vitro fertilization* means any fertilization of human ova which occurs outside the body of a female, either through admixture of donor human sperm and ova or by any other means.

[40 FR 33528, Aug. 8, 1975, as amended at 43 FR 1759, Jan. 11, 1978]

§ 46.204 **Ethical Advisory Boards.**

(a) One or more Ethical Advisory Boards shall be established by the Secretary. Members of these board(s) shall be so selected that the board(s) will be competent to deal with medical, legal, social, ethical, and related issues and may include, for example, research scientists, physicians, psychologists, sociologists, educators, lawyers, and ethicists, as well as representatives of the general public. No board member may be a regular, full-time employee of the Department of Health and Human Services.

(b) At the request of the Secretary, the Ethical Advisory Board shall render advice consistent with the policies and requirements of this part as to ethical issues, involving activities covered by this subpart, raised by individual applications or proposals. In addition, upon request by the Secretary, the Board shall render advice as to classes of applications or proposals and general policies, guidelines, and procedures.

(c) A Board may establish, with the approval of the Secretary, classes of applications or proposals which: (1) Must be submitted to the Board, or (2) need not be submitted to the Board. Where the Board so establishes a class of applications or proposals which must be submitted, no application or proposal within the class may be funded by the Department or any component thereof until the application or proposal has been reviewed by the Board and the Board has rendered advice as to its acceptability from an ethical standpoint.

[40 FR 33528, Aug. 8, 1975, as amended at 43 FR 1759, Jan. 11, 1978; 59 FR 28276, June 1, 1994]

§ 46.205 **Additional duties of the Institutional Review Boards in connection with activities involving fetuses, pregnant women, or human in vitro fertilization.**

(a) In addition to the responsibilities prescribed for Institutional Review Boards under Subpart A of this part,

Department of Health and Human Services   **§ 46.208**

the applicant's or offeror's Board shall, with respect to activities covered by this subpart, carry out the following additional duties:

(1) Determine that all aspects of the activity meet the requirements of this subpart;

(2) Determine that adequate consideration has been given to the manner in which potential subjects will be selected, and adequate provision has been made by the applicant or offeror for monitoring the actual informed consent process (e.g., through such mechanisms, when appropriate, as participation by the Institutional Review Board or subject advocates in: (i) Overseeing the actual process by which individual consents required by this subpart are secured either by approving induction of each individual into the activity or verifying, perhaps through sampling, that approved procedures for induction of individuals into the activity are being followed, and (ii) monitoring the progress of the activity and intervening as necessary through such steps as visits to the activity site and continuing evaluation to determine if any unanticipated risks have arisen);

(3) Carry out such other responsibilities as may be assigned by the Secretary.

(b) No award may be issued until the applicant or offeror has certified to the Secretary that the Institutional Review Board has made the determinations required under paragraph (a) of this section and the Secretary has approved these determinations, as provided in § 46.120 of Subpart A of this part.

(c) Applicants or offerors seeking support for activities covered by this subpart must provide for the designation of an Institutional Review Board, subject to approval by the Secretary, where no such Board has been established under Subpart A of this part.

[40 FR 33528, Aug. 8, 1975, as amended at 46 FR 8386, Jan. 26, 1981]

§ 46.206   General limitations.

(a) No activity to which this subpart is applicable may be undertaken unless:

(1) Appropriate studies on animals and nonpregnant individuals have been completed;

(2) Except where the purpose of the activity is to meet the health needs of the mother or the particular fetus, the risk to the fetus is minimal and, in all cases, is the least possible risk for achieving the objectives of the activity.

(3) Individuals engaged in the activity will have no part in: (i) Any decisions as to the timing, method, and procedures used to terminate the pregnancy, and (ii) determining the viability of the fetus at the termination of the pregnancy; and

(4) No procedural changes which may cause greater than minimal risk to the fetus or the pregnant woman will be introduced into the procedure for terminating the pregnancy solely in the interest of the activity.

(b) No inducements, monetary or otherwise, may be offered to terminate pregnancy for purposes of the activity.

[40 FR 33528, Aug. 8, 1975, as amended at 40 FR 51638, Nov. 6, 1975]

§ 46.207   Activities   directed   toward   pregnant women as subjects.

(a) No pregnant woman may be involved as a subject in an activity covered by this subpart unless: (1) The purpose of the activity is to meet the health needs of the mother and the fetus will be placed at risk only to the minimum extent necessary to meet such needs, or (2) the risk to the fetus is minimal.

(b) An activity permitted under paragraph (a) of this section may be conducted only if the mother and father are legally competent and have given their informed consent after having been fully informed regarding possible impact on the fetus, except that the father's informed consent need not be secured if: (1) The purpose of the activity is to meet the health needs of the mother; (2) his identity or whereabouts cannot reasonably be ascertained; (3) he is not reasonably available; or (4) the pregnancy resulted from rape.

§ 46.208   Activities   directed   toward   fetuses in utero as subjects.

(a) No fetus *in utero* may be involved as a subject in any activity covered by this subpart unless: (1) The purpose of the activity is to meet the health needs of the particular fetus and the fetus

119

§ 46.209

will be placed at risk only to the minimum extent necessary to meet such needs, or (2) the risk to the fetus imposed by the research is minimal and the purpose of the activity is the development of important biomedical knowledge which cannot be obtained by other means.

(b) An activity permitted under paragraph (a) of this section may be conducted only if the mother and father are legally competent and have given their informed consent, except that the father's consent need not be secured if: (1) His identity or whereabouts cannot reasonably be ascertained, (2) he is not reasonably available, or (3) the pregnancy resulted from rape.

§ 46.209 Activities directed toward fetuses ex utero, including nonviable fetuses, as subjects.

(a) Until it has been ascertained whether or not a fetus ex utero is viable, a fetus ex utero may not be involved as a subject in an activity covered by this subpart unless:

(1) There will be no added risk to the fetus resulting from the activity, and the purpose of the activity is the development of important biomedical knowledge which cannot be obtained by other means, or

(2) The purpose of the activity is to enhance the possibility of survival of the particular fetus to the point of viability.

(b) No nonviable fetus may be involved as a subject in an activity covered by this subpart unless:

(1) Vital functions of the fetus will not be artificially maintained,

(2) Experimental activities which of themselves would terminate the heartbeat or respiration of the fetus will not be employed, and

(3) The purpose of the activity is the development of important biomedical knowledge which cannot be obtained by other means.

(c) In the event the fetus ex utero is found to be viable, it may be included as a subject in the activity only to the extent permitted by and in accordance with the requirements of other subparts of this part.

(d) An activity permitted under paragraph (a) or (b) of this section may be conducted only if the mother and fa-

ther are legally competent and have given their informed consent, except that the father's informed consent need not be secured if: (1) His identity or whereabouts cannot reasonably be ascertained, (2) he is not reasonably available, or (3) the pregnancy resulted from rape.

[40 FR 33528, Aug. 8, 1975, as amended at 43 FR 1759, Jan. 11, 1978]

§ 46.210 Activities involving the dead fetus, fetal material, or the placenta.

Activities involving the dead fetus, mascerated fetal material, or cells, tissue, or organs excised from a dead fetus shall be conducted only in accordance with any applicable State or local laws regarding such activities.

§ 46.211 Modification or waiver of specific requirements.

Upon the request of an applicant or offeror (with the approval of its Institutional Review Board), the Secretary may modify or waive specific requirements of this subpart, with the approval of the Ethical Advisory Board after such opportunity for public comment as the Ethical Advisory Board considers appropriate in the particular instance. In making such decisions, the Secretary will consider whether the risks to the subject are so outweighed by the sum of the benefit to the subject and the importance of the knowledge to be gained as to warrant such modification or waiver and that such benefits cannot be gained except through a modification or waiver. Any such modifications or waivers will be published as notices in the FEDERAL REGISTER.

Subpart C—Additional Protections Pertaining to Biomedical and Behavioral Research Involving Prisoners as Subjects

SOURCE: 43 FR 53655, Nov. 16, 1978, unless otherwise noted.

§ 46.301 Applicability.

(a) The regulations in this subpart are applicable to all biomedical and behavioral research conducted or supported by the Department of Health

Department of Health and Human Services §46.305

and Human Services involving prisoners as subjects.

(b) Nothing in this subpart shall be construed as indicating that compliance with the procedures set forth herein will authorize research involving prisoners as subjects, to the extent such research is limited or barred by applicable State or local law.

(c) The requirements of this subpart are in addition to those imposed under the other subparts of this part.

§46.302 **Purpose.**

Inasmuch as prisoners may be under constraints because of their incarceration which could affect their ability to make a truly voluntary and uncoerced decision whether or not to participate as subjects in research, it is the purpose of this subpart to provide additional safeguards for the protection of prisoners involved in activities to which this subpart is applicable.

§46.303 **Definitions.**

As used in this subpart:

(a) *Secretary* means the Secretary of Health and Human Services and any other officer or employee of the Department of Health and Human Services to whom authority has been delegated.

(b) *DHHS* means the Department of Health and Human Services.

(c) *Prisoner* means any individual involuntarily confined or detained in a penal institution. The term is intended to encompass individuals sentenced to such an institution under a criminal or civil statute, individuals detained in other facilities by virtue of statutes or commitment procedures which provide alternatives to criminal prosecution or incarceration in a penal institution, and individuals detained pending arraignment, trial, or sentencing.

(d) *Minimal risk* is the probability and magnitude of physical or psychological harm that is normally encountered in the daily lives, or in the routine medical, dental, or psychological examination of healthy persons.

§46.304 **Composition of Institutional Review Boards where prisoners are involved.**

In addition to satisfying the requirements in §46.107 of this part, an Insti-

tutional Review Board, carrying out responsibilities under this part with respect to research covered by this subpart, shall also meet the following specific requirements:

(a) A majority of the Board (exclusive of prisoner members) shall have no association with the prison(s) involved, apart from their membership on the Board.

(b) At least one member of the Board shall be a prisoner, or a prisoner representative with appropriate background and experience to serve in that capacity, except that where a particular research project is reviewed by more than one Board only one Board need satisfy this requirement.

[43 FR 53655, Nov. 16, 1978, as amended at 46 FR 8386, Jan. 26, 1981]

§46.305 **Additional duties of the Institutional Review Boards where prisoners are involved.**

(a) In addition to all other responsibilities prescribed for Institutional Review Boards under this part, the Board shall review research covered by this subpart and approve such research only if it finds that:

(1) The research under review represents one of the categories of research permissible under §46.306(a)(2);

(2) Any possible advantages accruing to the prisoner through his or her participation in the research, when compared to the general living conditions, medical care, quality of food, amenities and opportunity for earnings in the prison, are not of such a magnitude that his or her ability to weigh the risks of the research against the value of such advantages in the limited choice environment of the prison is impaired;

(3) The risks involved in the research are commensurate with risks that would be accepted by nonprisoner volunteers;

(4) Procedures for the selection of subjects within the prison are fair to all prisoners and immune from arbitrary intervention by prison authorities or prisoners. Unless the principal investigator provides to the Board justification in writing for following some other procedures, control subjects must be selected randomly from the group of available prisoners who meet

§ 46.306

§ 46.306   45 CFR Subtitle A (10-1-00 Edition)

the characteristics needed for that particular research project;

(5) The information is presented in language which is understandable to the subject population;

(6) Adequate assurance exists that parole boards will not take into account a prisoner's participation in the research in making decisions regarding parole, and each prisoner is clearly informed in advance that participation in the research will have no effect on his or her parole; and

(7) Where the Board finds there may be a need for follow-up examination or care of participants after the end of their participation, adequate provision has been made for such examination or care, taking into account the varying lengths of individual prisoners' sentences, and for informing participants of this fact.

(b) The Board shall carry out such other duties as may be assigned by the Secretary.

(c) The institution shall certify to the Secretary, in such form and manner as the Secretary may require, that the duties of the Board under this section have been fulfilled.

§ 46.306 Permitted research involving prisoners.

(a) Biomedical or behavioral research conducted or supported by DHHS may involve prisoners as subjects only if:

(1) The institution responsible for the conduct of the research has certified to the Secretary that the Institutional Review Board has approved the research under § 46.305 of this subpart; and

(2) In the judgment of the Secretary the proposed research involves solely the following:

(i) Study of the possible causes, effects, and processes of incarceration, and of criminal behavior, provided that the study presents no more than minimal risk and no more than inconvenience to the subjects;

(ii) Study of prisons as institutional structures or of prisoners as incarcerated persons, provided that the study presents no more than minimal risk and no more than inconvenience to the subjects;

(iii) Research on conditions particularly affecting prisoners as a class (for example, vaccine trials and other research on hepatitis which is much more prevalent in prisons than elsewhere; and research on social and psychological problems such as alcoholism, drug addiction and sexual assaults) provided that the study may proceed only after the Secretary has consulted with appropriate experts including experts in penology medicine and ethics, and published notice, in the FEDERAL REGISTER, of his intent to approve such research; or

(iv) Research on practices, both innovative and accepted, which have the intent and reasonable probability of improving the health or well-being of the subject. In cases in which those studies require the assignment of prisoners in a manner consistent with protocols approved by the IRB to control groups which may not benefit from the research, the study may proceed only after the Secretary has consulted with appropriate experts, including experts in penology medicine and ethics, and published notice, in the FEDERAL REGISTER, of his intent to approve such research.

(b) Except as provided in paragraph (a) of this section, biomedical or behavioral research conducted or supported by DHHS shall not involve prisoners as subjects.

Subpart D—Additional Protections for Children Involved as Subjects in Research

SOURCE: 48 FR 9818, Mar. 8, 1983, unless otherwise noted.

§ 46.401 To what do these regulations apply?

(a) This subpart applies to all research involving children as subjects, conducted or supported by the Department of Health and Human Services.

(1) This includes research conducted by Department employees, except that each head of an Operating Division of the Department may adopt such nonsubstantive, procedural modifications as may be appropriate from an administrative standpoint.

(2) It also includes research conducted or supported by the Department of Health and Human Services outside the United States, but in appropriate

Department of Health and Human Services   **§ 46.406**

circumstances, the Secretary may, under paragraph (e) of § 46.101 of Subpart A, waive the applicability of some or all of the requirements of these regulations for research of this type.

(b) Exemptions at § 46.101(b)(1) and (b)(3) through (b)(6) are applicable to this subpart. The exemption at § 46.101(b)(2) regarding educational tests is also applicable to this subpart. However, the exemption at § 46.101(b)(2) for research involving survey or interview procedures or observations of public behavior does not apply to research covered by this subpart, except for research involving observation of public behavior when the investigator(s) do not participate in the activities being observed.

(c) The exceptions, additions, and provisions for waiver as they appear in paragraphs (c) through (i) of § 46.101 of Subpart A are applicable to this subpart.

[48 FR 9818, Mar. 8, 1983; 56 FR 28032, June 18, 1991; 56 FR 29757, June 28, 1991]

§ 46.402 Definitions.

The definitions in § 46.102 of Subpart A shall be applicable to this subpart as well. In addition, as used in this subpart:

(a) *Children* are persons who have not attained the legal age for consent to treatments or procedures involved in the research, under the applicable law of the jurisdiction in which the research will be conducted.

(b) *Assent* means a child's affirmative agreement to participate in research. Mere failure to object should not, absent affirmative agreement, be construed as assent.

(c) *Permission* means the agreement of parent(s) or guardian to the participation of their child or ward in research.

(d) *Parent* means a child's biological or adoptive parent.

(e) *Guardian* means an individual who is authorized under applicable State or local law to consent on behalf of a child to general medical care.

§ 46.403 IRB duties.

In addition to other responsibilities assigned to IRBs under this part, each IRB shall review research covered by this subpart and approve only research

which satisfies the conditions of all applicable sections of this subpart.

§ 46.404 Research not involving greater than minimal risk.

HHS will conduct or fund research in which the IRB finds that no greater than minimal risk to children is presented, only if the IRB finds that adequate provisions are made for soliciting the assent of the children and the permission of their parents or guardians, as set forth in § 46.408.

§ 46.405 Research involving greater than minimal risk but presenting the prospect of direct benefit to the individual subjects.

HHS will conduct or fund research in which the IRB finds that more than minimal risk to children is presented by an intervention or procedure that holds out the prospect of direct benefit for the individual subject, or by a monitoring procedure that is likely to contribute to the subject's well-being, only if the IRB finds that:

(a) The risk is justified by the anticipated benefit to the subjects;

(b) The relation of the anticipated benefit to the risk is at least as favorable to the subjects as that presented by available alternative approaches; and

(c) Adequate provisions are made for soliciting the assent of the children and permission of their parents or guardians, as set forth in § 46.408.

§ 46.406 Research involving greater than minimal risk and no prospect of direct benefit to individual subjects, but likely to yield generalizable knowledge about the subject's disorder or condition.

HHS will conduct or fund research in which the IRB finds that more than minimal risk to children is presented by an intervention or procedure that does not hold out the prospect of direct benefit for the individual subject, or by a monitoring procedure which is not likely to contribute to the well-being of the subject, only if the IRB finds that:

(a) The risk represents a minor increase over minimal risk;

(b) The intervention or procedure presents experiences to subjects that are reasonably commensurate with those inherent in their actual or expected medical, dental, psychological, social, or educational situations;

(c) The intervention or procedure is likely to yield generalizable knowledge about the subjects' disorder or condition which is of vital importance for the understanding or amelioration of the subjects' disorder or condition; and

(d) Adequate provisions are made for soliciting assent of the children and permission of their parents or guardians, as set forth in § 46.408.

§ 46.407 Research not otherwise approvable which presents an opportunity to understand, prevent, or alleviate a serious problem affecting the health or welfare of children.

HHS will conduct or fund research that the IRB does not believe meets the requirements of § 46.404, § 46.405, or § 46.406 only if:

(a) The IRB finds that the research presents a reasonable opportunity to further the understanding, prevention, or alleviation of a serious problem affecting the health or welfare of children; and

(b) The Secretary, after consultation with a panel of experts in pertinent disciplines (for example: science, medicine, education, ethics, law) and following opportunity for public review and comment, has determined either:

(1) That the research in fact satisfies the conditions of § 46.404, § 46.405, or § 46.406, as applicable, or

(2) The following:

(i) The research presents a reasonable opportunity to further the understanding, prevention, or alleviation of a serious problem affecting the health or welfare of children;

(ii) The research will be conducted in accordance with sound ethical principles;

(iii) Adequate provisions are made for soliciting the assent of children and the permission of their parents or guardians, as set forth in § 46.408.

§ 46.408 Requirements for permission by parents or guardians and for assent by children.

(a) In addition to the determinations required under other applicable sections of this subpart, the IRB shall determine that adequate provisions are made for soliciting the assent of the children, when in the judgment of the IRB the children are capable of providing assent. In determining whether children are capable of assenting, the IRB shall take into account the ages, maturity, and psychological state of the children involved. This judgment may be made for all children to be involved in research under a particular protocol, or for each child, as the IRB deems appropriate. If the IRB determines that the capability of some or all of the children is so limited that they cannot reasonably be consulted or that the intervention or procedure involved in the research holds out a prospect of direct benefit that is important to the health or well-being of the children and is available only in the context of the research, the assent of the children is not a necessary condition for proceeding with the research. Even where the IRB determines that the subjects are capable of assenting, the IRB may still waive the assent requirement under circumstances in which consent may be waived in accord with § 46.116 of Subpart A.

(b) In addition to the determinations required under other applicable sections of this subpart, the IRB shall determine, in accordance with and to the extent that consent is required by § 46.116 of Subpart A, that adequate provisions are made for soliciting the permission of each child's parents or guardian. Where parental permission is to be obtained, the IRB may find that the permission of one parent is sufficient for research to be conducted under § 46.404 or § 46.405. Where research is covered by §§ 46.406 and 46.407 and permission is to be obtained from parents, both parents must give their permission unless one parent is deceased, unknown, incompetent, or not reasonably available, or when only one parent has legal responsibility for the care and custody of the child.

(c) In addition to the provisions for waiver contained in § 46.116 of Subpart A, if the IRB determines that a research protocol is designed for conditions or for a subject population for which parental or guardian permission is not a reasonable requirement to protect the subjects (for example, neglected or abused children), it may

---

waive the consent requirements in Subpart A of this part and paragraph (b) of this section, provided an appropriate mechanism for protecting the children who will participate as subjects in the research is substituted, and provided further that the waiver is not inconsistent with Federal, state or local law. The choice of an appropriate mechanism would depend upon the nature and purpose of the activities described in the protocol, the risk and anticipated benefit to the research subjects, and their age, maturity, status, and condition.

(d) Permission by parents or guardians shall be documented in accordance with and to the extent required by §46.117 of Subpart A.

(e) When the IRB determines that assent is required, it shall also determine whether and how assent must be documented.

§46.409 Wards.

(a) Children who are wards of the state or any other agency, institution, or entity can be included in research approved under §46.406 or §46.407 only if such research is:

(1) Related to their status as wards; or

(2) Conducted in schools, camps, hospitals, institutions, or similar settings in which the majority of children involved as subjects are not wards.

(b) If the research is approved under paragraph (a) of this section, the IRB shall require appointment of an advocate for each child who is a ward, in addition to any other individual acting on behalf of the child as guardian or in loco parentis. One individual may serve as advocate for more than one child. The advocate shall be an individual who has the background and experience to act in, and agrees to act in, the best interests of the child for the duration of the child's participation in the research and who is not associated in any way (except in the role as advocate or member of the IRB) with the research, the investigator(s), or the guardian organization.

PART 50—U.S. EXCHANGE VISITOR PROGRAM—REQUEST FOR WAIVER OF THE TWO-YEAR FOREIGN RESIDENCE REQUIREMENT

Sec.
50.1 Authority.
50.2 Exchange Visitor Waiver Review Board.
50.3 Policy.
50.4 Procedures for submission of application to HHS.
50.5 Personal hardship, persecution and visa extension considerations.
50.6 Release from foreign government.

AUTHORITY: 75 Stat. 527 (22 U.S.C. 2451 et seq.); 84 Stat. 116 (8 U.S.C. 1182(e)).

SOURCE: 49 FR 9900, Mar. 16, 1984, unless otherwise noted.

§50.1 Authority.

Under the authority of Mutual Educational and Cultural Exchange Act of 1961 (75 Stat. 527) and the Immigration and Nationality Act as amended (84 Stat. 116), the Department of Health and Human Services is an "interested United States Government agency" with the authority to request the United States Information Agency to recommend to the Attorney General waiver of the two-year foreign residence requirement for exchange visitors under the Mutual Educational and Cultural Exchange Program.

§50.2 Exchange Visitor Waiver Review Board.

(a) *Establishment.* The Exchange Visitor Waiver Review Board is established to carry out the Department's responsibilities under the Exchange Visitor Program.

(b) *Functions.* The Exchange Visitor Waiver Review Board is responsible for making thorough and equitable evaluations of applications submitted by institutions, acting on behalf of exchange visitors, to the Department of HHS for a favorable recommendation to the United States Information Agency that the two-year foreign residence requirement for exchange visitors under the Exchanges Visitor Program be waived.

(c) *Membership.* The Exchange Visitor Waiver Review Board consists of no fewer than three members and two alternates, of whom no fewer than three

Appendix B

U.S. Food and Drug Administration

FDA INFORMATION SHEETS

Guidance for Institutional Review Boards and Clinical Investigation

1998 Update

This document represents the agency's current guidance on protection of human subjects of research. It is published as Level 2 guidance in accordance with the FDA "Good Guidance Practices." It does not create or confer any rights for or on any person and does not operate to bind FDA or the public. An alternative approach may be used if such approach satisfies the requirements of the applicable statute, regulations, or both. However, in many places throughout this document, a specific regulation is cited and the requirements of the regulation are reiterated. The regulations are enforceable.

This guidance document represents an update of the October 1995 revision of the Information Sheets. Comments and suggestions may be submitted at any time for Agency consideration. Comments received after publication may not be acted upon by the Agency until the document is next revised. For questions regarding the use or interpretation of this guidance, contact the Office of Health Affairs, HFY-20, 5600 Fishers Lane, Rockville, MD 20857, phone 301-827-1685, facsimile 301-443-0232.

Food and Drug Administration
Office of the Associate Commissioner for Health Affairs
5600 Fishers Lane
Rockville, Maryland, 20857

Revised 10/95
Updated 9/98

Appendix B
TABLE OF CONTENTS

IRB OPERATIONS AND CLINICAL INVESTIGATION REQUIREMENTS

General

Drugs and Biologics

Medical Devices

FDA Operations

Appendices

Frequently Asked Questions

The following is a compilation of answers to questions asked of FDA regarding the protection of human subjects of research. For ease of reference, the numbers assigned to the questions are consecutive throughout this section. These questions and answers are organized as follows:

I.   IRB Organization
II.   IRB Membership
III.   IRB Procedures
IV.   IRB Records
V.   Informed Consent Process
VI.   Informed Consent Document Content
VII.   Clinical Investigations
VIII.   General Questions

I. IRB Organization

1. What is an Institutional Review Board (IRB)?

Under FDA regulations, an IRB is an appropriately constituted group that has been formally designated to review and monitor biomedical research involving human subjects. In accordance with FDA regulations, an IRB has the authority to approve, require modifications in (to secure approval), or disapprove research. This group review serves an important role in the protection of the rights and welfare of human research subjects. The purpose of IRB review is to assure, both in advance and by periodic review, that appropriate steps are taken to protect the rights and welfare of humans participating as subjects in the research. To accomplish this purpose, IRBs use a group process to review research protocols and related materials (e.g., informed consent documents and investigator brochures) to ensure protection of the rights and welfare of human subjects of research.

2. Do IRBs have to be formally called by that name?

No, "IRB" is a generic term used by FDA (and HHS) to refer to a group whose function is to review research to assure the protection of the rights and welfare of the human subjects. Each institution may use whatever name it chooses. Regardless of the name chosen, the IRB is subject to the Agency's IRB regulations when studies of FDA regulated products are reviewed and approved.

3. Does an IRB need to register with FDA before approving studies?

Currently, FDA does not require IRB registration. The form FDA-1572 "Statement of Investigator" for a study conducted under an IND requires the name and address of the IRB that will be responsible for review of the study. IRBs that approve studies of FDA regulated products must be established and operated in compliance with 21 CFR part 56.

4. What is an "assurance" or a "multiple project assurance?"

An "assurance," is a document negotiated between an institution and the Department of Health and Human Services (HHS) in accordance with HHS regulations. For research involving human subjects conducted by HHS or supported in whole or in part by HHS, the HHS regulations require a written assurance from the performance-site institution that the institution will comply with the HHS protection of human subjects regulations [45 CFR part 46]. The assurance mechanism is described in 45 CFR 46.103. Once an institution's assurance has been approved by HHS, a number is assigned to the assurance. The assurance may be for a single grant or contract (a "single project assurance"); for multiple grants ("multiple project assurances" - formerly called "general assurances"); or for certain types of studies such as oncology group studies and AIDS research group studies ("cooperative project assurances"). The Office for Protection from Research Risks (OPRR), is responsible for implementing the HHS regulations. The address and telephone number for OPRR are: 6100 Executive Boulevard, Suite 3B01 (MSC-7507), Rockville, MD 20892-7507; (301) 496-7041.

5. Is an "assurance" required by FDA?

Currently, FDA regulations do not require an assurance. FDA regulations [21 CFR parts 50 and 56] apply to research involving products regulated by FDA - federal funds and/or support do not need to be involved for the FDA regulations to apply. When research studies involving products regulated by FDA are funded/supported by HHS, the research institution must comply with both the HHS and FDA regulations. Also, see the information sheet entitled "Significant Differences in HHS and FDA Regulations for the protection of Human Subjects."

6. Must an institution establish its own IRB?

No. Although institutions engaged in research involving human subjects will usually have their own IRBs to oversee research conducted within the institution or by the staff of the institution, FDA regulations permit an institution without an IRB to arrange for an "outside" IRB to be responsible for initial and continuing review of studies conducted at the non-IRB institution. Such arrangements should be documented in writing. Individuals conducting research in a non-institutional setting often use established IRBs (independent or institutional) rather than form their own IRBs. Also see the information sheets entitled "Non-local IRB Review" and "Cooperative Research."

7. May a hospital IRB review a study that will be conducted outside of the hospital?

Yes. IRBs may agree to review research from affiliated or unaffiliated investigators, however, FDA does not require IRBs to assume this responsibility. If the IRB routinely conducts these reviews, the IRB policies should authorize such reviews and the process should be described in the IRB's written procedures. A hospital IRB may review outside studies on an individual basis when the minutes clearly show the members are aware of where the study is to be conducted and when the IRB possesses appropriate knowledge about the study site(s).

8. May IRB members be paid for their services?

The FDA regulations do not preclude a member from being compensated for services rendered. Payment to IRB members should not be related to or dependent upon a favorable decision. Expenses, such as travel costs, may also be reimbursed.

9. What is the FDA role in IRB liability in malpractice suits?

FDA regulations do not address the question of IRB or institutional liability in the case of malpractice suits. FDA does not have authority to limit liability of IRBs or their members. Compliance with FDA regulations may help minimize an IRB's exposure to liability.

10. Is the purpose of the IRB review of informed consent to protect the institution or the subject?

The fundamental purpose of IRB review of informed consent is to assure that the rights and welfare of subjects are protected. A signed informed consent document is evidence that the document has been provided to a prospective subject (and presumably, explained) and that the subject has agreed to participate in the research. IRB review of informed consent documents also ensures that the institution has complied with applicable regulations.

11. Does an IRB or institution have to compensate subjects if injury occurs as a result of participation in a research study?

Institutional policy, not FDA regulation, determines whether compensation and medical treatment(s) will be offered and the conditions that might be placed on subject eligibility for compensation or treatment(s). The FDA informed consent regulation on compensation [21 CFR 50.25(a)(6)] requires that, for research involving more than minimal risk, the subject must be told whether any compensation and any medical treatment(s) are available if injury occurs and, if so, what they are, or where further information may be obtained. Any statement that compensation is not offered must avoid waiving or appearing to waive any of the subject's rights or releasing or appearing to release the investigator, sponsor, or institution from liability for negligence [21 CFR 50.20].

II. IRB Membership

12. May a clinical investigator be an IRB member?

Yes, however, the IRB regulations [21 CFR 56.107(e)] prohibit any member from participating in the IRB's initial or continuing review of any study in which the member has a conflicting interest, except to provide information requested by the IRB. When selecting IRB members, the potential for conflicts of interest should be considered. When members frequently have conflicts and must absent themselves from deliberation and abstain from voting, their contributions to the group review process may be diminished and could hinder the review procedure. Even greater disruptions may result if this person is chairperson of the IRB.

13. The IRB regulations require an IRB to have a diverse membership. May one member satisfy more than one membership category?

Yes. For example, one member could be otherwise unaffiliated with the institution and have a primary concern in a non-scientific area. This individual would satisfy two of the membership requirements of the regulations. IRBs should strive, however, for a membership that has a diversity of representative capacities and disciplines. In fact, the FDA regulations [21 CFR 56.107(a)] require that, as part of being qualified as an IRB, the IRB must have "... diversity of members, including consideration of race, gender, cultural backgrounds and sensitivity to such issues as community attitudes ...."

14. When IRB members cannot attend a convened meeting, may they send someone from their department to vote for them?

No. Alternates who are formally appointed and listed in the membership roster may substitute, but ad hoc substitutes are not permissible as members of an IRB. However, a member who is unable to be present at the convened meeting may participate by video-conference or conference telephone call, when the member has received a copy of the documents that are to be reviewed at the meeting. Such members may vote and be counted as part of the quorum. If allowed by IRB procedures, ad hoc substitutes may attend as consultants and gather information for the absent member, but they may not be counted toward the quorum or participate in either deliberation or voting with the board. The IRB may, of course, ask questions of this representative just as they could of any non-member consultant. Opinions of the absent members that are transmitted by mail, telephone, telefax or e-mail may be considered by the attending IRB members but may not be counted as votes or the quorum for convened meetings.

15. May the IRB use alternate members?

The use of formally appointed alternate IRB members is acceptable to the FDA, provided that the IRB's written procedures describe the appointment and function of alternate members. The IRB roster should identify the primary member(s) for whom each alternate member may substitute. To ensure maintaining an appropriate quorum, the alternate's qualifications should be comparable to the primary member to be replaced. The IRB minutes should document when an alternate member replaces a primary member. When alternates substitute for a primary member, the alternate member should have received and reviewed the same material that the primary member received or would have received.

16. Does a non-affiliated member need to attend every IRB meeting?

No. Although 21 CFR 56.108(c) does not specifically require the presence of a member not otherwise affiliated with the institution to constitute a quorum, FDA considers the presence of such members an important element of the IRB's diversity. Therefore, frequent absence of all non-affiliated members is not acceptable to FDA. Acknowledging their important role, many IRBs have appointed more than one member who is not otherwise affiliated with the institution. FDA encourages IRBs to appoint members in accordance with 21 CFR 56.107(a) who will be able to participate fully in the IRB process.

17. Which IRB members should be considered to be scientists and non-scientists?

21 CFR 56.107(c) requires at least one member of the IRB to have primary concerns in the scientific area and at least one to have primary concerns in the non-scientific area. Most IRBs include physicians and Ph.D. level physical or biological scientists. Such members satisfy the requirement for at least one scientist. When an IRB encounters studies involving science beyond the expertise of the members, the IRB may use a consultant to assist in the review, as provided by 21 CFR 56.107(f).

FDA believes the intent of the requirement for diversity of disciplines was to include members who had little or no scientific or medical training or experience. Therefore, nurses, pharmacists and other biomedical health professionals should not be regarded to have "primary concerns in the non-scientific area." In the past, lawyers, clergy and ethicists have been cited as examples of persons whose primary concerns would be in non-scientific areas.

Some members have training in both scientific and non-scientific disciplines, such as a J.D., R.N. While such members are of great value to an IRB, other members who are unambiguously non-scientific should be appointed to satisfy the non-scientist requirement.

III. IRB Procedures

18. The FDA regulations [21 CFR 56.104(c)] exempt an emergency use of a test article from prospective IRB review, however, "... any subsequent use of the test article at the institution is subject to IRB review." What does the phrase "subsequent use" mean?

FDA regulations allow for one emergency use of a test article in an institution without prospective IRB review, provided that such emergency use is reported to the IRB within five working days after such use. An emergency use is defined as a single use (or single course of treatment, e.g., multiple doses of antibiotic) with one subject. "Subsequent use" would be a second use with that subject or the use with another subject.

In its review of the emergency use, if it is anticipated that the test article may be used again, the IRB should request a protocol and consent document(s) be developed so that an approved protocol would be in place when the next need arises. In spite of the best efforts of the clinical investigator and the IRB, a situation may occur where a second emergency use needs to be considered. FDA believes it is inappropriate to deny emergency treatment to an individual when the only obstacle is lack of time for the IRB to convene, review the use and give approval.

19. Are there any regulations that require clinical investigators to report to the IRB when a study has been completed?

IRBs are required to function under written procedures. One of these procedural requirements [21 CFR 56.108(a)(3)] requires ensuring "prompt reporting to the IRB of changes in a research activity." The completion of the study is a change in activity and should be reported to the IRB. Although subjects will no longer be "at risk" under the study, a final report/notice to the IRB allows it to close its files as well as providing information that may be used by the IRB in the evaluation and approval of related studies.

20. What is expedited review?

Expedited review is a procedure through which certain kinds of research may be reviewed and approved without convening a meeting of the IRB. The Agency's IRB regulations [21 CFR 56.110] permit, but do not require, an IRB to review certain categories of research through an expedited procedure if the research involves no more than minimal risk. A list of categories was last published in the Federal Register on January 27, 1981 [46 FR 8980]. The list is reproduced as Appendix D of this document.

The IRB may also use the expedited review procedure to review minor changes in previously approved research during the period covered by the original approval. Under an expedited review procedure, review of research may be carried out by the IRB chairperson or by one or more experienced members of the IRB designated by the chairperson. The reviewer(s) may exercise all the authorities of the IRB, except disapproval. Research may only be disapproved following review by the full committee. The IRB is required to adopt a method of keeping all members advised of research studies that have been approved by expedited review.

On November 9, FDA published in the Federal Register concurrently with OPRR a new Expedited Review List. The entire Federal Register publication, including the FDA preamble, was published on pages 60353 - 60356 of the November 9, 1998 Federal Register and is available on the World Wide Web at the Dockets Management Page of the FDA home Page at http://www.fda.gov/ohrms/dockets/98fr/110998b.txt (or use suffix ".pdf" for Adobe Acrobat version) or alternatively at the Government Printing Office site at http://www.access.gpo.gov/su_docs/fedreg/a981109c.html and scroll down to Food and Drug Administration.

21. The number of studies we review has increased, and the size of the package of review materials we send to IRB members is becoming formidable. Must we send the full package to all IRB members?

The IRB system was designed to foster open discussion and debate at convened meetings of the full IRB membership. While it is preferable for every IRB member to have personal copies of all study materials, each member must be provided with sufficient information to be able to actively and constructively participate. Some institutions have developed a "primary reviewer" system to promote a thorough review. Under this system, studies are assigned to one or more IRB members for a full review of all materials. Then, at the convened IRB meeting the study is presented by the primary reviewer(s) and, after discussion by IRB members, a vote for an action is taken.

The "primary reviewer" procedure is acceptable to the FDA if each member receives, at a minimum; a copy of consent documents and a summary of the protocol in sufficient detail to determine the appropriateness of the study-specific statements in the consent documents. In addition, the complete documentation should be available to all members for their review, both before and at the meeting. The materials for review should be received by the membership sufficiently in advance of the meeting to allow for adequate review of the materials.

Some IRBs are also exploring the use of electronic submissions and computer access for IRB members. Whatever system the IRB develops and uses, it must ensure that each study receives an adequate review and that the rights and welfare of the subjects are protected.

22. Are sponsors allowed access to IRB written procedures, minutes and membership rosters?

The FDA regulations do not require public or sponsor access to IRB records. However, FDA does not prohibit the sponsor from requesting IRB records. The IRB and the institution may establish a policy on whether minutes or a pertinent portion of the minutes are provided to sponsors.

Because of variability, each IRB also needs to be aware of State and local laws regarding access to IRB records.

23. Must an investigator's brochure be included in the documentation when an IRB reviews an investigational drug study?

For studies conducted under an investigational new drug application, an investigator's brochure is usually required by FDA [21 CFR 312.23(a)(5) and 312.55]. Even though 21 CFR part 56does not mention the investigator's brochure by name, much of the information contained in such brochures is clearly required to be reviewed by the IRB. The regulations do outline the criteria for IRB approval of research. 21 CFR 56.111(a)(1) requires the IRB to assure that risks to the subjects are minimized. 21 CFR 56.111(a)(2) requires the IRB to assure that the risks to subjects are reasonable in relation to the anticipated benefits. The risks cannot be adequately evaluated without review of the results of previous animal and human studies, which are summarized in the investigator's brochure.

There is no specific regulatory requirement that the Investigator's Brochure be submitted to the IRB. There are regulatory requirements for submission of information which normally is included in the Investigator's Brochure. It is common that the Investigator's Brochure is submitted to the IRB, and the IRB may establish written procedures which require its submission. Investigator's Brochures may be part of the investigational plan that the IRB reviews when reviewing medical device studies.

24. To what extent is the IRB expected to actively audit and monitor the performance of the investigator with respect to human subject protection issues?

FDA does not expect IRBs to routinely observe consent interviews, observe the conduct of the study or review study records. However, 21 CFR 56.109(f) gives the IRB the authority to observe, or have a third party observe, the consent process and the research. When and if the IRB is concerned about the conduct of the study or the process for obtaining consent, the IRB may consider whether, as part of providing adequate oversight of the study, an active audit is warranted.

25. How can a sponsor know whether an IRB has been inspected by FDA, and the results of the inspection?

The Division of Scientific Investigations, Center for Drug Evaluation and Research, maintains an inventory of the IRBs that have been inspected, including dates of inspection and classification. The Division recently began including the results of inspections assigned by the Center for Biologics Evaluation and Research and the Center for Devices and Radiological Health. This information is available through Freedom of Information Act (FOIA) procedures. Once an investigational file has been closed, the correspondence between FDA and the IRB and the narrative inspectional report are also available under FOI.

26. If an IRB disapproves a study submitted to it, and it is subsequently sent to another IRB for review, should the second IRB be told of the disapproval?

Yes. When an IRB disapproves a study, it must provide a written statement of the reasons for its decision to the investigator and the institution [21 CFR 56.109(e)]. If the study is submitted to a second IRB, a copy of this written statement should be included with the study documentation so that it can make an informed decision about the study. 21 CFR 56.109(a) requires an IRB to "... review ... all research activities [emphasis added] ...." The FDA regulations do not prohibit submission of a study to another IRB following disapproval. However, all pertinent information about the study should be provided to the second IRB.

27. May an independent IRB review a study to be conducted in an institution with an IRB?

Generally, no. Most institutional IRB have jurisdiction over all studies conducted within that institution. An independent IRB may become the IRB of record for such studies only upon written agreement with the administration of the institution or the in-house IRB.

28. Could an IRB lose its quorum when members with a conflict of interest leave the room for deliberation and voting on a study?

Yes. "The quorum is the count of the number of members present. If the number present falls below a majority, the quorum fails. The regulations only require that a member who is conflicted not participate in the deliberations and voting on a study on which he or she is conflicted. The IRB may decide whether an individual should remain in the room."

29. Does FDA expect the IRB chair to sign the approval letters?

FDA does not specify the procedure that IRBs must use regarding signature of the IRB approval letter. The written operating procedures for the IRB should outline the procedure that is followed.

30. Does FDA prohibit direct communication between sponsors and IRBs?

It is important that a formal line of communication be established between the clinical investigator and the IRB. Clinical investigators should report adverse events directly to the responsible IRB, and should send progress reports directly to that IRB. However, FDA does not prohibit direct communication between the sponsor and the IRB, and recognizes that doing so could result in more efficient resolution of some problems.

FDA does require direct communication between the sponsors and the IRBs for certain studies of medical devices and when the 21 CFR 50.24 informed consent waiver has been invoked. Sponsors and IRBs are required to communicate directly for medical device studies under 21 CFR 812.2, 812.66 and 812.150(b). For informed consent waiver studies, direct communication between sponsors and IRBs is required under 21 CFR 50.24(e), 56.109(e), 56.109(g), 312.54(b), 312.130(d), 812.38(b)(4) and 812.47(b).

IV. IRB Records

31. Are annual IRB reviews required when all studies are reviewed by the IRB each quarter?

The IRB records for each study's initial and continuing review should note the frequency (not to exceed one year) for the next continuing review in either months or other conditions, such as after a particular number of subjects are enrolled.

An IRB may decide, to review all studies on a quarterly basis. If every quarterly report contains sufficient information for an adequate continuing review and is reviewed by the IRB under procedures that meet FDA requirements for continuing review, FDA would not require an additional "annual" review.

32. 21 CFR 56.115(a)(1) requires that the IRB maintain copies of "research proposals reviewed." Is the "research proposal" the same as the formal study protocol that the investigator receives from the sponsor of the research?

Yes. The IRB should receive and review all research activities [21 CFR 56.109(a)]. The documents reviewed should include the complete documents received from the clinical investigator, such as the protocol, the investigator's brochure, a sample consent document and any advertising intended to be seen or heard by prospective study subjects. Some IRBs also require the investigator to submit an institutionally-developed protocol summary form. A copy of all documentation reviewed is to be maintained for at least three years after completion of the research at that institution [21 CFR 56.115(b)]. However, when the IRB makes changes, such as in the wording of the informed consent document, only the finally approved copy needs to be retained in the IRB records.

33. What IRB records are required for studies that are approved but never started?

When an IRB approves a study, continuing review should be performed at least annually. All of the records listed in 21 CFR 56.115(a)(1) - (4) are required to be maintained. The clock starts on the date of approval, whether or not subjects have been enrolled. Written progress reports should be received from the clinical investigator for all studies that are in approved status prior to the date of expiration of IRB approval. If subjects were never enrolled, the clinical investigator's progress report would be brief. Such studies may receive continuing IRB review using expedited procedures. If the study is finally canceled without subject enrollment, records should be maintained for at least three years after cancellation [21 CFR 56.115(b)].

V. Informed Consent Process

34. Is getting the subject to sign a consent document all that is required by the regulations?

No. The consent document is a written summary of the information that should be provided to the subject. Many clinical investigators use the consent document as a guide for the verbal explanation of the study. The subject's signature provides documentation of agreement to participate in a study, but is only one part of the consent process. The entire informed consent process involves giving a subject adequate information concerning the study, providing adequate opportunity for the subject to consider all options, responding to the subject's questions, ensuring that the subject has comprehended this information, obtaining the subject's voluntary agreement to participate and, continuing to provide information as the subject or situation requires. To be effective, the process should provide ample opportunity for the investigator and the subject to exchange information and ask questions.

35. May informed consent be obtained by telephone from a legally authorized representative?

A verbal approval does not satisfy the 21 CFR 56.109(c) requirement for a signed consent document, as outlined in 21 CFR 50.27(a). However, it is acceptable to send the informed consent document to the legally authorized representative (LAR) by facsimile and conduct the consent interview by telephone when the LAR can read the consent as it is discussed. If the LAR agrees, he/she can sign the consent and return the signed document to the clinical investigator by facsimile.

36. 21 CFR 50.27(a) requires that a copy of the consent document be given to the person signing the form. Does this copy have to be a photocopy of the form with the subject's signature affixed?

No. The regulation does not require the copy of the form given to the subject to be a copy of the document with the subject's signature, although this is encouraged. It must, however, be a copy of the IRB approved document that was given to the subject to obtain consent [21 CFR 50.27(a) or 21 CFR 50.27(b)(2)]. One purpose of providing the person signing the form with a copy of the consent document is to allow the subject to review the information with others, both before and after making a decision to participate in the study, as well as providing a continuing reference for items such as scheduling of procedures and emergency contacts.

37. If an IRB uses a standard "fill-in-the-blank" consent format, does the IRB need to review the filled out form for each study?

Yes. A fill-in-the-blank format provides only some standard wording and a framework for organizing the relevant study information. The IRB should review a completed sample form, individualized for each study, to ensure that the consent document, in its entirety, contains all the information required by 21 CFR 50.25 in language the subject can understand. The completed sample form should be typed to enhance its readability by the subjects. The form finally approved by the IRB should be an exact copy of the form that will be presented to the research subjects. The IRB should also review the "process" for conducting the consent interviews, i.e., the circumstances under which consent will be obtained, who will obtain consent, and so forth.

38. The informed consent regulations [21 CFR 50.25 (a)(5)] require the consent document to include a statement that notes the possibility that FDA may inspect the records. Is this statement a waiver of the subject's legal right to privacy?

No. FDA does not require any subject to "waive" a legal right. Rather, FDA requires that subjects be informed that complete privacy does not apply in the context of research involving FDA regulated products. Under the authority of the Federal Food, Drug, and Cosmetic Act, FDA may inspect and copy clinical records to verify information submitted by a sponsor. FDA generally will not copy a subject's name during the inspection unless a more detailed study of the case is required or there is reason to believe that the records do not represent the actual cases studied or results obtained.

The consent document should not state or imply that FDA needs clearance or permission from the clinical investigator, the subject or the IRB for such access. When clinical investigators conduct studies for submission to FDA, they agree to allow FDA access to the study records, as outlined in 21 CFR 312.68 and 812.145. Informed consent documents should make it clear that, by participating in research, the subject's records automatically become part of the research database. Subjects do not have the option to keep their records from being audited/reviewed by FDA.

When an individually identifiable medical record (usually kept by the clinical investigator, not by the IRB) is copied and reviewed by the Agency, proper confidentiality procedures are followed within FDA. Consistent with laws relating to public disclosure of information and the law enforcement responsibilities of the Agency, however, absolute confidentiality cannot be guaranteed.

39. Who should be present when the informed consent interview is conducted?

FDA does not require a third person to witness the consent interview unless the subject or representative is not given the opportunity to read the consent document before it is signed, see 21 CFR 50.27(b). The person who conducts the consent interview should be knowledgeable about the study and able to answer questions. FDA does not specify who this individual should be. Some sponsors and some IRBs require the clinical investigator to personally conduct the consent interview. However, if someone other than the clinical investigator conducts the interview and obtains consent, this responsibility should be formally delegated by the clinical investigator and the person so delegated should have received appropriate training to perform this activity.

40. How do you obtain informed consent from someone who speaks and understands English but cannot read?

Illiterate persons who understand English may have the consent read to them and "make their mark," if appropriate under applicable state law. The 21 CFR 50.27(b)(2) requirements for signature of a witness to the consent process and signature of the person conducting consent interview must be followed, if a "short form" is used. Clinical investigators should be cautious when enrolling subjects who may not truly understand what they have agreed to do. The IRB should consider illiterate persons as likely to be vulnerable to coercion and undue influence and should determine that appropriate additional safeguards are in place when enrollment of such persons is anticipated, see 21 CFR 56.111(b).

41. Must a witness observe the entire consent interview or only the signature of the subject?

FDA does not require the signature of a witness when the subject reads and is capable of understanding the consent document, as outlined in 21 CFR 50.27(b)(1). The intended purpose is to have the witness present during the entire consent interview and to attest to the accuracy of the presentation and the apparent understanding of the subject. If the intent of the regulation were only to attest to the validity of the subject's signature, witnessing would also be required when the subject reads the consent.

42. Should the sponsor prepare a model informed consent document?

Although not required by the IND regulations, the sponsor provides a service to the clinical investigator and the IRB when it prepares suggested study-specific wording for the scientific and technical content of the consent document. However, the IRB has the responsibility and authority to determine the adequacy and appropriateness of all of the wording in the consent, see 21 CFR 56.109(a), 111(a)(4) and 111(a)(5). If an IRB insists on wording the sponsor cannot accept, the sponsor may decide not to conduct the study at that site. For medical device studies that are conducted under an IDE, copies of all forms and informational materials to be provided to subjects to obtain informed consent must be submitted to FDA as part of the IDE, see 21 CFR 812.25(g).

43 . Is the sponsor required to review the consent form approved by the IRB to make sure all FDA requirements are met?

For investigational devices, the informed consent is a required part of the IDE submission. It is, therefore, approved by FDA as part of the IDE application. When an IRB makes substantive changes in the document, FDA reapproval is required and the sponsor is necessarily involved in this process.

FDA regulations for other products do not specifically require the sponsor to review IRB approved consent documents. However, most sponsors do conduct such reviews to assure the wording is acceptable to the sponsor.

44. Are there alternatives to obtaining informed consent from a subject?

The regulations generally require that the investigator obtain informed consent from subjects. Investigators also may obtain informed consent from a legally authorized representative of the subject. FDA recognizes that a durable power of attorney might suffice as identifying a legally authorized representative under some state and local laws. For example, a subject might have designated an individual to provide consent with regard to health care decisions through a durable power of attorney and have specified that the individual also has the power to make decisions on entry into research. FDA defers to state and local laws regarding who is a legally authorized representative. Therefore, the IRB should assure that the consent procedures comply with state and local laws, including assurance that the law applies to obtaining informed consent for subjects participating in research as well as for patients who require health care decisions."

Alternatives 1 and 2 are provided for in the regulations and are appropriate. Alternative 3 allows a designated individual to provide consent for a patient with regard to health care decisions and is appropriate when it specifically includes entry into research. FDA defers to state and local laws regarding substituted consent. Therefore, the IRB must assure itself that the substituted consent procedures comply with state and local law, including assurance the law applies to obtaining informed consent for subjects participating in research as well as for patients who require health care decisions.

45. When should study subjects be informed of changes in the study?

Protocol amendments must receive IRB review and approval before they are implemented, unless an immediate change is necessary to eliminate an apparent hazard to the subjects (21 CFR 56.108(a)(4)). Those subjects who are presently enrolled and actively participating in the study should be informed of the change if it might relate to the subjects' willingness to continue their participation in the study (21 CFR 50.25(b)(5)). FDA does not require reconsenting of subjects that have completed their active participation in the study, or of subjects who are still actively participating when the change will not affect their participation, for example when the change will be implemented only for subsequently enrolled subjects.

VI. Informed Consent Document Content

46. May an IRB require that the sponsor of the study and/or the clinical investigator be identified on the study's consent document?

Yes. The FDA requirements for informed consent are the minimum basic elements of informed consent that must be presented to a research subject [21 CFR 50.25]. An IRB may require inclusion of any additional information which it considers important to a subject's decision to participate in a research study [21 CFR 56.109(b)].

47. Does FDA require the informed consent document to contain a space for assent by children?

No, however, many investigators and IRBs consider it standard practice to obtain the agreement of older children who can understand the circumstances before enrolling them in research. While the FDA regulations do not specifically address enrollment of children (other than to include them as a class of vulnerable subjects), the basic requirement of 21 CFR 50.20 applies, i.e., the legally effective informed consent of the subject or the subject's legally authorized representative must be obtained before enrollment. Parents, legal guardians and/or others may have the ability to give permission to enroll children in research, depending on applicable state and local law of the jurisdiction in which the research is conducted. (Note: permission to enroll in research is not the same as permission to provide medical treatment.) IRBs generally require investigators to obtain the permission of one or both of the parents or guardian (as appropriate) and the assent of children who possess the intellectual and emotional ability to comprehend the concepts involved. Some IRBs require two documents, a fully detailed explanation for parents and older children to read and sign, and a shorter, simpler one for younger children. [For research supported by DHHS, the additional protections at 45 CFR 46 Subpart D are also required. The Subpart D regulations provide appropriate guidance for all other pediatric studies.]

48. Does FDA require the signature of children on informed consent documents?

As indicated above, researchers may seek assent of children of various ages. Older children may be well acquainted with signing documents through prior experience with testing, licensing and/or other procedures normally encountered in their lives. Signing a form to give their assent for research would not be perceived as unusual and would be reasonable. Younger children, however, may never have had the experience of signing a document. For these children requiring a signature may not be appropriate, and some other technique to verify assent could be used. For example, a third party may verify, by signature, that the assent of the child was obtained.

49. Who should be listed on the consent as the contact to answer questions?

21 CFR 50.25(a)(7) requires contacts for questions about the research, the research subject's rights and in case of a research-related injury. It does not specify whom to contact. The same person may be listed for all three. However, FDA and most IRBs believe it is better to name a knowledgeable person other than the clinical investigator as the contact for study subject rights. Having the clinical investigator as the only contact may inhibit subjects from reporting concerns and/or possible abuses.

50. May the "compensation" for participation in a trial offered by a sponsor include a coupon good for a discount on the purchase price of the product once it has been approved for marketing?

No. This presumes, and inappropriately conveys to the subjects, a certainty of favorable outcome of the study and prompt approval for marketing. Also, if the product is approved, the coupon may financially coerce the subject to insist on that product, even though it may not be the most appropriate medically.

51. Must informed consent documents be translated into the written language native to study subjects who do not understand English?

The signed informed consent document is the written record of the consent interview. Study subjects are given a copy of the consent to be used as a reference document to reinforce their understanding of the study and, if desired, to consult with their physician or family members about the study.

In order to meet the requirements of 21 CFR 50.20, the consent document must be in language understandable to the subject. When the prospective subject is fluent in English, and the consent interview is conducted in English, the consent document should be in English. However, when the study subject population includes non-English speaking people so that the clinical investigator or the IRB anticipates that the consent interviews are likely to be conducted in a language other than English, the IRB should assure that a translated consent form is prepared and that the translation is accurate.

A consultant may be utilized to assure that the translation is correct. A copy of the translated consent document must be given to each appropriate subject. While a translator may be used to facilitate conversation with the subject, routine ad hoc translation of the consent document may not be substituted for a written translation.

Also see FDA Information Sheets: "A Guide to Informed Consent Documents" and "Informed Consent and the Clinical Investigator"

52. Is it acceptable for the consent document to say specimens are "donated"?

What about a separate donation statement? It would be acceptable for the consent to say that specimens are to be used for research purposes. However, the word "donation" implies abandonment of rights to the "property". 21 CFR 50.20 prohibits requiring subjects to waive or appear to waive any rights as a condition for participation in the study. Whether or not the wording is contained in "the actual consent form" is immaterial. All study-related documents must be submitted to the IRB for review. Any separate "donation" agreement is regarded to be part of the informed consent documentation, and must be in compliance with 21 CFR 50.

53. Do informed consent forms have to justify fees charged to study subjects?

FDA does not require the consent to contain justification of charges.

VII. Clinical Investigations

54. Does a physician, in private practice, conducting research with an FDA regulated product, need to obtain IRB approval?

Yes. The FDA regulations require IRB review and approval of regulated clinical investigations, whether or not the study involves institutionalized subjects. FDA has included non-institutionalized subjects because it is inappropriate to apply a double standard for the protection of research subjects based on whether or not they are institutionalized.

An investigator may be able to obtain IRB review by submitting the research proposal to a community hospital, a university/medical school, an independent IRB, a local or state government health agency or other organizations. If IRB review cannot be accomplished by one of these means, investigators may contact the FDA for assistance (Health Assessment Policy Staff 301-827-1685).

55. Does a clinical investigation involving a marketed product require IRB review and approval?

Yes, if the investigation is governed by FDA regulations [see 21 CFR 56.101, 56.102(c), 312.2(b)(1), 361.1, 601.2, and 812.2]. Also, see the information sheet entitled " 'Off-label' and Investigational Use of Marketed Drugs and Biologics" for more information.

VIII. General Questions

56. Which FDA office may an IRB contact to determine whether an investigational new drug application (IND) or investigational device exemption (IDE) is required for a study of a test article?

For drugs, the IRB may contact the Drug Information Branch, Center for Drug Evaluation and Research (CDER), at (301) 827-4573.

For a biological blood product, contact the Office of Blood Research and Review, Center for Biologics Evaluation and Research (CBER), at 301-827-3518. For a biological vaccine product, contact the Office of Vaccines Research and Review at 301-827-0648. For a biological Therapeutic product, contact the Office of Therapeutics Research and Review, CBER, at 301-594-2860.

For a medical device, contact the Program Operation Staff, Office of Device Evaluation, Center for Devices and Radiological Health (CDRH), at (301) 594-1190.

If the IRB is unsure about whether a test article is a "drug," a "biologic" or a "device," the IRB may contact the Health Assessment Policy Staff, Office of Health Affairs, at (301) 827-1685.

57. What happens during an FDA inspection of an IRB?

FDA field investigators interview institutional officials and examine the IRB records to determine compliance with FDA regulations. Also, see the information sheet entitled "FDA Institutional Review Board Inspections" for a complete description of the inspection process.

58. Does a treatment IND/IDE [21 CFR 312.34/812.36] require prior IRB approval?

Test articles given to human subjects under a treatment IND/IDE require prior IRB approval, with two exceptions. If a life-threatening emergency exists, as defined by 21 CFR 56.102(d), the procedures described in 56.104(c) ("Exemptions from IRB Requirement") may be followed. In addition, FDA may grant the sponsor or sponsor/investigator a waiver of the IRB requirement in accord with 21 CFR 56.105. An IRB may still choose to review a study even if FDA has granted a waiver. For further information see the information sheets entitled "Emergency Use of an Investigational Drug or Biologic," "Emergency Use of Unapproved Medical Devices," "Waiver of IRB Requirements" and "Treatment use of Investigational Drugs and Biologics."

59. How have the FDA policies on enrollment of special populations changed?

On July 22, 1993, the FDA published the Guideline for the Study and Evaluation of Gender Differences in the Clinical Evaluation of Drugs, in the Federal Register [58 FR 39406]. The guideline was developed to ensure that the drug development process provides adequate information about the effects of drugs and biological products in women. For further information, see the information sheet entitled "Evaluation of Gender Differences in Clinical Investigations."

On December 13, 1994, FDA published a final rule on the labeling of prescription drugs for pediatric populations [59 FR 64240]. The rule [21 CFR 201.57] encourages sponsors to include pediatric subjects in clinical trials so that more complete information about the use of drugs and biological products in the pediatric population can be developed.

60. What is a medical device?

A medical device is any instrument, apparatus, or other similar or related article, including component, part, or accessory, which is: (a) recognized in the official National Formulary, or the United States Pharmacopeia, or any supplement to them; (b) intended for use in the diagnosis of disease or other conditions, or in the cure, mitigation, treatment, or prevention of disease, in humans or other animals; or (c) intended to affect the structure or any function of the human body or in animals; and does not achieve any of its principal intended purposes through chemical action within or on the human body or in animals and is not dependent upon being metabolized for the achievement of its principal intended purposes.

Approximately 1,700 types of medical devices are regulated by FDA. The range of devices is broad and diverse, including bandages, thermometers, ECG electrodes, IUDs, cardiac pacemakers, and hemodialysis machines. For further information, see the information sheets entitled "Medical Devices," "Frequently Asked Questions about IRB Review of Medical Devices" and "Significant Risk and Nonsignificant Risk Medical Device Studies."

61. Are in vitro diagnostic products medical devices?

Yes. The definition of a "device" includes in vitro diagnostic products - devices that aid in the diagnosis of disease or medical/physiological conditions (e.g., pregnancy) by using human or animal components to cause chemical reactions, fermentation, and the like. A few diagnostic products are intended for use in controlling other regulated products (such as those used to screen the blood supply for transfusion-transmitted diseases) and are regulated as biological products.

62. What are the IRB's general obligations towards intraocular lens (IOL) clinical investigations?

An IRB is responsible for the initial and continuing review of all IOL clinical investigations. Each individual IOL style is subject to a separate review by the IRB. This does not, however, preclude the IRB from using prior experience with other IOL investigations in considering the comparative merits of a new lens style. All IOL studies are also subject to FDA approval.

63. Considering the large number of IOL studies, how does an IRB approach the review of a new IOL style?

Full IRB review is required for all new IOLs that exhibit major departures from available lenses. Minor changes to existing lenses may be approved through expedited review. FDA designates new IOL styles as either major or minor changes based upon a predetermined classification scheme and advises the sponsor of its determination. The sponsor, through the investigator, should provide the IRB with the investigational plan which indicates the FDA study requirements, as well as the informed consent document and other comparative information on the proposed lens that describes its characteristics. It is the IRB's prerogative to request any relevant information on a new IOL to arrive at a decision or to be more rigorous in its evaluation than FDA considers minimally required.

64. Must a manufacturer comply with 21 CFR 50 and 56 when conducting trials within its own facility using employees as subjects?

Yes. This situation represents a prime example of a vulnerable subject population.

65. Do Radioactive Drug Research Committees (RDRCs) have authority to approve initial clinical studies in lieu of an IND?

No. An IND is required when the purpose of the study is to determine safety and efficacy of the drug or for immediate therapeutic, diagnostic or similar purposes. RDRCs are provided for in 21 CFR 361.1 Radioactive Drugs for Certain Research Uses. Radioactive drugs (as defined in 21 CFR 310.3(n)) may be administered to human research subjects without obtaining an IND when the purpose of the research project is to obtain basic information regarding the metabolism (including kinetics, distribution, and localization) of a radioactively labelled drug or regarding human physiology, pathophysiology, or biochemistry. Certain basic research studies, e.g., studies to determine whether a drug localizes in a particular organ or fluid space and to describe the kinetics of that localization, may have eventual therapeutic or diagnostic implications, but the initial studies are considered to be basic research within the meaning of 21 CFR 361.1. Such basic research studies must be conducted under the conditions set forth in 21 CFR 361.1(b).

All RDRC approved studies must also be approved by an IRB prior to initiation of the studies.

66. Does FDA approve RDRCs?

Yes. An RDRC must obtain and maintain approval by the Food and Drug Administration, as outlined in 21 CFR 361.1(c). RDRCs must register with the Division of Medical Imaging and Radiopharmaceutical Drug Products, (HFD-160), Center for Drug Evaluation and Research, FDA, 5600 Fishers Lane, Rockville, Maryland 20857. The FDA contact for compliance issues is the Human Subject Protection Team (HFD-343), CDER, FDA, 7520 Standish Place, Rockville, MD 20855.

Cooperative Research

Cooperative research studies involve more than one institution. The Food and Drug Administration (FDA) and Department of Health and Human Services (HHS) regulations permit institutions involved in multi-institutional studies to use reasonable methods of joint or cooperative review [21 CFR 56.114 and 45 CFR 46.114, respectively]. While the IRB assumes responsibility for oversight and continuing review, the clinical investigator and the research site retain the responsibility for the conduct of the study.

Scope of Cooperative Research Activities

The regulatory provision for cooperative review arrangements may be applied to different types of cooperative clinical investigations. Examples include research coordinated by cooperative oncology groups and participation by investigators and subjects in a clinical study primarily conducted at or administered by another institution. Often, one institution has the primary responsibility for the conduct of the study and the responsibility for administrative or coordinating functions. At other times, multi center trials may be coordinated by an office or organization that does not actually conduct the clinical study or have an IRB.

Written Cooperative Review Agreements

The cooperative research arrangements between institutions may apply to the review of one study, to certain specific categories of studies or to all studies. A single cooperative IRB may provide review for several participating institutions, but the respective responsibilities of the IRB and each institution should be agreed to in writing.

An institution may agree to delegate the responsibility for initial and continuing review to another institution's IRB. In turn, the IRB agrees to assume responsibility for initial and continuing review. The institution delegating the responsibility for review should understand that it is agreeing to abide by the reviewing IRB's decisions. The delegating institution remains responsible for ensuring that the research conducted within its own institution is in full accordance with the determinations of the IRB providing the review and oversight.

The IRB which agrees to review studies conducted at another institution has responsibility for initial and continuing review of the research. Such an IRB, in initially reviewing the study, should take into account the required criteria for approval, the facilities and capabilities of the other institution, and the measures taken by the other institution to ensure compliance with the IRB's determinations. The reviewing IRB needs to be sensitive to factors such as community attitudes.

The agreement for IRB review of cooperative research should be documented. Depending upon the scope of the agreement, documentation may be simple, in the form of a letter, or more complex such as a formal memorandum of understanding. In the case of studies supported or conducted by HHS, arrangements or agreements may be subject to approval by HHS through the Office for Protection from Research Risks (OPRR) and should be executed in accordance with OPRR's instructions. Whatever form of documentation is used, copies should be furnished to all parties to the agreement, and to those responsible for ensuring compliance with the regulations and the IRB's determinations. The IRB's records should include documentation of such agreements.

When an IRB approves a study, it notifies (in writing) the clinical investigator and the institution at each location for which the IRB has assumed responsibility [21 CFR 56.109(d)]. All required reports from the clinical investigators should be sent directly to the responsible IRB with copies to the investigator's institution, as appropriate.

Multi-institutional IRB

Another form of cooperative research activity is a multi-institutional IRB, that oversees the research activities of more than one institution in a defined area, such as a city or county. Such an IRB is formed by separate but cooperating institutions and eliminates the need for each facility to organize and staff its own IRB. A variation of this is an IRB that is established by a corporate entity to oversee research at its operating components, for example, a hospital system with facilities at several locations.

Also see FDA Information Sheet: "Non-Local IRB Review"

Non-Local IRB Review

Under certain circumstances, local review by an Institutional Review Board (IRB) may not be available, e.g., research conducted by investigators unaffiliated with an institution with an IRB. Although conceptually modeled for local IRB review, the Food and Drug Administration (FDA) regulations do not prohibit review of research by IRBs in locations other than where the research is to be performed (e.g., independent or non-institutional IRB). Therefore, an IRB may review studies that are not performed on-site as long as the 21 CFR parts 50 and 56 requirements are met.

When non-local IRB review takes place, the reviewing IRB must document its role and responsibility. A written agreement should be executed between the performance site where the research is to be conducted (e.g., private practitioner's office, clinic, etc.) and the IRB or its institution. The agreement should confirm the authority of the IRB to oversee the study. While the IRB assumes responsibility for oversight and continuing review, the clinical investigator and the research site retain the responsibility for the conduct of the study.

Community Attitudes

The non-local IRB should have adequate knowledge of community attitudes, information on conditions surrounding the conduct of the research, and the continuing status of the research to assure fulfilling the requirements of 21 CFR 56.107, 56.111(a)(3), (a)(7) and (b) for each study site. The non-local IRB needs to ensure these requirements are met for each location for which it has assumed IRB oversight responsibility.

The FDA regulations require all IRBs to have membership sufficiently qualified to promote respect for the IRB's advice and counsel in safeguarding the rights and welfare of human subjects [21 CFR 56.107]. IRBs conducting non-local review need to be knowledgeable about the community from which the subjects are drawn to ensure that subject rights will be protected and that the consent process is appropriate for the subject population involved. The IRB should be sensitive to community laws and mores because state and local laws and community attitudes pertaining to research may be more restrictive than Federal regulations or the prevailing standards of the community where the IRB is located.

IRBs can obtain knowledge of community attitudes with a site visit by a representative of the IRB, by appointing an IRB member from that community, or by having a consultant from the community advise the IRB, either prior to or during the deliberations. If travel is not feasible, participation in the IRB meeting can be by video-conference or conference telephone call, or by using other technologies that allow for real-time conversational interaction between the remote member and the members at the convened location. All IRB members should receive an advance copy of the documents that are to be reviewed at the meeting. The minutes of the meeting, during which non-local research is reviewed, should document the procedures used to assure that community attitudes were adequately taken into consideration.

IRB Information Needs

IRBs should have access to a variety of information to properly conduct initial and continuing reviews. Knowledge of the conditions surrounding the conduct of the research is needed to ensure that risks to subjects are minimized [21 CFR 56.111]. An IRB should have sufficient information to judge the qualifications of the researcher conducting the study in question. The researcher's curriculum vitae, a listing of other studies conducted, letters of reference, information from the sponsor of the research, and information from licensing boards and professional societies are examples of information a non-local IRB may want to review. If the research is to be conducted in an institution, the clinical investigator should provide a description of that institution and associated medical facilities. The acknowledgment and/or the permission of the institution should also be provided. If the research is to be conducted outside an institutional setting, the IRB may request a plan for emergency medical care. Depending upon the degree of risk inherent in the study, a hospital should certify that its facilities are available.

The IRB should explicitly detail the information it needs in written reports from the researcher. In addition to scheduled continuing review of progress reports, an IRB may use other methods of obtaining information on the conduct of the study. All IRBs should have procedures that assure the IRB becomes aware of unexpected problems in ongoing studies in a timely manner. Fulfilling this requirement may call for additional efforts for non-local IRBs, such as visiting the study site, contacting the sponsor's research monitor for information on the monitor's site visits, or arranging for other oversight of the study.

IRB Contact

The FDA informed consent regulations [21 CFR 50.25(a)(7)] require that the subject be given the name of a person to contact "... for answers to pertinent questions about the research and research subjects' rights, and whom to contact in the event of a research-related injury to the subject." Non-local IRBs should include, in the consent document, an IRB contact person and a telephone number (toll-free if long-distance). The non-local IRB may also designate an individual at the research site to be the contact and to relay reports to the IRB.

IRB Jurisdiction

When an institution has a local IRB, the written procedures of that IRB or of the institution should define the scope of studies subject to review by that IRB. A non-local IRB may not become the IRB of record for studies within that defined scope unless the local IRB or the administration of the institution agree. Any agreement to allow review by a non-local IRB should be in writing.

Also see FDA Information Sheet: "Cooperative Research."

Continuing Review After Study Approval

Institutional Review Boards (IRBs) are responsible for continuing review of ongoing research to ensure that the rights and welfare of human subjects are protected. The Food and Drug Administration (FDA) regulations regarding continuing review require an IRB to develop and follow written procedures for:

- conducting continuing review of research at intervals appropriate to the degree of risk, but not less than once per year [21 CFR 56.108(a)(1) and 56.109(f)];
- determining which studies need verification from sources other than the investigator that no material changes in the research have occurred since the previous IRB review [21 CFR 56.108(a)(2)];
- ensuring that changes in approved research are promptly reported to, and approved by, the IRB [21 CFR 56.108(a)(3-4)]; and
- suspending or terminating approval of research that is not being conducted in accordance with the IRB's requirements [21 CFR 56.108(b)(2) and 56.113].

The FDA continuing review regulations outline minimum requirements; they do not provide specific instructions to IRBs on how to set up their own rules for continuing review within the framework of the regulations. Therefore, the regulations allow institutions or IRBs to impose greater and more detailed standards of protection for human subjects than those specified by the regulations and permit each IRB to develop procedures appropriate to its needs. By regulation, the IRB has the authority and the responsibility to take appropriate steps such as terminating or suspending approval of research that is not being conducted in accordance with the IRB's requirements.

1. Criteria for Conducting Continuing Review

FDA regulations set forth the criteria to be satisfied if an IRB is to approve research [21 CFR 56.111]. These criteria are the same for initial review and continuing review and include a determination by the IRB that

- risks to subjects are minimized;
- risks to subjects are reasonable in relation to anticipated benefits;
- selection of subjects is equitable;
- informed consent is adequate and appropriately documented;
- where appropriate, the research plan makes adequate provision for monitoring the data collected to ensure the safety of subjects;
- where appropriate, there are adequate provisions to protect the privacy of subjects and to maintain the confidentiality of data; and
- appropriate safeguards have been included to protect vulnerable subjects.

2. Process for Conducting Continuing Review

Routine continuing review should include IRB review of a written progress report(s) from the clinical investigator. Progress reports include information such as: the number of subjects entered into the research study; a summary description of subject experiences (benefits, adverse reactions); numbers of withdrawals from the research; reasons for withdrawals; the research results obtained thus far; a current risk-benefit assessment based on study results; and any new information since the IRB's last review. Special attention should be paid to determining whether new information or unanticipated risks were discovered since the previous IRB review.

Any significant new findings which may relate to the subjects' willingness to continue participation should be provided to the subjects in accordance with 21 CFR 50.25(b)(5).

The IRB should obtain a copy of the consent document currently in use and determine whether the information contained in it is still accurate and complete, including whether new information that may have been obtained during the course of the study needs to be added. Obtaining the consent document also provides a check on whether the document being used by the clinical investigator has current IRB approval.

The purpose of continuing review is to review the progress of the entire study, not just changes in it. Continuing review of a study may not be conducted through an expedited review procedure, unless 1) the study was eligible for, and initially reviewed by, an expedited review procedure, or 2) the study has changed such that the only activities remaining are eligible for expedited review.

The IRB should determine that the frequency and extent of continuing review for each study is adequate to ensure the continued protection of the rights and welfare of research subjects. The factors considered in setting the frequency of review may include: the nature of the study; the degree of risk involved; and the vulnerability of the study subject population. Note that 21 CFR 56.108(a)(2) requires IRBs to follow written procedures for determining the frequency and extent of continuing review.

The continuation of research after expiration of IRB approval is a violation of the regulations [21 CFR 56.103(a)]. If the IRB has not reviewed and approved a research study by the study's current expiration date, i.e., IRB approval has expired, research activities should stop. No new subjects may be enrolled in the study. However, if the investigator is actively pursuing renewal with the IRB and the IRB believes that an over-riding safety concern or ethical issue is involved, the IRB may permit the study to continue for the brief time required to complete the review process.

When study approval is terminated by the IRB, in addition to stopping all research activities, any subjects currently participating should be notified that the study has been terminated. Procedures for withdrawal of enrolled subjects should consider the rights and welfare of subjects. If follow-up of subjects for safety reasons is permitted/required by the IRB, the subjects should be so informed and any adverse events/outcomes should be reported to the IRB and the sponsor.

3. Process for Dealing with Reports of Adverse Reactions and Unexpected Events

a. Written Procedures

IRB continuing review responsibilities include reviewing reports of adverse reactions and unexpected events involving risks to subjects or others. The IRB should establish a procedure for receiving and reviewing these reports. The level and promptness of review may depend upon factors such as the seriousness of the event, whether the event is described in the study protocol and consent and whether the event occurred at a location for which the IRB is the IRB of record. The written procedures may include a brief form to be completed by the principal investigator when an adverse event occurs, asking for his/her opinion as to whether the event was related to the study and other information to aid the IRB in an appropriate and efficient review of the event.

Researchers should be made aware of the IRB's policies and procedures concerning reporting and continuing review requirements. This can be accomplished by notifying the investigator, in the IRB's letter of approval, of the requirement to report changes and unanticipated problems in research activities. The IRB's written procedures pertaining to continuing review and reporting requirements should be distributed to ensure that all individuals involved in research activities understand their obligations.

b. Process

Unanticipated risks are sometimes discovered during the course of research. Information that may impact on the risk/benefit ratio should be promptly reported to, and reviewed by, the IRB to ensure adequate protection of the welfare of the subjects. Based upon such information, the IRB may need to reconsider its approval of the study, require modifications to the study or, revise the continuing review timetable.

IRBs are also responsible for ensuring that reports of unanticipated problems involving risks to human subjects or others are reported to the FDA [21 CFR 56.108(b)(1)]. Usually, this reporting is accomplished through the normal reporting channel, i.e., the investigator to the sponsor to FDA.

4. Process for Reviewing Changes in Ongoing Research During the Approval Period

In accord with 21 CFR 56.110(b), an IRB may use expedited review procedures to review minor changes in ongoing previously-approved research during the period for which approval is authorized. An expedited review may be carried out by the IRB chairperson or by one or more experienced reviewers designated by the chairperson from among members of the IRB.

When a proposed change in a research study is not minor (e.g., procedures involving increased risk or discomfort are to be added), then the IRB must review and approve the proposed change at a convened meeting before the change can be implemented. The only exception is a change necessary to eliminate apparent immediate hazards to the research subjects [21 CFR 56.108(a)(4)]. In such a case, the IRB should be promptly informed of the change following its implementation and should review the change to determine that it is consistent with ensuring the subjects' continued welfare.

THIS PAGE INTENTIONALLY LEFT BLANK

Sponsor–Investigator–IRB Interrelationship

The interrelationship and interaction between the research sponsor (e.g., drug, biologic and device manufacturers), the clinical investigator and the Institutional Review Board (IRB) may be very complex. The regulations do not prohibit direct sponsor-IRB contacts, although, the sponsor-IRB interaction customarily occurs through the investigator who conducts the clinical study. The clinical investigator generally provides the communication link between the IRB and the sponsor. Such linkage is agreed to by the sponsors and investigators when they sign forms FDA-1571 and FDA-1572, respectively, for drug and biologic studies or an investigator agreement for device studies. There are occasions when direct communication between the IRB and the sponsor may facilitate resolution of concerns about study procedures or specific wording in an informed consent document. The clinical investigator should be kept apprised of the discussion.

Sponsor Assurance that IRBs Operate in Compliance with 21 CFR Part 56

FDA regulations [21 CFR 312.23(a)(1)(iv)] require that a sponsor assure the FDA that a study will be conducted in compliance with the informed consent and IRB regulations [21 CFR parts 50 and 56]. This requirement has been misinterpreted to mean that it is a sponsor's obligation to determine IRB compliance with the regulations. This is not the case. Sponsors should rely on the clinical investigator, who assures the sponsor on form FDA-1572 for drugs and biologics or the investigator agreement for devices that the study will be reviewed by an IRB. Because clinical investigators work directly with IRBs, it is appropriate that they assure the sponsor that the IRB is functioning in compliance with the regulations.

An IRB must notify an investigator in writing of its decision to approve, disapprove or request modifications in a proposed research activity [21 CFR 56.109(e)]. This correspondence should be made available to the sponsor by the clinical investigator. In the Agency's view, this required documentation provides the sponsor with reasonable assurance that an IRB complies with 21 CFR part 56 and that it will be responsible for initial and continuing review of the study. Also, the sponsor and, in fact, anyone who is interested, may obtain an Establishment Inspection Report from an FDA inspection of an IRB. These reports summarize the conditions observed during the IRB inspection. FDA, however, does not certify IRBs.

Sponsor Access to Medical Records

The IRB is responsible for ensuring that informed consent documents include the extent to which the confidentiality of medical records will be maintained [21 CFR 50.25(a)(5)]. FDA requires sponsors (or research monitors hired by them) to monitor the accuracy of the data submitted to FDA in accordance with regulatory requirements. These data are generally in the possession of the clinical investigator. Each subject must be advised during the informed consent process of the extent to which confidentiality of records identifying the subject will be maintained and of the possibility that the FDA may inspect the records. While FDA access to medical records is a regulatory requirement, subject names are not usually requested by FDA unless the records of particular individuals require a more detailed study of the cases, or unless there is reason to believe that the records do not represent actual cases studied or actual results obtained. The consent document should list all other entities (e.g., the sponsor) who will have access to records identifying the subject. The extent to which confidentiality will be maintained may affect a subject's decision to participate in a clinical investigation.

Confidentiality of Sponsor Information

The IRB's primary responsibility with respect to protecting confidentiality is to the research subject. IRBs should, however, respect the sponsor's need to maintain confidentiality of certain information about products under development. IRB members and staff should be aware that information submitted for review may be confidential, trade secret, and of commercial interest and should recognize the need for maintaining the confidentiality of the review materials and IRB records. It is advisable for IRBs to have policies that address this issue.

Nonsignificant Risk Device Studies

"A sponsor's preliminary determination that a medical device study presents an NSR is subject to IRB approval." The effect of the IRB's NSR decision is important to research sponsors and investigators because significant risk (SR) studies require sponsors to file an Investigational Device Exemption (IDE) with FDA before they may begin. NSR studies, however, may begin as soon as the IRB approves the study. The sponsor, usually through the clinical investigator, provides the IRB with information necessary to make a judgment on the risk of a device study. While the investigational plan and supporting materials usually contain sufficient information to make a determination, the IRB can request additional information if needed [21 CFR 812.150(b)(10)]. If the IRB believes that additional information is needed, it may contact the sponsor directly, but it should keep the clinical investigator apprised of the request. While making the SR/NSR determination, any of the three parties may ask FDA to provide a risk assessment. See FDA Information Sheet: "Significant Risk and Nonsignificant Risk Medical Device Studies" for further information.

Disagreements

The sponsor may choose not to conduct, to terminate, or to discontinue studies that do not conform with the sponsor's wishes. For example, the sponsor, clinical investigator, and IRB may reach an impasse about study procedures or specific wording in an informed consent document. The FDA will not mediate such disagreements. The Agency's policy of decentralized ethical review of clinical investigations allows such decisions to be made by local IRBs, and any disagreements between a sponsor, IRB, and clinical investigator should be resolved through appropriate communication among those parties.

Acceptance of Foreign Clinical Studies

The Food and Drug Administration (FDA) may accept clinical studies conducted outside the United States in support of safety and efficacy claims for drugs, biological products and medical devices.

All drug, biologic and device studies conducted under an Investigational New Drug (IND) or Investigational Device Exemption (IDE) are governed by the FDA informed consent and IRB requirements. [See 21 CFR part 312 IND regulations and 21 CFR part 812 IDE regulations.]

Under 21 CFR 312.120(c)(1), FDA will accept a foreign clinical study involving a drug or biological product not conducted under an IND only if the study conforms to whichever of the following provides greater protection of the human subjects:

- the ethical principles contained in the 1989 version of the Declaration of Helsinki, or
- the laws and regulations of the country in which the research was conducted.

Under 21 CFR 814.15(a) and (b), FDA will accept a foreign clinical study involving a medical device not conducted under an IDE only if the study conforms to whichever of the following provides greater protection of the human subjects:

- the ethical principles contained in the 1983 version of the Declaration of Helsinki, or
- the laws and regulations of the country in which the research was conducted.

Also see these FDA Information Sheets:
"Non-Local IRB Review"
"Waiver of IRB Requirements for Drug and Biologic Studies"
"Informed Consent and the Clinical Investigator"
Declaration of Helsinki—the 1983 and 1989 versions

Charging for Investigational Products

This information sheet discusses FDA policy on allowing charges for the test articles in clinical investigations.

Decisions concerning charging subjects for investigational products are guided by professional ethics, institutional policies, and FDA regulations. The FDA informed consent regulations require the consent document to include a description of any additional costs to the subject that may result from participation in the research [21 CFR 50.25(b)(3)]. IRBs should ensure that the informed consent documents outline any additional costs that will be billed to study subjects or their insurance company as a result of participation in the study. IRBs should also ensure that any such charges are appropriate and equitable.

Because the regulations governing drugs and biologics vary from those governing medical devices, the Agency's position on charging for the test articles will be discussed separately. FDA does not prohibit charging the subjects for related treatment or for services.

1. Charging for Investigational Medical Devices and Radiological Health Products

The Investigational Device Exemption (IDE) regulations allow sponsors to charge for an investigational device, however, the charge should not exceed an amount necessary to recover the costs of manufacture, research, development, and handling of the investigational device [21 CFR 812.7(b)]. A sponsor justifies the proposed charges for the device in the IDE application, states the amount to be charged, and explains why the charge does not constitute commercialization [21 CFR 812.20(b)(8)]. FDA generally allows sponsors to charge investigators for investigational devices, and this cost usually is passed on to the subjects.

2. Charging for Investigational Drugs and Biologics

The Investigational New Drug (IND) regulations [21 CFR 312.7(d)] permit a sponsor to charge for an investigational drug or biologic that has not been approved for marketing, only under the conditions outlined below. In both a clinical trial and a treatment IND, the charge should not exceed an amount that is necessary to recover the costs associated with the manufacture, research, development, and handling of the investigational drug or biologic. FDA may withdraw authorization to charge if the Agency finds that the conditions underlying the authorization are no longer satisfied.

(i) Clinical Trials Under an IND

A sponsor may not charge for an investigational drug or biologic in a clinical trial under an IND without the Agency's prior written approval. In requesting such approval, the sponsor must explain why a charge is necessary, i.e., why providing the product without charge should not be considered part of the normal cost of conducting a clinical trial [21 CFR 312.7(d)(1)].

(ii) Treatment Protocol or Treatment IND

A sponsor or investigator may charge for an investigational drug or biologic for a treatment use under a treatment protocol or treatment IND, as outlined in 21 CFR 312.34 and 312.35, provided: (1) there is adequate enrollment in the ongoing clinical investigations under the authorized IND; (2) charging does not constitute commercial marketing of a new drug for which a marketing application has not been approved; (3) the drug or biologic is not being commercially promoted or advertised; and (4) the sponsor is actively pursuing marketing approval with due diligence. FDA must be notified in writing prior to commencing any such charges. Authorization for charging goes into effect automatically 30 days after receipt of the information by FDA, unless FDA notifies the sponsor to the contrary [21 CFR 312.7(d)(2)].

There is no specific regulatory requirement that the Investigator's Brochure be submitted to the IRB. There are regulatory requirements for submission of information which normally is included in the Investigator's Brochure. It is common that the Investigator's Brochure is submitted to the IRB, and the IRB may establish written procedures which require its submission. Investigator's Brochures may be part of the investigational plan that the IRB reviews when reviewing medical device studies.

Recruiting Study Subjects

FDA requires that an Institutional Review Board (IRB) review and have authority to approve, require modifications in, or disapprove all research activities covered by the IRB regulations [21 CFR 56.109(a)]. An IRB is required to ensure that appropriate safeguards exist to protect the rights and welfare of research subjects [21 CFR 56.107(a) and 56.111]. In fulfilling these responsibilities, an IRB is expected to review all the research documents and activities that bear directly on the rights and welfare of the subjects of proposed research. The protocol, the consent document and, for studies conducted under the Investigational New Drug (IND) regulations, the investigator's brochure are examples of documents that the IRB should review. The IRB should also review the methods and material that investigators propose to use to recruit subjects.

A. Media Advertising:

Direct advertising for research subjects, i.e., advertising that is intended to be seen or heard by prospective subjects to solicit their participation in a study, is not in and of itself, an objectionable practice. Direct advertising includes, but is not necessarily limited to: newspaper, radio, TV, bulletin boards, posters, and flyers that are intended for prospective subjects. Not included are: (1) communications intended to be seen or heard by health professionals, such as "dear doctor" letters and doctor-to-doctor letters (even when soliciting for study subjects), (2) news stories and (3) publicity intended for other audiences, such as financial page advertisements directed toward prospective investors.

IRB review and approval of listings of clinical trials on the internet would provide no additional safeguard and is not required when the system format limits the information provided to basic trial information, such as: the title; purpose of the study; protocol summary;

basic eligibility criteria; study site location(s); and how to contact the site for further information. Examples of clinical trial listing services that do not require prospective IRB approval include the National Cancer Institute's cancer clinical trial listing (PDQ) and the government-sponsored AIDS Clinical Trials Information Service (ACTIS). However, when the opportunity to add additional descriptive information is not precluded by the data base system, IRB review and approval may assure that the additional information does not promise or imply a certainty of cure or other benefit beyond what is contained in the protocol and the informed consent document.

FDA considers direct advertising for study subjects to be the start of the informed consent and subject selection process. Advertisements should be reviewed and approved by the IRB as part of the package for initial review. However, when the clinical investigator decides at a later date to advertise for subjects, the advertising may be considered an amendment to the ongoing study. When such advertisements are easily compared to the approved consent document, the IRB chair, or other designated IRB member, may review and approve by expedited means, as provided by 21 CFR 56.110(b)(2). When the IRB reviewer has doubts or other complicating issues are involved, the advertising should be reviewed at a convened meeting of the IRB.

FDA expects IRBs to review the advertising to assure that it is not unduly coercive and does not promise a certainty of cure beyond what is outlined in the consent and the protocol. This is especially critical when a study may involve subjects who are likely to be vulnerable to undue influence. [21 CFR 50.20, 50.25, 56.111(a)(3), 56.111(b) and 812.20(b)(11).]

When direct advertising is to be used, the IRB should review the information contained in the advertisement and the mode of its communication, to determine that the procedure for recruiting subjects is not coercive and does not state or imply a certainty of favorable outcome or other benefits beyond what is outlined in the consent document and the protocol. The IRB should review the final copy of printed advertisements to evaluate the relative size of type used and other visual effects. When advertisements are to be taped for broadcast, the IRB should review the final audio/video tape. The IRB may review and approve the wording of the advertisement prior to taping to preclude re-taping because of inappropriate wording. The review of the final taped message prepared from IRB-approved text may be accomplished through expedited procedures. The IRB may wish to caution the clinical investigators to obtain IRB approval of message text prior to taping, in order to avoid re-taping because of inappropriate wording.

No claims should be made, either explicitly or implicitly, that the drug, biologic or device is safe or effective for the purposes under investigation, or that the test article is known to be equivalent or superior to any other drug, biologic or device. Such representation would not only be misleading to subjects but would also be a violation of the Agency's regulations concerning the promotion of investigational drugs [21 CFR 312.7(a)] and of investigational devices [21 CFR 812.7(d)].

Advertising for recruitment into investigational drug, biologic or device studies should not use terms such as "new treatment," "new medication" or "new drug" without explaining that the test article is investigational. A phrase such as "receive new treatments" leads study subjects to believe they will be receiving newly improved products of proven worth.

Advertisements should not promise "free medical treatment," when the intent is only to say subjects will not be charged for taking part in the investigation. Advertisements may state that subjects will be paid, but should not emphasize the payment or the amount to be paid, by such means as larger or bold type.

Generally, FDA believes that any advertisement to recruit subjects should be limited to the information the prospective subjects need to determine their eligibility and interest. When appropriately worded, the following items may be included in advertisements. It should be noted, however, that FDA does not require inclusion of all of the listed items.

1. the name and address of the clinical investigator and/or research facility;
2. the condition under study and/or the purpose of the research;
3. in summary form, the criteria that will be used to determine eligibility for the study;
4. a brief list of participation benefits, if any (e.g., a no-cost health examination);
5. the time or other commitment required of the subjects; and
6. the location of the research and the person or office to contact for further information.

B. Receptionist Scripts.

The first contact prospective study subjects make is often with a receptionist who follows a script to determine basic eligibility for the specific study. The IRB should assure the procedures followed adequately protect the rights and welfare of the prospective subjects. In some cases personal and sensitive information is gathered about the individual. The IRB should have assurance that the information will be appropriately handled. A simple statement such as "confidentiality will be maintained" does not adequately inform the IRB of the procedures that will be used.

Examples of issues that are appropriate for IRB review: What happens to personal information if the caller ends the interview or simply hangs up? Are the data gathered by a marketing company? If so, are names, etc. sold to others? Are names of non-eligibles maintained in case they would qualify for another study? Are paper copies of records shredded or are readable copies put out as trash? The acceptability of the procedures would depend on the sensitivity of the data gathered, including; personal, medical and financial.

Also see these FDA Information Sheets:
"A Guide to Informed Consent Documents"
"Payment to Research Subjects"

Payment to Research Subjects

The Institutional Review Board (IRB) should determine that the risks to subjects are reasonable in relation to anticipated benefits [21 CFR 56.111(a)(2)] and that the consent document contains an adequate description of the study procedures [21 CFR 50.25(a)(1)] as well as the risks [21 CFR 50.25(a)(2)] and benefits [21 CFR 50.25(a)(3)]. It is not uncommon for subjects to be paid for their participation in research, especially in the early phases of investigational drug, biologic or device development. Payment to research subjects for participation in studies is not considered a benefit, it is a recruitment incentive. Financial incentives are often used when health benefits to subjects are remote or non-existent. The amount and schedule of all payments should be presented to the IRB at the time of initial review. The IRB should review both the amount of payment and the proposed method and timing of disbursement to assure that neither are coercive or present undue influence [21 CFR 50.20].

Any credit for payment should accrue as the study progresses and not be contingent upon the subject completing the entire study. Unless it creates undue inconvenience or a coercive practice, payment to subjects who withdraw from the study may be made at the time they would have completed the study (or completed a phase of the study) had they not withdrawn. For example, in a study lasting only a few days, an IRB may find it permissible to allow a single payment date at the end of the study, even to subjects who had withdrawn before that date.

While the entire payment should not be contingent upon completion of the entire study, payment of a small proportion as an incentive for completion of the study is acceptable to FDA, providing that such incentive is not coercive. The IRB should determine that the amount paid as a bonus for completion is reasonable and not so large as to unduly induce subjects to stay in the study when they would otherwise have withdrawn. All information concerning payment, including the amount and schedule of payment(s), should be set forth in the informed consent document.

Also see these FDA Information Sheets:
"A Guide to Informed Consent Documents"
"Recruiting Study Subjects."

Screening Tests Prior to Study Enrollment

For some studies, the use of screening tests to assess whether prospective subjects are appropriate candidates for inclusion in studies is an appropriate pre-entry activity. While an investigator may discuss availability of studies and the possibility of entry into a study with a prospective subject without first obtaining consent, informed consent must be obtained prior to initiation of any clinical procedures that are performed solely for the purpose of determining eligibility for research, including withdrawal from medication (wash-out). When wash-out is done in anticipation of or in preparation for the research, it is part of the research.

Procedures that are to be performed as part of the practice of medicine and which would be done whether or not study entry was contemplated, such as for diagnosis or treatment of a disease or medical condition, may be performed and the results subsequently used for determining study eligibility without first obtaining consent. On the other hand, informed consent must be obtained prior to initiation of any clinical screening procedures that is performed solely for the purpose of determining eligibility for research. When a doctor-patient relationship exists, prospective subjects may not realize that clinical tests performed solely for determining eligibility for research enrollment are not required for their medical care. Physician-investigators should take extra care to clarify with their patient-subjects why certain tests are being conducted.

Clinical screening procedures for research eligibility are considered part of the subject selection and recruitment process and, therefore, require IRB oversight. If the screening qualifies as a minimal risk procedure [21 CFR 56.102(i)], the IRB may choose to use expedited review procedures [21 CFR 56.110]. The IRB should receive a written outline of the screening procedure to be followed and how consent for screening will be obtained. The IRB may find it appropriate to limit the scope of the screening consent to a description of the screening tests and to the reasons for performing the tests including a brief summary description of the study in which they may be asked to participate. Unless the screening tests involve more than minimal risk or involve a procedure for which written consent is normally required outside the research context, the IRB may decide that prospective study subjects need not sign a consent document [21 CFR 56.109(c)]. If the screening indicates that the prospective subject is eligible, the informed consent procedures for the study, as approved by the IRB, would then be followed.

Certain clinical tests, such as for HIV infection, may have State requirements regarding (1) the information that must be provided to the participant, (2) which organizations have access to the test results and (3) whether a positive result has to be reported to the health department. Prospective subjects should be informed of any such requirements and how an unfavorable test result could affect employment or insurance before the test is conducted. The IRB may wish to confirm that such tests are required by the protocol of the study.

Also see this FDA Information Sheet:
"Recruiting Study Subjects"

THIS PAGE INTENTIONALLY LEFT BLANK

A Guide to Informed Consent

Consent Document Content

For studies that are subject to the requirements of the FDA regulations, the informed consent documents should meet the requirements of 21 CFR 50.20 and contain the information required by each of the eight basic elements of 21 CFR 50.25(a), and each of the six elements of 21 CFR 50.25(b) that is appropriate to the study. IRBs have the final authority for ensuring the adequacy of the information in the informed consent document.

IRB standard format

Many IRBs have developed standard language and/or a standard format to be used in portions of all consent documents. Standard language is typically developed for those elements that deal with confidentiality, compensation, answers to questions, and the voluntary nature of participation. Each investigator should determine the local IRB's requirements before submitting a study for initial review. Where changes are needed from the standard paragraphs or format, the investigator can save time by anticipating the local IRB's concerns and explaining in the submission to the IRB why the changes are necessary.

Sponsor-prepared sample consent documents

Sample or draft consent documents may be developed by a sponsor or cooperative study group. However, the IRB of record is the final authority on the content of the consent documents that is presented to the prospective study subjects.

Investigational New Drug Applications (IND) submitted to FDA are not required to contain a copy of the consent document. If the sponsor submits a copy, or if FDA requests a copy, the Agency will review the document and may comment on the document's adequacy.

For significant risk medical devices, the consent document is considered to be a part of the investigational plan in the Application for an Investigational Device Exemption (IDE). FDA always reviews these consent documents. The Agency's review is generally limited to ensuring the presence of the required elements of informed consent and the absence of exculpatory language. Any substantive changes to the document made by an IRB must be submitted to FDA (by the sponsor) for review and approval.

Revision of Consent Documents during the study

Study protocols are often changed during the course of the study. When these changes require revision of the informed consent document, the IRB should have a system that identifies the revised consent document, in order to preclude continued use of the older version and to identify file copies. While not required by FDA regulations, some IRBs stamp the final copy of the consent document with the approval date. The investigator then photocopies the consent document for use. [Note: the wording of the regulations is provided in italics, followed by explanatory comments.]

21 CFR 50.20 General requirements for informed consent

Except as provided in ß50.23, no investigator may involve a human being as a subject in research covered by these regulations unless the investigator has obtained the legally effective informed consent of the subject or the subject's legally authorized representative. An investigator shall seek such consent only under circumstances that provide the prospective subject or the representative sufficient opportunity to consider whether or not to participate and that minimize the possibility of coercion or undue influence. The information that is given to the subject or the representative shall be in language understandable to the subject or the representative. No informed consent, whether oral or written, may include any exculpatory language through which the subject or the representative is made to waive or appear to waive any of the subject's rights, or releases or appears to release the investigator, the sponsor, the institution, or its agents from liability for negligence.

The IRB should ensure that technical and scientific terms are adequately explained or that common terms are substituted. The IRB should ensure that the informed consent document properly translates complex scientific concepts into simple concepts that the typical subject can read and comprehend.

Although not prohibited by the FDA regulations, use of the wording, "I understand..." in informed consent documents may be inappropriate as many prospective subjects will not "understand" the scientific and medical significance of all the statements. Consent documents are more understandable if they are written just as the clinical investigator would give an oral explanation to the subject, that is, the subject is addressed as "you" and the clinical investigator as "I/we." This second person writing style also helps to communicate that there is a choice to be made by the prospective subject. Use of first person may be interpreted as presumption of subject consent, i.e., the subject has no choice. Also, the tone of the first person "I understand" style seems to misplace emphasis on legal statements rather than on explanatory wording enhancing the subject's comprehension.

Subjects are not in a position to judge whether the information provided is complete. Subjects may certify that they understand the statements in the consent document and are satisfied with the explanation provided by the consent process (e.g., "I understand the statements in this informed consent document)." They should not be required to certify completeness of disclosure (e.g., "This study has been fully explained to me," or, "I fully understand the study.")

Consent documents should not contain unproven claims of effectiveness or certainty of benefit, either explicit or implicit, that may unduly influence potential subjects. Overly optimistic representations are misleading and violate FDA regulations concerning the promotion of investigational drugs [21 CFR 312.7] or investigational devices [21 CFR 812.7(d)] as well as the requirement to minimize the possibility of coercion or undue influence [21 CFR 50.20].

FDA approval of studies

Investigational drug and biologic studies are not officially approved by FDA. When a sponsor submits a study to FDA as part of the initial application for an investigational new drug (IND), FDA has thirty days to review the application and place the study on "hold" if there are any obvious reasons why the proposed study should not be conducted. Therefore, subjects are likely to impute a greater involvement by the Agency in a research study than actually exists if phrases such as, "FDA has given permission..." or "FDA has approved..." are used in consent documents. If FDA does not place the study on hold within the thirty day period, the study may begin (with IRB approval).

FDA also believes that an explicit statement that an IRB has approved solicitation of subjects to participate in research could mislead or unduly induce subjects. Subjects might think that, because the IRB had approved the research, there is no need to evaluate the study for themselves to determine whether or not they should participate.

Non-English Speaking Subjects

To meet the requirements of 21 CFR 50.20, the informed consent document should be in language understandable to the subject (or authorized representative). When the consent interview is conducted in English, the consent document should be in English. When the study subject population includes non-English speaking people or the clinical investigator or the IRB anticipates that the consent interviews will be conducted in a language other than English, the IRB should require a translated consent document to be prepared and assure that the translation is accurate. As required by 21 CFR 50.27, a copy of the consent document must be given to each subject. In the case of non-English speaking subjects, this would be the translated document. While a translator may be helpful in facilitating conversation with a non-English speaking subject, routine ad hoc translation of the consent document should not be substituted for a written translation.

If a non-English speaking subject is unexpectedly encountered, investigators will not have a written translation of the consent document and must rely on oral translation. Investigators should carefully consider the ethical/legal ramifications of enrolling subjects when a language barrier exists. If the subject does not clearly understand the information presented, the subject's consent will not truly be informed and may not be legally effective. If investigators enroll subjects without an IRB approved written translation, a "short form" written consent document, in a language the subject understands, should be used to document that the elements of informed consent required by 21 CFR 50.25 were presented orally. The required signatures on a short form are stated in 21 CFR 50.27(b)(2).

Illiterate English-Speaking Subjects

A person who speaks and understands English, but does not read and write, can be enrolled in a study by "making their mark" on the consent document, when consistent with applicable state law.

A person who can understand and comprehend spoken English, but is physically unable to talk or write, can be entered into a study if they are competent and able to indicate approval or disapproval by other means. If (1) the person retains the ability to understand the concepts of the study and evaluate the risk and benefit of being in the study when it is explained verbally (still competent) and (2) is able to indicate approval or disapproval to study entry, they may be entered into the study. The consent form should document the method used for communication with the prospective subject and the specific means by which the prospective subject communicated agreement to participate in the study. An impartial third party should witness the entire consent process and sign the consent document. A video tape recording of the consent interview is recommended.

Assent of children

Although not addressed in the regulations, FDA believes that IRBs should consider whether to require the approval of older children before they are enrolled in a research study. For research with children, some IRBs have required that two consent documents be developed. One for obtaining the parents permission and one, which outlines the study in simplified language, for obtaining the assent of children who can understand the concepts involved. Although not required by FDA regulations, the HHS regulations for conduct of studies in children may be used as guidance [45 CFR 46, Subpart D].

21 CFR 50.25 Elements of Informed Consent

(a) Basic elements of informed consent. In seeking informed consent, the following information shall be provided to each subject:

(1) A statement that the study involves research, an explanation of the purposes of the research and the expected duration of the subject's participation, a description of the procedures to be followed, and identification of any procedures which are experimental.

The statement that the study involves research is important because the relationship between patient-physician is different than that between subject-investigator. Any procedures relating solely to research (e.g., randomization, placebo control, additional tests) should be explained to the subjects. The procedures subjects will encounter should be outlined in the consent document, or an explanation of the procedures, such as a treatment chart, may be attached to and referenced in the consent document.

Consent documents for studies of investigational articles should include a statement that a purpose of the study includes an evaluation of the safety of the test article. Statements that test articles are safe or statements that the safety has been established in other studies, are not appropriate when the purpose of the study includes determination of safety. In studies that also evaluate the effectiveness of the test article, consent documents should include that purpose, but should not contain claims of effectiveness.

(2) A description of any reasonably foreseeable risks or discomforts to the subject.

The risks of procedures relating solely to research should be explained in the consent document. The risks of the tests required in the study protocol should be explained, especially for tests that carry significant risk of morbidity/mortality themselves. The explanation of risks should be reasonable and should not minimize reported adverse effects.

The explanation of risks of the test article should be based upon information presented in documents such as the protocol and/or investigator's brochure, package labeling, and previous research study reports. For IND studies, the IRB should assure that the clinical investigator submits the investigator's brochure (when one exists) with the other study materials for review.

(3) A description of any benefits to the subject or to others which may reasonably be expected from the research.

The description of benefits to the subject should be clear and not overstated. If no direct benefit is anticipated, that should be stated. The IRB should be aware that this element includes a description not only of the benefits to the subject, but to "others" as well. This may be an issue when benefits accruing to the investigator, the sponsor, or others are different than that normally expected to result from conducting research. Thus, if these benefits may be materially relevant to the subject's decision to participate, they should be disclosed in the informed consent document.

(4) A disclosure of appropriate alternative procedures or courses of treatment, if any, that might be advantageous to the subject.

To enable a rational choice about participating in the research study, subjects should be aware of the full range of options available to them. Consent documents should briefly explain any pertinent alternatives to entering the study including, when appropriate, the alternative of supportive care with no additional disease-directed therapy. While this should be more than just a list of alternatives, a full risk/benefit explanation of alternatives may not be appropriate to include in the written document. The person(s) obtaining the subjects' consent, however, should be able to discuss available alternatives and answer questions that the subject may raise about them. As with other required elements, the consent document should contain sufficient information to ensure an informed decision.

> *(5) A statement describing the extent, if any, to which confidentiality of records identifying the subject will be maintained and that notes the possibility that the Food and Drug Administration may inspect the records.*

Study subjects should be informed of the extent to which the institution intends to maintain confidentiality of records identifying the subjects. In addition, they should be informed that FDA may inspect study records (which include individual medical records). If any other entity, such as the sponsor of the study, may gain access to the study records, the subjects should be so informed. The consent document may, at the option of the IRB, state that subjects' names are not routinely required to be divulged to FDA. When FDA requires subject names, FDA will treat such information as confidential, but on rare occasions, disclosure to third parties may be required. Therefore, absolute protection of confidentiality by FDA should not be promised or implied. Also, consent documents should not state or imply that FDA needs clearance or permission from the subject for access. When clinical investigators conduct a study for submission to FDA, they agree to allow FDA access to the study records. Informed consent documents should make it clear that, by participating in research, the subject's records automatically become part of the research database. Subjects do not have the option to keep their records from being audited/reviewed by FDA.

> *(6) For research involving more than minimal risk, an explanation as to whether any compensation and an explanation as to whether any medical treatments are available if injury occurs and, if so, what they consist of, or where further information may be obtained.*

Informed consent documents should describe any compensation or medical treatments that will be provided if injury occurs. If specific statements cannot be made (e.g., each case is likely to require a different response), the subjects should be informed where further information may be obtained. The consent should also indicate whether subjects will be billed for the cost of such medical treatments. When costs will be billed, statements such as "will be billed to you or your insurer in the ordinary manner," "the sponsor has set some funds aside for medical costs related to.... Here's how to apply for reimbursement if you think you might be eligible" or "no funds have been set aside..." are preferred. Statements such as: "will be the responsibility of you or your insurance company" or "compensation is not available," could appear to relieve the sponsor or investigator of liability for negligence, see 21 CFR 50.20.

Compensation v. Waiver of Subject's Rights

The consent document must explain whether there is compensation available in case of injury but must not waive or appear to waive the rights of the subject or release or appear to release those conducting the study from liability for negligence. When no system has been set up to provide funds, the preferred wording is: "no funds have been set aside for" "[the cost] will be billed to you or your insurance," or similar wording that explains the provisions or the process. Wording such as: "will be your responsibility or that of your third-party payor" has been erroneously interpreted by some subjects to mean the insurance company is required to pay.

(7) An explanation of whom to contact for answers to pertinent questions about the research and research subjects' rights, and whom to contact in the event of a research-related injury to the subject.

This requirement contains three components, each of which should be specifically addressed. The consent document should provide the name of a specific office or person and the telephone number to contact for answers to questions about: 1) the research subjects' rights; 2) a research-related injury; and 3) the research study itself. It is as important for the subject to know why an individual should be contacted as it is for the subject to know whom to contact. Although a single contact might be able to fulfill this requirement, IRBs should consider requiring that the person(s) named for questions about research subjects' rights not be part of the research team as this may tend to inhibit subjects from reporting concerns and discovering possible problems.

(8) A statement that participation is voluntary, that refusal to participate will involve no penalty or loss of benefits to which the subject is otherwise entitled, and that the subject may discontinue participation at any time without penalty or loss of benefits to which the subject is otherwise entitled.

This element requires that subjects be informed that they may decline to participate or to discontinue participation at any time without penalty or loss of benefits. Language limiting the subject's right to withdraw from the study should not be permitted in consent documents. If the subjects who withdraw will be asked to permit follow-up of their condition by the researchers, the process and option should be outlined in the consent document.

(b) Additional elements of informed consent. When appropriate, one or more of the following elements of information shall also be provided to each subject:

(1) A statement that the particular treatment or procedure may involve risks to the subject (or to the embryo or fetus, if the subject is or may become pregnant) which are currently unforeseeable.

A statement that there may be unforeseen risks to the embryo or fetus may not be sufficient if animal data are not available to help predict the risk to a human fetus. Informed consent documents should explain that mutagenicity (the capability to induce genetic mutations) and teratogenicity (the capability to induce fetal malformations) studies have not yet been conducted/completed in animals. [Note: The lack of animal data does not constitute a valid reason for restricting entry of women of childbearing potential into a clinical trial.] Subjects, both women and men, need to understand the danger of taking a drug whose effects on the fetus are unknown. If relevant animal data are available, however, the significance should be explained to potential subjects. Investigators should ensure that the potential risks that the study poses are adequately explained to subjects who are asked to enter a study. If measures to prevent pregnancy should be taken while in the study, that should be explained.

FDA guidance on the inclusion of women in clinical trials [58 FR 39406] now gives IRBs broader discretion to encourage the entry of a wide range of individuals into the early phases of clinical trials. FDA urges IRBs to question any study that appears to limit enrollment based on gender and/or minority status. Statements such as, "you may not participate in this research study if you are a woman who could become pregnant" should not routinely be included in informed consent documents.

(2) Anticipated circumstances under which the subject's participation may be terminated by the investigator without regard to the subject's consent.

When applicable, subjects should be informed of circumstances under which their participation may be terminated by the investigator without the subject's consent. An unexplained statement that the investigator and/or sponsor may withdraw subjects at any time, does not adequately inform the subjects of anticipated circumstances for such withdrawal.

A statement that the investigator may withdraw subjects if they do not "follow study procedures" is not appropriate. Subjects are not in a position to know all the study procedures. Subjects may be informed, however, that they may be withdrawn if they do not follow the instructions given to them by the investigator.

> *(3) Any additional costs to the subject that may result from participation in the research.*

If the subjects may incur an additional expense because they are participating in the research, the costs should be explained. IRBs should consider that some insurance and/or other reimbursement mechanisms may not fund care that is delivered in a research context.

> *(4) The consequences of a subjects' decision to withdraw from the research and procedures for orderly termination of participation by the subject.*

When withdrawal from a research study may have deleterious effects on the subject's health or welfare, the informed consent should explain any withdrawal procedures that are necessary for the subject's safety and specifically state why they are important to the subject's welfare. An unexplained statement that the subject will be asked to submit to tests prior to withdrawal, does not adequately inform the subjects why the tests are necessary for the subject's welfare.

> *(5) A statement that significant new findings developed during the course of the research which may relate to the subject's willingness to continue participation will be provided to the subject.*

When it is anticipated that significant new findings that would be pertinent to the subject's continued participation are likely to occur during the subject's participation in the study, the IRB should determine that a system, or a reasonable plan, exists to make such notification to subjects.

> *(6) The approximate number of subjects involved in the study.*

If the IRB determines that the numbers of subjects in a study is material to the subjects' decision to participate, the informed consent document should state the approximate number of subjects involved in the study.

The Consent Process

Informed consent is more than just a signature on a form, it is a process of information exchange that may include, in addition to reading and signing the informed consent document, subject recruitment materials, verbal instructions, question/answer sessions and measures of subject understanding. Institutional Review Boards (IRBs), clinical investigators, and research sponsors all share responsibility for ensuring that the informed consent process is adequate. Thus, rather than an endpoint, the consent document should be the basis for a meaningful exchange between the investigator and the subject.

The clinical investigator is responsible for ensuring that informed consent is obtained from each research subject before that subject participates in the research study. FDA does not require the investigator to personally conduct the consent interview. The investigator remains ultimately responsible, even when delegating the task of obtaining informed consent to another individual knowledgeable about the research.

In addition to signing the consent, the subject/representative should enter the date of signature on the consent document, to permit verification that consent was actually obtained before the subject began participation in the study. If consent is obtained the same day that the subject's involvement in the study begins, the subject's medical records/case report form should document that consent was obtained prior to participation in the research. A copy of the consent document must be provided to the subject and the original signed consent document should be retained in the study records. Note that the FDA regulations do not require the subject's copy to be a signed copy, although a photocopy with signature(s) is preferred.

The IRB should be aware of who will conduct the consent interview. The IRB should also be informed of such matters as the timing of obtaining informed consent and of any waiting period (between informing the subject and obtaining the consent) that will be observed.

The consent process begins when a potential research subject is initially contacted. Although an investigator may not recruit subjects to participate in a research study before the IRB reviews and approves the study, an investigator may query potential subjects to determine if an adequate number of potentially eligible subjects is available.

21 CFR 50.27 Documentation of Informed Consent

(a) *Except as provided in 56.109(c), informed consent shall be documented by the use of a written consent form approved by the IRB and signed and dated by the subject or the subject's legally authorized representative at the time of consent. A copy shall be given to the person signing the form.*

(b) *Except as provided in 56.109(c), the consent form may be either of the following:*

(1) *A written consent document that embodies the elements of informed consent required by 50.25. This form may be read to the subject or the subject's legally authorized representative, but , in any event, the investigator shall give either the subject or the representative adequate opportunity to read it before it is signed.*

(2) *A short form written consent document stating that the elements of informed consent required by 50.25 have been presented orally to the subject or the subject's legally authorized representative. When this method is used, there shall be a witness to the oral presentation. Also, the IRB shall approve a written summary of what is to be said to the subject or the representative. Only the short form itself is to be signed by the subject or the representative. However, the witness shall sign both the short form and a copy of the summary, and the person actually obtaining the consent shall sign a copy of the summary. A copy of the summary shall be given to the subject or the representative in addition to a copy of the short form.*

The informed consent documentation requirements [21 CFR 50.27] permit the use of either a written consent document that embodies the elements of informed consent or a "short form" stating that the elements of informed consent have been presented orally to the subject. Whichever document is used, a copy must be given to the person signing the document.

When a short form consent document is to be used [21 CFR 50.27(b)(2)], the IRB should review and approve the written summary of the full information to be presented orally to the subjects. A witness is required to attest to the adequacy of the consent process and to the subject's voluntary consent. Therefore, the witness must be present during the entire consent interview, not just for signing the documents. The subject or the subject's legally authorized representative must sign and date the short form. The witness must sign both the short form and a copy of the summary, and the person actually obtaining the consent must sign a copy of the summary. The subject or the representative must be given a copy of the summary as well as a copy of the short form. While the regulations do not prohibit the use of multiple consent documents, FDA suggests that they be used with caution. Multiple consent documents may be confusing to a research subject and if, inadvertently, one document is not presented, critical information may not be relayed to the research subject. For some studies, however, the use of multiple documents may improve subject understanding by "staging" information in the consent process. This process may be useful for studies with separate and distinct, but linked, phases through which the subject may proceed. If this technique is used, the initial document should explain that subjects will be asked to participate in the additional phases. It should be clear whether the phases are steps in one study or separate but interrelated studies. For certain types of studies, the Agency encourages the process of renewing the consent of subjects.

Also see these FDA information sheets:
"Sponsor-Investigator-IRB Interrelationship"
"Acceptance of Foreign Clinical Studies"
"Emergency Use of an Investigational Drug or Biologic"
"Emergency Use of Unapproved Medical Devices"
"Screening Tests Prior to Study Enrollment"
"Recruiting Study Subjects"
"Payment to Research Subjects"
"Evaluation of Gender Differences in Clinical Investigations"
"Significant Differences in HHS and FDA Regulations for the Protection of Human Subject"

THIS PAGE INTENTIONALLY LEFT BLANK

Use of Investigational Products When Subjects Enter a Second Institution

Several issues are raised when a subject who is participating in a research study at one institution is admitted to another facility. To help illustrate, the following will serve as the model for this information sheet: Regional Medical Center (RMC) has developed a research protocol; the study has been reviewed and approved by the RMC institutional review board (RMC-IRB); each subject receives a test drug for a 16 week period (4 weeks inpatient, 12 weeks outpatient); some research subjects will live in a distant town with a local health care facility, Memorial Hospital (MH). For these subjects, participation at RMC will involve considerable travel time and costs. While several examples can be imagined, the three scenarios below may help to illustrate some key points.

1. The least complex (first) scenario is when a subject's treatment/hospitalization is not related to the research. Procedures should be in place for rapidly identifying test drugs and devices (e.g., an emergency contact number and unblinding procedure). For this example, we will assume that hospitalization at MH is medically necessary and that the local physician has determined that it is appropriate to continue the subject (now patient) on the test drug. In this case, MH is providing incidental medical care and is not participating as a research site. Therefore, MH staff are not investigators and the MH-IRB does not need to review the protocol. The usual procedures for dealing with drugs prescribed out-of-facility would be followed (often, this is a pharmacy department policy). The investigator at RMC remains responsible for test drug administration and follow-up and therefore, should be aware of the hospitalization. The RMC investigator may need to report the event as an unexpected adverse incident, if it is possibly related to use of the test article. The RMC-IRB remains the IRB of record.

2. For the second scenario, the involvement of MH is reasonably foreseen and is an anticipated part of the study protocol (e.g., the need for inpatient care is anticipated for the condition under study, or the need for subjects to return home and receive medical follow-up). The RMC-IRB should be aware that other institutions and/or providers will be providing medical care/follow-up and should ensure that adequate reporting and safety systems are in place before approving the study. In this example, the protocol allows the test drug to be sent to the subjects' regular health care providers. Even though the test article is being given at MH, only routine medical monitoring is conducted by the local provider with little or no reporting to the RMC investigator, who remains responsible for the test drug administration and collects research data when the subject returns to RMC. The involvement of MH is incidental to the study (i.e., research data are not collected) and thus, it is not participating as a research site. In the first two scenarios, prior to continuing the investigational drug, the local physician should obtain from the clinical investigator the information necessary to safely continue the investigational drug. The information conveyed might include a description of treatment procedures, warnings of possible adverse reactions, emergency procedures, a copy of the signed informed consent document (which is a research summary as well as documentation of consent).

3. For the third scenario, MH is designated as an extension of the research milieu. In this instance, the second institution (MH) is responsible for a portion of the research protocol. For this example, a physician at MH has been identified in the protocol as a sub-investigator for subjects residing in that local catchment area. As sub-investigator, this physician is responsible for conducting examinations of subjects to monitor status and measure effects of the test drug (data collection). These research data are systematically reported to the RMC investigator.

Because MH is conducting research, it is responsible for complying with the applicable research regulations. The MH-IRB may review, approve and be responsible for monitoring the portion of the research conducted at MH just as it would for any other research in the facility or, MH may agree to accept the RMC-IRB as the responsible IRB. If the RMC-IRB is to accept responsibility for other sites, it should consider the rationale for transferring or referring subjects to another institution; the circumstances under which responsibility will be shared; the instructions that will be given to the sub-investigators; the monitoring procedures that will be followed; and the informed consent process.

Informed Consent

Although not specifically discussed in the FDA regulations, requiring the subject to sign a second research consent document for the secondary facility should be avoided when feasible. In the first and second scenarios, research is not being conducted at MH and therefore, noresearch consent is needed for the second facility (however, consent for medical treatment may be required). Since the medical need in the first scenario is unexpected, the informed consent document would not describe such involvement. In the second scenario, because MH involvement is planned, the informed consent document should describe the activities to be carried out at MH. When some of the research activities are carried out at a secondary location, the investigator and the IRB should consider whether any additional information, such as a local emergency contact number, needs to be included in the informed consent document.

The third scenario is the most complex. Because MH is involved in research, the informed consent process should include a description of this activity. As appropriate, this could be included in the consent document presented to all subjects, or a separate informed consent document could be prepared for those subjects entering MH. If the RMC-IRB is accepting responsibility for other sites, it would review and approve the informed consent document(s). If MH does not agree to cooperative review, however, MH-IRB may accept the RMC informed consent document if it adequately describes the involvement of MH (i.e., not require a second document). MH-IRB may also decide to develop its own informed consent document. In this case it is important that the subject not receive conflicting information and the two IRBs should work to resolve such issues. If there are two consent documents, generally the RMC document would cover the overall study and the MH document would only detail the specific procedures involved while at that facility.

Also see this FDA Information Sheet:
"Cooperative Research"

Personal Importation and Use of Unapproved Products

FDA permits individuals to bring into the U.S., for their personal use, up to a three months supply of FDA-regulated products sold abroad but not approved in the U.S. Importation may be in personal baggage or by mail. All of the following four conditions must be met in order to permit importation:

1. The product was purchased for personal use.
2. The product is not for commercial distribution and the amount of product is not excessive (i.e., 3 month supply or less).
3. The intended use of the product is appropriately identified.
4. The patient seeking to import the product affirms in writing that it is for the patient's own use and provides the name and address of the licensed physician in the U.S. responsible for his or her treatment with the product.

This importation policy applies to most drugs, biologics and medical devices intended for personal import, provided they are not fraudulently promoted and do not present an unreasonable risk. Importation by a physician for use by his/her patients does not meet the requirements for personal importation.

Since the person using the product initiates the importation, that person is presumed to be knowledgeable about the product and its use. Therefore, such personal importation is not regarded by FDA to be research and an IND/IDE is not required. Also, neither IRB review nor informed consent is required by FDA for such personal importation and use.

The policy on personal importation and use of unapproved products is undergoing review and is subject to change.

THIS PAGE INTENTIONALLY LEFT BLANK

Exception from Informed Consent For Studies Conducted in Emergency Settings: Regulatory Language and Excerpts from Preamble

The federal regulations for the protection of human subjects in research require informed consent, with a few narrow exceptions. Following industry-FDA meetings (1993); a Congressional hearing (May 1994); a Coalition-conference of academic, medical and research organizations (October 1994); and an FDA-sponsored public forum (January 1995), FDA published in the Federal Register in September 1995, [60 FR 49086] a proposal to amend its regulations to permit a limited class of research in emergency settings without consent. Following a careful review of the comments received on the proposal, a final regulation was published in the Federal Register on October 2, 1996, [61 FR 51498]. The Department of Health and Human Services published, in the same issue, its waiver criteria which match the FDA requirements [61 FR 51531]. These documents establish a single standard for this class of research.

The new FDA regulation (21 CFR 50.24) provides a narrow exception to the requirement for informed consent from each human subject, or his or her legally authorized representative, prior to initiation of an experimental intervention. The exception would apply to a limited class of research activities involving human subjects who are in need of emergency medical intervention but who cannot give informed consent because of their life-threatening medical condition, and who do not have a legally authorized person to represent them. The intent of the new regulation is to allow research on life-threatening conditions for which available treatments are unproven or unsatisfactory and where it is not possible to obtain informed consent, while establishing additional protections to provide for safe and ethical studies.

FDA recognizes that persons with life-threatening conditions who can neither give informed consent nor refuse enrollment are a vulnerable population. FDA recognizes that the lack of autonomy and inability of subjects to give informed consent requires additional protective procedures in the review, approval, and operation of this research. The exception from the informed consent requirement permitted by the rule is conditional upon documented findings by an Institutional Review Board (IRB).

The regulation specifically requires the concurrence of a licensed physician "who is a member of or consultant to the IRB and who is not otherwise participating in the clinical investigation" (Sec. 50.24(a)). This requirement is similar to 21 CFR 50.23 which requires an independent assessment by a physician not otherwise participating in the research when an investigational product is to be used in a life-threatening situation. Because 21 CFR 50.24 permits an exception from the requirement for informed consent for a group of subjects, the case-by-case independent determination is replaced by the general concurrence of a licensed physician. The option for use of a consultant to the IRB is to provide flexibility, for example, when the physician member(s) cannot participate in the deliberation and voting due to conflict of interest. Because the documented concurrence of the physician is required for approval of these studies, IRBs should ensure that meeting minutes specifically record this affirmative vote.

[Note: The numbering system used below follows the regulation numbers. The regulatory language is in italics. Readers are referred to the full text of the regulation and the preamble for additional guidance.]

According to 21 CFR 50.24, the IRB must find and document each of the following. It is clear from the regulations's wording that it is the IRB's responsibility to make decisions as to whether the criteria of the rule are met.

50.24(a)(1) *The human subjects are in a life-threatening situation,*

> The criteria contained in the rule do not require the condition to be immediately life-threatening or to immediately result in death. Rather, the subjects must be in a life-threatening situation requiring intervention before consent from a legally authorized representative is feasible. Life-threatening includes diseases or conditions where the likelihood of death is high unless the course of the disease or condition is interrupted. (See Sec. 312.81.) People with the conditions cited in the examples provided in the comments—e.g., long-term or permanent coma, stroke, and head injury—may survive for long periods but the likelihood of survival is not known during the therapeutic window of treatment. People with these conditions are clearly at increased risk of death due to infection, pulmonary embolism, progression of disease, etc. The rule would apply in such situations if the intervention must be given before consent is feasible in order to be successful. The informed consent waiver provision is not intended to apply to persons who are not in an emergent situation, e.g., individuals who have been in a coma for a long period of time and for whom the research intervention should await the availability of a legally authorized representative of the subject.

available treatments are unproven or unsatisfactory, and

> "Clinical equipoise" must exist. "When the relative benefits and risks of the proposed intervention, as compared to standard therapy, are unknown, or thought to be equivalent or better, there is clinical equipoise between the historic intervention and the proposed test intervention." (60 FR 49086 at 49093, September 21, 1995.)

the collection of valid scientific evidence, which may include evidence obtained through randomized placebo-controlled investigations, is necessary to determine the safety and effectiveness of particular interventions.

> Although the regulation specifically references placebo controlled trials, this was done to indicate that such trials may be conducted when appropriate. Other controls, e.g, active controls and historical controls, may also be used when they are appropriate and adequate to the task of providing evidence that the drug or device will have the effect claimed. In virtually all cases, when a placebo is used, standard care, if any, would be given to all subjects, with subjects randomized to receive, in addition, the test treatment or a placebo. An exception to this would be the situation in which the test is to determine whether standard treatment is in fact useful. In that case, there must be a group that does not receive it.

(2) Obtaining informed consent is not feasible because:
(i) the subjects will not be able to give their informed consent as a result of their medical condition;

> Subjects do not have to be comatose, but the medical condition under study must prevent obtaining valid informed consent. The agency expects the IRB to determine, based on the specific details of the individual clinical investigation (including the window of opportunity for treatment), the procedures the investigator must follow to attempt to obtain informed consent before enrolling a subject in an investigation without such consent. IRBs also should be knowledgeable about an institution's procedures regarding the use of advance medical directives and assess whether the proposed clinical investigation is consistent with those procedures.

(ii) the intervention under investigation must be administered before consent from the subjects' legally authorized representatives is feasible; and

The agency expects the IRB to determine, based on the specific details of the individual clinical investigation (including the window of opportunity for treatment), the procedures the investigator must follow to attempt to obtain informed consent before enrolling a subject in an investigation without such consent.

(iii) There is no reasonable way to identify prospectively the individuals likely to become eligible for participation in the clinical investigation.

If an IRB determines that it is not appropriate to waive the requirement for informed consent because there is a reasonable way to identify prospectively the individuals likely to become eligible for the study, then this exception would not apply. In that case, only those subjects with the condition who gave prior consent may be enrolled in the study. Those individuals who either did not make a decision or who refused participation would be excluded from participation in the study. While an exception would not be allowed under this rule, the individual exception allowed under 21 CFR 50.23 might be applicable in some circumstances.

(3) Participation in the research holds out the prospect of direct benefit to the subjects because: (i) subjects are facing a life-threatening situation that necessitates intervention;

(ii) appropriate animal and other preclinical studies have been conducted, and the information derived from those studies and related evidence support the potential for the intervention to provide a direct benefit to the individual subjects; and

(iii) risks associated with the investigation are reasonable in relation to what is known about the medical condition of the potential class of subjects, the risks and benefits of standard therapy, if any, and what is known about the risks and benefits of the proposed intervention or activity.

(4) The clinical investigation could not practicably be carried out without the waiver.

If scientifically sound research can be practicably carried out using only consenting subjects (directly, or in most cases for the research contemplated in the rule, with legally authorized representatives), then the agency thinks it should be carried out without involving nonconsenting subjects. By practicable, the agency means, for example, (1) that recruitment of consenting subjects does not bias the science and the science is no less rigorous as a result of restricting it to consenting subjects; or (2) that the research is not unduly delayed by restricting it to consenting subjects.

(5) The proposed investigational plan defines the length of the potential therapeutic window based on scientific evidence, and the investigator has committed to attempting to contact a legally authorized representative for each subject within that window of time and, if feasible, to asking the legally authorized representative contacted for consent within that window rather than proceeding without consent. The investigator will summarize efforts made to contact legally authorized representatives and make this information available to the IRB at the time ofcontinuing review.

The agency believes that these procedures will ensure that appropriate efforts are made by the investigator to obtain consent from subjects prior to enrollment. The agency expects these procedures to be documented in the protocol and/or by the IRB, and the efforts made by investigators to be documented in the material presented to the IRB for its continuing review.

(6) The IRB has reviewed and approved informed consent procedures and an informed consent document consistent with Sec. 50.25. These procedures and the informed consent document are to be used with subjects or their legally authorized representatives in situations where use of such procedures and documents is feasible. The IRB has reviewed and approved procedures and information to be used when providing an opportunity for a family member to object to a subject's participation in the clinical investigation consistent with paragraph (a)(7)(v) of this section.

IRBs need to be aware of state and local laws. Some states have laws which prohibit entry of subjects into research without their express consent. This new rule does not preempt state/or local law.

The agency has specifically included family members under this rule because the opportunity for an available family member to object to a potential subject's participation in such a clinical investigation provides an additional and an important protection to these individuals. Otherwise, if consent from a subject or the subject's legally authorized representative were not feasible, the eligible individual could be enrolled into the investigation. Thus, by permitting a family member (even one who is not a legally authorized representative) to object to an individual's inclusion in the investigation, a further protection is provided to that individual. A family member must be provided an opportunity to object to the potential subject's participation, if feasible within the therapeutic window, when obtaining informed consent from the subject is not feasible and a legally authorized representative is not available. The agency recognizes that this may not constitute legally effective informed consent if the family member is not a legally authorized representative under State law. FDA is not establishing a hierarchy of family members although an IRB may consider the need for creating a hierarchy in reviewing individual investigations. Under this rule only one family member would need to be consulted and agree or object to the patient's participation in the research. If family members were to disagree, the researcher and family members would need to work out the disagreement.

(7) Additional protections of the rights and welfare of subjects will be provided, including, at least: (i) consultation (including, where appropriate, consultation carried out by the IRB) with representatives of the communities in which the clinical investigation will be conducted and from which the subjects will be drawn

While an IRB may appropriately decide to supplement its members with consultants from the community, broader consultation with the community is needed for this type of research. The agency expects the IRB to provide an opportunity for the community from which research subjects may be drawn to understand the proposed clinical investigation and its risks and benefits and to discuss the investigation. The IRB should consider this community discussion in reviewing the investigation. Based on this community consultation, the IRB may decide, among other things, that it is appropriate to attempt to exclude certain groups from participation in the investigation, or that wider community consultation and discussion is needed. As described in the preamble to the proposed rule (60 FR 49086, September 21, 1995), IRBs should consider, for example, having a public meeting in the community to discuss the protocol; establishing a separate panel of members of the community from which the subjects will be drawn; including consultants to the IRB from the community from which the subjects will be drawn; enhancing the membership of the IRB by adding members who are not affiliated with the institution and are representative of the community; or developing other mechanisms to ensure community involvement and input into the IRB's decisionmaking process. It is likely that multiple methods may be needed in order to provide the supplemental information that the IRB will need from the community to review this research.

(ii) Public disclosure to the communities in which the clinical investigation will be conducted and from which the subjects will be drawn, prior to initiation of the clinical investigation, of plans for the investigation and its risks and expected benefits;

It is the IRB's responsibility to determine the information to be disclosed. This information could include, but may not necessarily be limited to, the information that is found in the informed consent document, the investigator's brochure, and the research protocol. The IRB should consider how best to publicly disclose, prior to commencement of the clinical investigation, sufficient information to describe the investigation's risks and benefits, e.g., relevant information from the investigator's brochure, the informed consent document, and the investigational protocol. Initial disclosure of information will occur during the community consultation process. Disclosure of this information to the community will inform individuals within the community about the clinical investigation and permit them to raise concerns and objections.

(iii) Public disclosure of sufficient information following completion of the clinical investigation to apprise the community and researchers of the study, including the demographic characteristics of the research population, and its results;

It is necessary to provide comprehensive summary data from the completed trial to the research community in order to permit other researchers to assess the results of the clinical investigation. The agency thinks that there must be a scientific need to conduct clinical investigations involving subjects who are unable to consent; if previous investigations have already provided the scientific answer, this should be shared broadly with the research community. Sufficient information may be contained in a scientific publication of the results of the completed investigation; in other instances, a publication may need to be supplemented by additional information. The agency has modified Sec. 50.24(a)(7)(iii) to clarify that the information to be disclosed is to include the demographic characteristics (age, gender, and race) of the research population. For a multicenter investigation, the agency anticipates that the sponsor and/or lead investigators will be responsible for analyzing the results of the overall investigation, including the demographic characteristics of the research population, and that these results will be published (or reported in the lay press) within a reasonable period of time following completion of the investigation. Publication in a scientific journal or reports of the results by lay press, that would be supplemented upon request by comprehensive summary data, will enable the research community, e.g., researchers not connected to the clinical investigation, to learn of the research's results. Following publication, the IRB will be responsible for determining appropriate mechanisms for providing this information, possibly supplemented by a lay description, to the community from which research subjects were drawn. The usual rules of marketing and promotion apply to the disclosure of this information. The agency notes that it is common for the results of research to be reported in the lay press and published in peer reviewed journals.

(iv) Establishment of an independent data monitoring committee to exercise oversight of the clinical investigation; and

A data monitoring committee will help ensure that if it becomes clear that the benefits of the investigational intervention are established, or that risks are greater than anticipated, or that the benefits do not justify the risks of the research, the investigation can be modified to minimize those risks or the clinical investigation can be halted. The data monitoring committee is established by the sponsor of the research, as an advisory body to the sponsor. An independent committee is constituted of individuals not otherwise connected with the particular clinical investigation. A variety of expertise is required for an effective data monitoring committee. Typically included are clinicians specializing in the relevant medical field(s), biostatisticians, and bioethicists. The data monitoring committee receives study data on an ongoing basis on a schedule generally defined in the investigational protocol; based on its review of the data it may recommend to the sponsor that the clinical investigation be modified or stopped. In effect, it is responsible for making sure that continuing the investigation in its current format remains appropriate, on both safety and scientific grounds. A number of reasonable models for establishment and function of these committees are described and discussed in S. Ellenberg, N. Geller, R. Simon, S. Yusuf (editors), Practical issues in data monitoring of clinical trials (Proceeding of an International Workshop) Statistics in Medicine, vol. 12; 1993. If a sponsor accepts a data monitoring committee's recommendation to stop the investigation or to institute a major modification of the trial, the sponsor is required to notify FDA and all participating investigators and IRBs in a written IND or IDE safety report within 10 working days after the sponsor's initial receipt of the information. (See Secs. 312.32, 312.56(d), and 812.150(b)(1)).

If an IRB, a subcommittee of the IRB, or some other preexisting institutional committee were to serve as a data monitoring committee, it would need to be constituted as a data monitoring committee when it functions in that capacity. The agency thinks that the duties and scope of activities of an IRB and a data monitoring committee are quite different and that it is important for separate entities to be established. The agency would not object, however, to an already established committee, such as an IRB, serving as a data monitoring committee as long as that committee was constituted to perform the duties of a data monitoring committee and operated as such separately and distinctly from its IRB activities.

(v) If obtaining informed consent is not feasible and a legally authorized representative is not reasonably available, the investigator has committed, if feasible, to attempting to contact within the therapeutic window the subject's family member who is not a legally authorized representative, and asking whether he or she objects to the subject's participation in the clinical investigation. The investigator will summarize efforts made to contact family members and make this information available to the IRB at the time of continuing review.

See previous explanation under Sec. 50.24(a)(6).

(b) The IRB is responsible for ensuring that procedures are in place to inform, at the earliest feasible opportunity, each subject, or if the subject remains incapacitated, a legally authorized representative of the subject, or if such a representative is not reasonably available, a family member, of the subject's inclusion in the clinical investigation, the details of the investigation and other information contained in the informed consent document. The IRB shall also ensure that there is a procedure to inform the subject, or if the subject remains incapacitated, a legally authorized representative of the subject, or if such a representative is not reasonably available, a family member, that he or she may discontinue the subject's participation at any time without penalty or loss of benefits to which the subject is otherwise entitled. If a legally authorized representative or family member is told about the clinical investigation and the subject's condition improves, the subject is also to be informed as soon as feasible. If a subject is entered into a clinical investigation with waived consent and the subject dies before a legally authorized representative or family member can be contacted, information about the clinical investigation is to be provided to the subject's legally authorized representative or family member, if feasible.

The agency thinks that it may not always be possible to develop a meaningful informed consent document for continued participation in the research, because the relevant information may vary significantly depending upon when it becomes feasible to provide the information to the subject or legally authorized representative. It is up to the IRB to determine whether it is possible or desirable, given the nature of the clinical investigation, to have an actual document that could be signed for continued participation in the investigation. The agency notes that such a document, that would be signed after entry into an investigation, would not constitute consent for what had already occurred; it could, however, serve to document that the subject consented to continued participation in the investigation. The agency notes that Secs. 312.60 and 812.140 require the clinical investigator to document data pertinent to each individual in the investigation. This documentation should include information that the subject, legally authorized representative, or family member was informed of the subject's inclusion in the clinical investigation, the details of the investigation, and other information contained in the informed consent document. Like other IRB records, records of the determinations above must be kept for a minimum of three years after the completion of the clinical investigation (21 CFR 50.24(c)). Again, like other IRB records, these are subject to inspection and copying by FDA.

(d) Protocols involving an exception to the informed consent requirement under this section must be performed under a separate investigational new drug application (IND) or investigational device exemption (IDE) that clearly identifies such protocols as protocols that may include subjects who are unable to consent. The submission of those protocols in a separate IND/IDE is required even if an IND for the same drug product or an IDE for the same device already exists. Applications for investigations under this section may not be submitted as amendments under Secs. 312.30 or 812.35 of this chapter.

The submission of a separate IND or IDE will ensure that FDA reviews the application before the study may proceed. FDA review of the application will enable the agency to assess whether the available treatments for the condition are unproven or unsatisfactory, whether the intervention is reasonable, whether the study design will provide the information sought, and whether other conditions of the regulations are met. The amount of information needed in the application will differ depending upon the particular intervention. If an IND or IDE exists, the separate application does not need to duplicate, and the sponsor does not need to resubmit, information that is contained in the existing IND or IDE; the separate application will need to reference the existing IND or IDE, contain a protocol for the clinical investigation that includes a description of how the investigation proposes to meet the conditions of this regulation, and contain only the study-specific information required by Secs. 312.23, 812.20, and 812.25, as appropriate. If the investigation involves a product that has received marketing approval and the use is within the product's approved labeling, and without dosage or schedule change if for a drug product, the protocol may simply need to be accompanied by the product's approved labeling and a description of how the investigation proposes to meet the conditions of this regulation; no toxicology or manufacturing controls or chemistry information may need to be submitted. By submitting this information to the agency for review, the dual review by both FDA and an IRB will provide additional protections to the subjects of this research.

If the clinical investigation involves a product that has received marketing approval, but involves a route of administration or dosage level or use in a subject population or other factor that significantly increases the risks (or decreases the acceptability of the risks) associated with the use of the product, or if the investigation involves an investigational product for which an IND or IDE does not exist, then the IND or IDE would need to include information to support the altered conditions of use, including toxicology, chemistry, and clinical information, as appropriate.

(e) If an IRB determines that it cannot approve a clinical investigation because the investigation does not meet the criteria in the exception provided under paragraph (a) of this section or because of other relevant ethical concerns, the IRB must document its findings and provide these findings promptly in writing to the clinical investigator and to the sponsor of the clinical investigation. The sponsor of the clinical investigation must promptly disclose this information to FDA and to the sponsor's clinical investigators who are participating or are asked to participate in this or a substantially equivalent clinical investigation of the sponsor, and to other IRBs that have been, or are, asked to review this or a substantially equivalent investigation by that sponsor.

By "substantially equivalent" the agency means other clinical investigations that propose to invoke this exception from informed consent and that involve basically the same medical conditions and investigational treatments. The agency intends this requirement to refer to clinical investigations conducted by the same sponsor.

It is the sponsor's responsibility to determine that a study is "substantially equivalent." If a protocol invoking this exception is modified by the sponsor in order to respond to IRB concerns that it does not meet the criteria in Sec. 50.24(a) of the exception or because of other relevant ethical concerns, and it is a multicenter study, then the IRB's written findings are to be disclosed to other centers that either are, or may be, participating in the study. If there is a change in a protocol in a multicenter trial, there is re-review of the protocol by all the IRBs of the institutions participating in the multicenter trial. If the change is minor, it may be eligible for expedited review under Sec. 56.110, which permits the IRB to use an expedited review procedure to review minor changes in previously approved research during the period for which approval is authorized. If the change is significant, it would need to be reviewed by the full committee. It is the sponsor's responsibility to determine if it has a substantially similar protocol necessitating information dissemination.

"Off-Label" and Investigational Use Of Marketed Drugs, Biologics, and Medical Devices

"Off-Label" Use of Marketed Drugs, Biologics and Medical Devices

Good medical practice and the best interests of the patient require that physicians use legally available drugs, biologics and devices according to their best knowledge and judgement. If physicians use a product for an indication not in the approved labeling, they have the responsibility to be well informed about the product, to base its use on firm scientific rationale and on sound medical evidence, and to maintain records of the product's use and effects. Use of a marketed product in this manner when the intent is the "practice of medicine" does not require the submission of an Investigational New Drug Application (IND), Investigational Device Exemption (IDE) or review by an Institutional Review Board (IRB). However, the institution at which the product will be used may, under its own authority, require IRB review or other institutional oversight.

Investigational Use of Marketed Drugs, Biologics and Medical Devices

The investigational use of approved, marketed products differs from the situation described above. "Investigational use" suggests the use of an approved product in the context of a clinical study protocol [see 21 CFR 312.3(b)]. When the principal intent of the investigational use of a test article is to develop information about the product's safety or efficacy, submission of an IND or IDE may be required. However, according to 21 CFR 312.2(b)(1), the clinical investigation of a marketed drug or biologic does not require submission of an IND if all six of the following conditions are met:

(i)   it is not intended to be reported to FDA in support of a new indication for use or to support any other significant change in the labeling for the drug;

(ii) it is not intended to support a significant change in the advertising for the product;

(iii) it does not involve a route of administration or dosage level, use in a subject population, or other factor that significantly increases the risks (or decreases the acceptability of the risks) associated with the use of the drug product;

(iv) it is conducted in compliance with the requirements for IRB review and informed consent [21 CFR parts 56 and 50, respectively];

(v)   it is conducted in compliance with the requirements concerning the promotion and sale of drugs [21 CFR 312.7]; and

(vi) it does not intend to invoke 21 CFR 50.24.

Contact information is listed on the following page

For additional information on whether or not an IND or IDE is required in a specific situation, contact:

For DRUG PRODUCTS contact:
Drug Information Branch (HFD-210)
Center for Drug Evaluation and Research
Food and Drug Administration
5600 Fishers Lane
Rockville, Maryland 20857
301-827-4573

For a BIOLOGICAL BLOOD product, contact:
Office of Blood Research and Review (HFM-300)
Center for Biologic Evaluation and Research
Food and Drug Administration
1401 Rockville Pike
Rockville, Maryland 20852
301-827-3518

For a BIOLOGICAL VACCINE product, contact:
Office of Vaccines Research and Review (HFM-400)
Food and Drug Administration
8800 Rockville Pike
Bethesda, Maryland 20892-0001
301-827-0648

For a BIOLOGICAL THERAPEUTIC product, contact:
Office of Therapeutics Research and Review (HFM-500)
Food and Drug Administration
1451 Rockville Pike
Rockville, Maryland 20852-1420
301-594-2860

For a MEDICAL DEVICE product, contact:
Program Operations Staff (HFZ-403)
Office of Device Evaluation
Center for Devices and Radiological Health
Food and Drug Administration
9200 Corporate Blvd.
Rockville, Maryland 20850
301-594-1190

<u>Drugs and Biologics</u>

Emergency Use of an Investigational Drug or Biologic

- Obtaining an Emergency IND
- Emergency Exemption from Prospective IRB Approval
- Exception From Informed Consent Requirement
- Planned Emergency Research, Informed Consent Exception

Treatment Use of Investigational Drugs
Waiver of IRB Requirements
Drug Study Designs
Evaluation of Gender Differences

Emergency Use of an Investigation Drug or Biologic

The emergency use of test articles frequently prompts questions from Institutional Review Boards (IRBs) and investigators. This information sheet addresses three areas of concern: emergency Investigational New Drug (IND) requirements; IRB procedures; and informed consent requirements.

Obtaining an Emergency IND

The emergency use of an unapproved investigational drug or biologic requires an IND. If the intended subject does not meet the criteria of an existing study protocol, or if an approved study protocol does not exist, the usual procedure is to contact the manufacturer and determine if the drug or biologic can be made available for the emergency use under the company's IND.

The need for an investigational drug or biologic may arise in an emergency situation that does not allow time for submission of an IND. In such a case, FDA may authorize shipment of the test article in advance of the IND submission. Requests for such authorization may be made by telephone or other rapid communication means [21 CFR 312.36].

FDA Contacts for Obtaining an Emergency IND

Product	Office/Divison to Contact
Drug Products	**Drug Information Branch** **(HFD-210)** 301-827-4573
Biological Blood Products	**Office of Blood Research and Review** **(HFM-300)** 301-827-3518
Biological Vaccine Products	**Office of Vaccines Research and Review** **(HFM-400)** 301-827-0648
Biological Therapeutic Products	**Office of Therapeutics Research and Review** **(HFM-500)** 301-594-2860
On Nights and Weekends	**Division of Emergency and Epidemiological Operations** **(HFC-160)** 301-443-1240

Emergency Exemption from Prospective IRB Approval

Emergency use is defined as the use of an investigational drug or biological product with a human subject in a life-threatening situation in which no standard acceptable treatment is available and in which there is not sufficient time to obtain IRB approval [21 CFR 56.102(d)]. The emergency use provision in the FDA regulations [21 CFR 56.104(c)] is an exemption from prior review and approval by the IRB. The exemption, which may not be used unless all of the conditions described in 21 CFR 56.102(d) exist, allows for one emergency use of a test article without prospective IRB review. FDA regulations require that any subsequent use of the investigational product at the institution have prospective IRB review and approval. FDA acknowledges, however, that it would be inappropriate to deny emergency treatment to a second individual if the only obstacle is that the IRB has not had sufficient time to convene a meeting to review the issue.

Life-threatening, for the purposes of section 56.102(d), includes the scope of both life-threatening and severely debilitating, as defined below.

> **Life-threatening** means diseases or conditions where the likelihood of death is high unless the course of the disease is interrupted and diseases or conditions with potentially fatal outcomes, where the end point of clinical trial analysis is survival. The criteria for life-threatening do not require the condition to be immediately life-threatening or to immediately result in death. Rather, the subjects must be in a life-threatening situation requiring intervention before review at a convened meeting of the IRB is feasible.

> **Severely debilitating** means diseases or conditions that cause major irreversible morbidity. Examples of severely debilitating conditions include blindness, loss of arm, leg, hand or foot, loss of hearing, paralysis or stroke.

Institutional procedures may require that the IRB be notified prior to such use, however, this notification should not be construed as an IRB approval. Notification should be used by the IRB to initiate tracking to ensure that the investigator files a report within the five day time-frame required by 21 CFR 56.104(c). The FDA regulations do not provide for expedited IRB approval in emergency situations. Therefore, "interim," "compassionate," "temporary" or other terms for an expedited approval process are not authorized. An IRB must either convene and give "full board" approval of the emergency use or, if the conditions of 21 CFR 56.102(d) are met and it is not possible to convene a quorum within the time available, the use may proceed without any IRB approval.

Some manufacturers will agree to allow the use of the test article, but their policy requires "an IRB approval letter" before the test article will be shipped. If it is not possible to convene a quorum of the IRB within the time available, some IRBs have sent to the sponsor a written statement that the IRB is aware of the proposed use and considers the use to meet the requirements of 21 CFR 56.104(c). Although, this is not an "IRB approval," the acknowledgment letter has been acceptable to manufacturers and has allowed the shipment to proceed.

This policy is undergoing review and is subject to change.

Exception From Informed Consent Requirement

Even for an emergency use, the investigator is required to obtain informed consent of the subject or the subject's legally authorized representative unless both the investigator and a physician who is not otherwise participating in the clinical investigation certify in writing all of the following [21 CFR 50.23(a)]:

(1) The subject is confronted by a life-threatening situation necessitating the use of the test article.
(2) Informed consent cannot be obtained because of an inability to communicate with, or obtain legally effective consent from, the subject.
(3) Time is not sufficient to obtain consent from the subject's legal representative.
(4) No alternative method of approved or generally recognized therapy is available that provides an equal or greater likelihood of saving the subject's life.

If, in the investigator's opinion, immediate use of the test article is required to preserve the subject's life, and if time is not sufficient to obtain an independent physician's determination that the four conditions above apply, the clinical investigator should make the determination and, within 5 working days after the use of the article, have the determination reviewed and evaluated in writing by a physician who is not participating in the clinical investigation. The investigator must notify the IRB within 5 working days after the use of the test article [21 CFR 50.23(c)].

Exception from Informed Consent for Planned Emergency Research

The conduct of planned research in life-threatening emergent situations where obtaining prospective informed consent has been waived, is provided by 21 CFR 50.24. The research plan must be approved in advance by FDA and the IRB, and publicly disclosed to the community in which the research will be conducted. Such studies are usually not eligible for the emergency approvals described above. The information sheet "Exception from Informed Consent for Studies Conducted in Emergency Settings: Regulatory Language and Excerpts from Preamble," is a compilation of the wording of 21 CFR 50.24 and pertinent portions of the preamble from the October 2, 1996 Federal Register.

Also see these FDA Information Sheets:
"Exception from Informed Consent for Studies Conducted in Emergency
Settings: Regulatory Language and Excerpts from Preamble"
"Emergency Use of Unapproved Medical Devices"
"Treatment Use of Investigational Drugs"

THIS PAGE INTENTIONALLY LEFT BLANK

Treatment Use of Investigational Drugs

Expanded Access of Investigational Drugs

Investigational products are sometimes used for treatment of serious or life-threatening conditions either for a single subject or for a group of subjects. The procedures that have evolved for an investigational new drug (IND) used for these purposes reflect the recognition by the Food and Drug Administration (FDA) that, when no satisfactory alternative treatment exists, subjects are generally willing to accept greater risks from test articles that may treat life-threatening and debilitating illnesses. The following mechanisms expand access to promising therapeutic agents without compromising the protection afforded to human subjects or the thoroughness and scientific integrity of product development and marketing approval.

OPEN LABEL PROTOCOL OR OPEN PROTOCOL IND

These are usually uncontrolled studies, carried out to obtain additional safety data (Phase 3 studies). They are typically used when the controlled trial has ended and treatment is continued so that the subjects and the controls may continue to receive the benefits of the investigational drug until marketing approval is obtained. These studies require prospective Institutional Review Board (IRB) review and informed consent.

TREATMENT IND

The treatment IND [21 CFR 312.34 and 312.35] is a mechanism for providing eligible subjects with investigational drugs for the treatment of serious and life-threatening illnesses for which there are no satisfactory alternative treatments. A treatment IND may be granted after sufficient data have been collected to show that the drug "may be effective" and does not have unreasonable risks. Because data related to safety and side effects are collected, treatment INDs also serve to expand the body of knowledge about the drug.

There are four requirements that must be met before a treatment IND can be issued: 1) the drug is intended to treat a serious or immediately life-threatening disease; 2) there is no satisfactory alternative treatment available; 3) the drug is already under investigation, or trials have been completed; and 4) the trial sponsor is actively pursuing marketing approval.

Treatment IND studies require prospective IRB review and informed consent. A sponsor may apply for a waiver of local IRB review under a treatment IND if it can be shown to be in the best interest of the subjects, and if a satisfactory alternate mechanism for assuring the protection of human subjects is available, e.g., review by a central IRB. Such a waiver does not apply to the informed consent requirement. An IRB may still opt to review a study even if FDA has granted a waiver.

GROUP C TREATMENT IND

The "Group C" treatment IND was established by agreement between FDA and the National Cancer Institute (NCI). The Group C program is a means for the distribution of investigational agents to oncologists for the treatment of cancer under protocols outside the controlled clinical trial. Group C drugs are generally Phase 3 study drugs that have shown evidence of relative and reproducible efficacy in a specific tumor type. They can generally be administered by properly trained physicians without the need for specialized supportive care facilities. Group C drugs are distributed only by the National Institutes of Health under NCI protocols. Although treatment is the primary objective and patients treated under Group C guidelines are not part of a clinical trial, safety and effectiveness data are collected. Because administration of Group C drugs is not done with research intent, FDA has generally granted a waiver from the IRB review requirements [21 CFR 56.105]. Even though FDA has granted a waiver for these drugs, an IRB may still choose to conduct a review under its policies and procedures. The usage of a Group C drug is described in its accompanying "Guideline Protocol" document. The Guideline Protocol contains an FDA-approved informed consent document which must be used if there has been no local IRB review.

PARALLEL TRACK

The Agency's Parallel Track policy [57 FR 13250] permits wider access to promising new drugs for AIDS/HIV related diseases under a separate "expanded access" protocol that "parallels" the controlled clinical trials that are essential to establish the safety and effectiveness of new drugs. It provides an administrative system that expands the availability of drugs for treating AIDS/HIV. These studies require prospective IRB review and informed consent.

EMERGENCY USE IND

The need for an investigational drug may arise in an emergency situation that does not allow time for submission of an IND in the usual manner. In such cases, FDA may authorize shipment of the drug for a specified use [21 CFR 312.36]. Such authorization is usually conditioned upon the sponsor filing an appropriate application as soon as practicable. Prospective IRB review is required unless the conditions for exemption are met [21 CFR 56.104(c) and 56.102(d)]. Informed consent is required unless the conditions for exception are met [21 CFR 50.23].

Also see this FDA Information Sheet:
"Emergency Use of an Investigational Drug or Biologic"

Waiver of IRB Requirements for Drug and Biologics Studies

In accordance with 21 CFR 56.105, FDA may waive any of the requirements contained in the Institutional Review Board (IRB) regulations [21 CFR part 56] if requested by the sponsor or sponsor-investigator. A waiver can be granted for specific research activities or for classes of research activities otherwise covered by the IRB regulations. Note that the waiver provision does not apply to the informed consent requirements [21 CFR part 50]. An institution may still require IRB review on the local level even if a waiver from FDA is granted.

FDA uses the waiver provision only where it would be in the best interest of the subjects and where alternative mechanisms for assuring the protection of the subjects are adequate. Circumstances which FDA will consider for a waiver include "treatment INDs," i.e., the use of an investigational drug or biologic primarily for the treatment of a subject with a serious or immediately life-threatening disease for whom comparable or satisfactory alternate therapy is unavailable. [See 21 CFR 312.34.] The waiver provision is not needed for an emergency use because the regulations contain a provision for exemption from prospective IRB review in an emergency, provided that such use is reported to the IRB within 5 working days [21 CFR 56.104(c)].

FDA will handle waiver requests expeditiously. A request for waiver should contain the following information:

(1) The specific requirement or requirements in the IRB regulations for which a waiver is requested.
(2) The specific research activity for which the waiver will be applied and why this is a special situation.
(3) Why a waiver would be in the interest of subjects.
(4) What alternate mechanism(s) for assuring the protection of human subjects is available and would be utilized.
(5) A copy of the proposed consent document.

The sponsor or sponsor-investigator should submit a request for a waiver associated with an IND to the Review Division in the Center for Drug Evaluation and Research (CDER) or to the Review Division in the Center for Biologic Evaluation and Research (CBER) responsible for reviewing the IND. If the identity of the responsible Review Division is unknown, the waiver request may be sent to:

For DRUG PRODUCTS:
Drug Information Branch (HFD-211)
Center for Drug Evaluation and Research
Food and Drug Administration
5600 Fishers Lane
Rockville, Maryland 20857
(301) 827- 4573

For a BIOLOGICAL BLOOD product, contact:
Office of Blood Research and Review (HFM-300)
Center for Biologic Evaluation and Research
Food and Drug Administration
1401 Rockville Pike
Rockville, Maryland 20852
301-827-3518

For a BIOLOGICAL VACCINE product, contact:
Office of Vaccines Research and Review (HFM-400)
Food and Drug Administration
8800 Rockville Pike
Bethesda, Maryland 20892-0001
301-827-0648

For a BIOLOGICAL THERAPEUTIC product, contact:
Office of Therapeutics Research and Review (HFM-500)
Food and Drug Administration
1451 Rockville Pike
Rockville, Maryland 20852-1420
301-594-2860

Also see these FDA Information Sheets:
"Emergency Use of an Investigational Drug or Biologic"
"Treatment Use of Investigational Drugs"

Drug Study Designs

Before a new drug or biologic can be marketed, its sponsor must show, through adequate and well-controlled clinical studies, that it is effective. A well-controlled study permits a comparison of subjects treated with the new agent with a suitable control population, so that the effect of the new agent can be determined and distinguished from other influences, such as spontaneous change, "placebo" effects, concomitant therapy, or observer expectations. FDA regulations [21 CFR 312.126] cite five different kinds of controls that can be useful in particular circumstances:

(1) placebo concurrent control
(2) dose-comparison concurrent control
(3) no-treatment concurrent control
(4) active-treatment concurrent control, and
(5) historical control

No general preference is expressed for any one type, but the study design chosen must be adequate to the task. Thus, in discussing historical controls, the regulation notes that, because it is relatively difficult to be sure that historical control groups are comparable to the treated subjects with respect to variables that could effect outcome, use of historical control studies has been reserved for special circumstances, notably cases where the disease treated has high and predictable mortality (a large difference from this usual course would be easy to detect) and those in which the effect is self-evident (e.g., a general anesthetic).

Placebo control, no-treatment control (suitable where objective measurements are felt to make blinding unnecessary), and dose-comparison control studies are all study designs in which a difference is intended to be shown between the test article and some control. The alternative study design generally proposed to these kinds of studies is an active-treatment concurrent control in which a finding of no difference between the test article and the recognized effective agent (active-control) would be considered evidence of effectiveness of the new agent. There are circumstances in which this is a fully valid design. Active-controls are usually used in antibiotic trials, for example, because it is easy to tell the difference between antibiotics that have the expected effect on specific infections and those that do not. In many cases, however, the active-control design may be simply incapable of allowing any conclusion as to whether or not the test article is having an effect.

There are three principal difficulties in interpreting active-control trials. First, active-control trials are often too small to show that a clinically meaningful difference between the two treatments, if present, could have been detected with reasonable assurance; i.e., the trials have a high "beta-error." In part, this can be overcome by increasing sample size, but two other problems remain even if studies are large. One problem is that there are numerous ways of conducting a study that can obscure differences between treatments, such as poor diagnostic criteria, poor methods of measurement, poor compliance, medication errors, or poor training of observers. As a general statement, carelessness of all kinds will tend to obscure differences between treatments. Where the objective of a study is to show a difference, investigators have powerful stimuli toward assuring study excellence. Active-control studies, however, which are intended to show no significant difference between treatments, do not provide the same incentives toward study excellence, and it is difficult to detect or assess the kinds of poor study quality that can arise. The other problem is that a finding of no difference between a test article and an effective treatment may not be meaningful. Even where all the incentives toward study excellence are present, i.e., in placebo-controlled trials, effective drugs are not necessarily demonstrably effective (i.e., superior to placebo) every time they are studied. In the absence of a placebo group, a finding of no difference in an active-control study therefore can mean that both agents are effective, that neither agent was effective in that study, or that the study was simply unable to tell effective from ineffective agents. In other words, to draw the conclusion that the test article was effective, one has to know with assurance that the active-control would have shown superior results to a placebo, had a placebo group been included in the study.

For certain drug classes, such as analgesics, antidepressants or antianxiety drugs, failure to show superiority to placebo in a given study is common. This is also often seen with antihypertensives, anti-angina drugs, anti-heart failure treatments, antihistamines, and drugs for asthma prophylaxis. In these situations, active-control trials showing no difference between the new drug and control are of little value as primary evidence of effectiveness and the active-control design (the study design most often proposed as an alternative to use of a placebo) is not credible.

In many situations, deciding whether an active-control design is likely to be a useful basis for providing data for marketing approval is a matter of judgment influenced by available evidence. If, for example, examination of prior studies of a proposed active-control reveals that the test article can very regularly (almost always) be distinguished from placebo in a particular setting (subject population, dose, and other defined parameters), an active-control design may be reasonable if it reproduces the setting in which the active-control has been regularly effective.

It is often possible to design a successful placebo-controlled trial that does not cause investigator discomfort nor raise ethical issues. Treatment periods can be kept short; early "escape" mechanisms can be built into the study so that subjects will not undergo prolonged placebo-treatment if they are not doing well. In some cases randomized placebo-controlled therapy withdrawal studies have been used to minimize exposure to placebo or unsuccessful therapy; in such studies apparent responders to a treatment in an open study are randomly assigned to continued treatment or to placebo. Subjects who fail (e.g., blood pressure rises, angina worsens) can be removed promptly, with such failure representing a study endpoint.

IRBs may face difficult issues in deciding on the acceptability of placebo-controlled and active-control trials. Placebo-controlled trials, regardless of any advantages in interpretation of results, are obviously not ethically acceptable where existing treatment is life-prolonging. A placebo-controlled study that exposes subjects to a documented serious risk is not acceptable, but it is critical to review the evidence that harm would result from denial of active treatment, because alternative study designs, especially active-control studies, may not be informative, exposing subjects to risk but without being able to collect useful information.

For additional information, contact:

For DRUG PRODUCTS:
Drug Information Branch (HFD-211)
Center for Drug Evaluation and Research
Food and Drug Administration
5600 Fishers Lane
Rockville, Maryland 20857
(301) 827- 4573

For a BIOLOGICAL BLOOD product, contact:
Office of Blood Research and Review (HFM-300)
Center for Biologic Evaluation and Research
Food and Drug Administration
1401 Rockville Pike
Rockville, Maryland 20852
301-827-3518

For a BIOLOGICAL VACCINE product, contact:
Office of Vaccines Research and Review (HFM-400)
Food and Drug Administration
8800 Rockville Pike
Bethesda, Maryland 20892-0001
301-827-0648

For a BIOLOGICAL THERAPEUTIC product, contact:
Office of Therapeutics Research and Review (HFM-500)
Food and Drug Administration
1451 Rockville Pike
Rockville, Maryland 20852-1420
301-594-2860

THIS PAGE INTENTIONALLY LEFT BLANK

Evaluation of Gender Differences in Clinical Investigations

FDA Guideline

On July 22, 1993, the FDA published the Guideline for the Study and Evaluation of Gender Differences in the Clinical Evaluation of Drugs, in the Federal Register [58 FR 39406]. The guideline was developed amidst growing concerns that the drug development process did not provide adequate information about the effects of drugs or biological products in women and a general consensus that women should be allowed to determine for themselves the appropriateness of participating in early clinical trials.

Many aspects of the guideline may be important to an Institutional Review Board (IRB) as part of its initial deliberations about protocols and ongoing surveillance of research. While the guideline specifically addresses drug and biologic testing, the Agency suggests that when reviewing medical device studies, IRBs consider whether the principles of the guideline apply to the device under investigation and, if so, whether to include these principles in their review of the protocol. IRBs should be aware that the FDA guideline represents current policy and describes the Agency's expectations regarding the inclusion of subjects in drug development.

The guideline presents the following critical changes that should be reflected in drug and biologic product protocols presented to IRBs:

- First, the guideline lifts a restriction on participation by most women with childbearing potential from entering Phase 1 and early Phase 2 trials, and now encourages their participation. FDA believes that early drug and biologic trials can be safely conducted in women even before completion of all animal reproduction studies through protocol designs that include monitoring for pregnancy as well as measures to prevent pregnancy during exposure to investigational agents. Pregnancy testing is recommended, and women must be counseled about the reliable use of contraception or abstinence from intercourse while participating in the clinical trial. The guideline does not, however, specify the type of contraception to be used because FDA believes that decisions of this nature are best left to the woman in consultation with her health care provider. It is important that investigators have access to gynecologic consultants who can provide information about contraceptives and advice for study participants.

- Second, the guideline states that sponsors should collect gender-related data during research and development and should analyze the data for gender effects in addition to other variables such as age and race. FDA requires sponsors to include a fair representation of both genders as participants in clinical trials so that clinically significant gender-related differences in response can be detected. The guideline also underscores the importance of collecting pharmacokinetics data on demographic differences beginning in the Phase 1 and 2 studies, so that relevant study designs are developed for later trials.

- In addition, the guideline identifies three specific pharmacokinetics issues to be considered when feasible: (1) effect of the stages of the menstrual cycle; (2) effect of exogenous hormonal therapy including oral contraceptives; and (3) effect of the drug or biologic on the pharmacokinetics of oral contraceptives.

Informed Consent Issues

A critical responsibility of the investigator and the IRB has always included ensuring that there is an adequate informed consent process for study subjects. When preclinical teratology and reproductive toxicology studies are not completed prior to the initial studies in humans, male and female study subjects should be informed about lack of full characterization of the test article and the potential effects of the test agent on conception and fetal development. All study subjects should be provided with new pertinent information arising from preclinical studies as it becomes available, and informed consent documents should be updated when appropriate. Study subjects should also be informed about any new clinical data that emerge regarding general safety and effectiveness, including relevant gender effects.

Summary

IRBs now have broader discretion to encourage the entry of a wide range of individuals into the early phases of clinical trials. FDA appreciates the cooperation of IRBs in assisting the Agency to foster changes in product development that will promote the overall health of all people. FDA urges IRBs not to needlessly exclude women or other groups.

Medical Devices

Medical Devices

A medical device is defined, in part, as any health care product that does not achieve its primary intended purposes by chemical action or by being metabolized. Medical devices include, among other things, surgical lasers, wheelchairs, sutures, pacemakers, vascular grafts, intraocular lenses, and orthopedic pins. Medical devices also include diagnostic aids such as reagents and test kits for in vitro diagnosis (IVD) of disease and other medical conditions such as pregnancy.

Clinical investigations of medical devices must comply with the Food and Drug Administration (FDA) informed consent and Institutional Review Board (IRB) regulations [21 CFR parts 50 and 56, respectively]. Federal requirements governing investigations involving medical devices were enacted as part of the Medical Device Amendments of 1976 and the Safe Medical Devices Act of 1990. These amendments to the Federal Food, Drug, and Cosmetic Act (the Act) define the regulatory framework for medical device development, testing, approval, and marketing.

Except for certain low risk devices, each manufacturer who wishes to introduce a new medical device to the market must submit a premarket notification to FDA. FDA reviews these notifications to determine if the new device is "substantially equivalent" to a device that was marketed prior to passage of the Amendments (i.e., a "pre-amendments device"). If the new device is deemed substantially equivalent to a pre-amendments device, it may be marketed immediately and is regulated in the same regulatory class as the pre-amendments device to which it is equivalent. (The premarket notification requirement for new devices and devices that are significant modifications of already marketed devices is set forth in section 510(k) of the Act. Devices determined by FDA to be "substantially equivalent" are often referred to as "510(k) devices". If the new device is deemed not to be substantially equivalent to a pre-amendments device, it must undergo clinical testing and premarket approval before it can be marketed unless it is reclassified into a lower regulatory class.

Investigational Device Exemption (IDE)

An investigational device is a medical device which is the subject of a clinical study designed to evaluate the effectiveness and/or safety of the device. Clinical investigations undertaken to develop safety and effectiveness data for medical devices must be conducted according to the requirements of the IDE regulations [21 CFR part 812]. An IDE study may not necessarily commence 30 days after an IDE submission to FDA. Certain clinical investigations of devices (e.g., certain studies of lawfully marketed devices) may be exempt from the IDE regulations [21 CFR 812.2(c)]. Unless exempt from the IDE regulations, an investigational device must be
categorized as either "significant risk" (SR) or "nonsignificant risk" (NSR). The determination that a device presents a nonsignificant or significant risk is initially made by the sponsor. The proposed study is then submitted either to FDA (for SR studies) or to an IRB (for NSR studies).

The IRB's SR/NSR determination has significant consequences for the study sponsor, FDA, and prospective research subjects. SR device studies must be conducted in accordance with the full IDE requirements [21 CFR part 812], and may not commence until 30 days following the sponsor's submission of an IDE application to FDA. Submission of the IDE application

enables FDA to review information about the technical characteristics of the device, the results of any prior studies (laboratory, animal and human) involving the device, and the proposed study protocol and consent documents. Based upon the review of this information, FDA may impose restrictions on the study to ensure that risks to subjects are minimized and do not outweigh the anticipated benefits to the subjects and the importance of the knowledge to be gained. The study may not commence until FDA has approved the IDE application and the IRB has approved the study.

In contrast, NSR device studies do not require submission of an IDE application to FDA. Instead, the sponsor is required to conduct the study in accordance with the "abbreviated requirements" of the IDE regulations [21 CFR 812.2(b)]. Unless otherwise notified by FDA, an NSR study is considered to have an approved IDE if the sponsor fulfills the abbreviated requirements. The abbreviated requirements address, among other things, the requirements for IRB approval and informed consent, recordkeeping, labeling, promotion, and study monitoring. NSR studies may commence immediately following IRB approval.

IRB Review of the Protocol and Informed Consent

Once the final SR/NSR decision has been rendered by the IRB (or FDA), the IRB must consider whether or not the study should be approved. In considering whether a study should be approved, the IRB should use the same criteria it would use in considering approval of any research involving an FDA regulated product [21 CFR 56.111]. Some NSR studies may also qualify as "minimal risk" studies, and thus may be reviewed through an expedited review procedure [21 CFR 56.110]. FDA considers all SR studies to present more than minimal risk, and thus, full IRB review is necessary. In making its determination on approval, the IRB should consider the risks and benefits of the medical device compared to the risks and benefits of alternative devices or procedures.

Also see these FDA Information Sheets:
"Significant Risk and Nonsignificant Risk Medical Device Studies"
"Sponsor-Investigator-IRB Interrelationship"

Frequently Asked Questions About IRB Review of Medical Devices

1. What is meant by Class I, II and III devices?

The class distinction is made primarily on the level of risk to users/patients and, therefore, the level of FDA oversight needed to ensure that the device is safe and effective as labeled. Generally, but not always, this corresponds to logical risk evaluations.

Class	Controls	Products
Class I	General controls	crutches, band aids
Class II	Special controls	wheelchairs, tampons
Class III	PreMarket Approval	heart valves (known to present hazards requiring clinical demonstration of safety and effectiveness) - OR - not enough known about safety or effectiveness to assign to Class I or II

2. What is the difference between marketing approval under a 510(k) and under a PMA?

A 510(k) application demonstrates that a new device is substantially equivalent to another device that is legally on the market without a PMA. If FDA agrees that the new device is substantially equivalent, it can be marketed. Clinical data are not required in most 510(k) applications; however if clinical data are necessary to demonstrate substantial equivalence, the clinical studies need to be conducted in compliance withthe requirements of the IDE regulations, IRB review and informed consent (21 CFR parts 812, 56 and 50, respectively).

3. Why should an IRB decide whether a device is non-significant risk (NSR)?

The sponsors (usually the manufacturer of the device) makes the initial decision whether a device imparts significant risk (SR) to study subjects or others. If so, the sponsor obtains an Investigational Device Exemption (IDE) from FDA. If the sponsor believes the device does not impart significant risk, IRB approval of a study as an NSR device can be sought. The NSR category was created to avoid delay and expense where the anticipated risk to human subjects did not justify the involvement of FDA. If the IRB agrees that the study is NSR, no submission to or review by FDA is necessary before starting studies in humans. If the IRB considers the study to be SR, the sponsor must obtain an IDE from FDA before proceeding with clinical studies.

4. What does FDA know about an NSR study?

"There is no requirement to report to FDA when an NSR study starts." The requirements for IRB review, informed consent, adverse event reporting and labeling still apply. In addition, the sponsor should understand that proceeding with an NSR study is at their risk (meaning that the FDA can later disagree) and they may voluntarily seek advice or inform FDA about the decision to proceed without filing an IDE with FDA.

5. How does an IRB decide whether a device is SR or NSR?

The IRB uses its best abilities, the information in the regulations and the guidelines, and the risk evaluation provided by the applicant. It can, as always, seek outside assistance. The IRB should have written policies and procedures regarding device review. The information sheet "Significant Risk and Non-Significant Risk Medical Device Studies" provides additional guidance.

6. Does an IRB that reviews medical device studies need written procedures for determining whether the device is SR or NSR?

When the IRB determines that an investigation presented for approval as involving an NSR device actually involves an SR device, 21 CFR 812.66 requires the IRB to so notify the investigator and, where appropriate, the sponsor. 21 CFR 56.108(a)(1) requires the IRB to follow written procedures for conducting its initial review of research and for reporting its findings and actions to the investigator. The procedures followed in determining whether a study is SR or NSR should be included among those written procedures.

7. Does FDA require IRB review of the off-label use of a marketed device?

YES, if the off-label use is part of a research project involving human subjects. NO, if the off-label use is intended to be solely the practice of medicine, i.e., for a physician treating a patient and no research is being done.

8. What is the meaning of exemption in 21 CFR 812.2(c)(2)?

The exemption applies only to investigations in which 510(k)'d products are being used in accordance with the labeling cleared by FDA. Investigation of an off-label use of a 510(k) product takes it outside this exemption. A device subject to 510(k) remains "investigational" until the 510(k) is cleared by FDA and the investigational use is subject to the requirements of the IDE regulation, informed consent and IRB review (21 CFR 812, 50 and 56, respectively).

9. Must an IRB review a clinical investigation being done after submission of a 510(k)?

YES, if it's research the 21 CFR 50 and 56 regulations apply, and an IRB should review it. A 510(k) allows commercial distribution; it doesn't address research use. A 510(k) application can take time to process during which it remains an investigational product. It cannot be distributed except for investigational use until FDA clears the 510(k) application.

Also see these FDA Information Sheets:
"Medical Devices"
"Significant Risk and Nonsignificant Risk Medical Device Studies"
"Emergency Use of Unapproved Medical Devices."

Significant Risk and Nonsignificant Risk Medical Device Studies

The Investigational Device Exemption (IDE) regulations [21 CFR part 812] describe two types of device studies, "significant risk" (SR) and "nonsignificant risk" (NSR). An SR device study is defined [21 CFR 812.3(m)] as a study of a device that presents a potential for serious risk to the health, safety, or welfare of a subject and (1) is intended as an implant; or (2) is used in supporting or sustaining human life; or (3) is of substantial importance in diagnosing, curing, mitigating or treating disease, or otherwise prevents impairment of human health; or (4) otherwise presents a potential for serious risk to the health, safety, or welfare of a subject. An NSR device investigation is one that does not meet the definition for a significant risk study. NSR device studies, however, should not be confused with the concept of "minimal risk," a term utilized in the Institutional Review Board (IRB) regulations [21 CFR part 56] to identify certain studies that may be approved through an "expedited review" procedure. For both SR and NSR device studies, IRB approval prior to conducting clinical trials and continuing review by the IRB are required. In addition, informed consent must be obtained for either type of study [21 CFR part 50].

Distinguishing Between SR and NSR Device Studies

The effect of the SR/NSR decision is very important to research sponsors and investigators. SR device studies are governed by the IDE regulations [21 CFR part 812]. NSR device studies have fewer regulatory controls than SR studies and are governed by the abbreviated requirements [21 CFR 812.2(b)]. The major differences are in the approval process and in the record keeping and reporting requirements. The SR/NSR decision is also important to FDA because the IRB serves, in a sense, as the Agency's surrogate with respect to review and approval of NSR studies. FDA is usually not apprised of the existence of approved NSR studies because sponsors and IRBs are not required to report NSR device study approvals to FDA. If an investigator or a sponsor proposes the initiation of a claimed NSR investigation to an IRB, and if the IRB agrees that the device study is NSR and approves the study, the investigation may begin at that institution immediately, without submission of an IDE application to FDA.

If an IRB believes that a device study is SR, the investigation may not begin until both the IRB and FDA approve the investigation. To help in the determination of the risk status of the device, IRBs should review information such as reports of prior investigations conducted with the device, the proposed investigational plan, a description of subject selection criteria, and monitoring procedures. The sponsor should provide the IRB with a risk assessment and the rationale used in making its risk determination [21 CFR 812.150(b)(10)].

SR/NSR Studies and the IRB
The NSR/SR Decision

The assessment of whether or not a device study presents a NSR is initially made by the sponsor. If the sponsor considers that a study is NSR, the sponsor provides the reviewing IRB an explanation of its determination and any other information that may assist the IRB in evaluating the risk of the study. The sponsor should provide the IRB with a description of the device, reports of prior investigations with the device, the proposed investigational plan, a description of patient selection criteria and monitoring procedures, as well as any other information that the IRB deems necessary to make its decision. The sponsor should inform the IRB whether other IRBs have reviewed the proposed study and what determination was made. The sponsor must inform the IRB of the Agency's assessment of the device's risk if such an assessment has been made. The IRB may also consult with FDA for its opinion.

The IRB may agree or disagree with the sponsor's initial NSR assessment. If the IRB agrees with the sponsor's initial NSR assessment and approves the study, the study may begin without submission of an IDE application to FDA. If the IRB disagrees, the sponsor should notify FDA that an SR determination has been made. The study can be conducted as an SR investigation following FDA approval of an IDE application.

The risk determination should be based on the proposed use of a device in an investigation, and not on the device alone. In deciding if a study poses an SR, an IRB must consider the nature of the harm that may result from use of the device. Studies where the potential harm to subjects could be life-threatening, could result in permanent impairment of a body function or permanent damage to body structure, or could necessitate medical or surgical intervention to preclude permanent impairment of a body function or permanent damage to body structure should be considered SR. Also, if the subject must undergo a procedure as part of the investigational study, e.g., a surgical procedure, the IRB must consider the potential harm that could be caused by the procedure in addition to the potential harm caused by the device. Two examples follow:

> The study of a pacemaker that is a modification of a commercially—available pacemaker poses a SR because the use of any pacemaker presents a potential for serious harm to the subjects. This is true even though the modified pacemaker may pose less risk, or only slightly greater risk, in comparison to the commercially-available model. The amount of potential reduced or increased risk associated with the investigational pacemaker should only be considered (in relation to possible decreased or increased benefits) when assessing whether the study can be approved. The study of an extended wear contact lens is considered SR because wearing the lens continuously overnight while sleeping presents a potential for injuries not normally seen with daily wear lenses, which are considered NSR.

FDA has the ultimate decision in determining if a device study is SR or NSR. If the Agency does not agree with an IRB's decision that a device study presents an NSR, an IDE application must be submitted to FDA. On the other hand, if a sponsor files an IDE with FDA because it is presumed to be an SR study, but FDA classifies the device study as NSR, the Agency will return the IDE application to the sponsor and the study would be presented to IRBs as an NSR investigation.

IRB and Sponsor Responsibilities Following SR/NSR Determination

If the IRB decides the study is Significant Risk:

1. IRB Responsibilities:

Notify sponsor and investigator of SR decision After IDE obtained by sponsor, proceed to review study applying requisite criteria [21 CFR 56.111]

2. Sponsor Responsibilities:

Submit IDE to FDA or, if electing not to proceed with study, notify FDA (CDRH Program Operations Staff 301-594-1190) of the SR determination; Study may not begin until FDA approves IDE and IRB approves the study.Sponsor and investigator(s) must comply with IDE regulations [21 CFR part 812], as well as informed consent and IRB regulations [21 CFR parts 50 and 56].

If the IRB decides the study is Nonsignificant Risk:

1. IRB proceeds to review study applying requisite criteria [21 CFR 56.111]

2. If the study is approved by the IRB, the sponsor and investigator must comply with "abbreviated IDE requirements" [21 CFR 812.2(b)], and informed consent and IRB regulations [21 CFR parts 50 and 56].

The Decision to Approve or Disapprove

Once the SR/NSR decision has been reached, the IRB should consider whether the study should be approved or not. The criteria for deciding if SR and NSR studies should be approved are the same as for any other FDA regulated study [21 CFR 56.111]. The IRB should assure that risks to subjects are minimized and are reasonable in relation to anticipated benefits and knowledge to be gained, subject selection is equitable, informed consent materials and procedures are adequate, and provisions for monitoring the study and protecting the privacy of subjects are acceptable. To assure that the risks to the subject are reasonable in relation to the anticipated benefits, the risks and benefits of the investigation should be compared to the risks and benefits of alternative devices or procedures. This differs from the judgment about whether a study poses a SR or NSR which is based solely upon the seriousness of the harm that may result from the use of the device. Minutes of IRB meetings must document the rationale for SR/NSR and subsequent approval or disapproval decisions for the clinical investigation.

FDA considers studies of all significant risk devices to present more than minimal risk; thus, full IRB review for all studies involving significant risk devices is necessary. Generally, IRB review at a convened meeting is also required when reviewing NSR studies. Some NSR studies, however, may qualify as minimal risk [21 CFR 56.102(i)] and the IRB may choose to review those studies under its expedited review procedures [21 CFR 56.110].

Examples of NSR/SR Devices

The following examples are provided to assist sponsors and IRBs in making SR/NSR determinations. The list includes many commonly used medical devices. Inclusion of a device in the NSR category should not be viewed as a conclusive determination, because the proposed use of a device in a study is the ultimate determinant of the potential risk to subjects. It is unlikely that a device included in the SR category could be deemed NSR due to the inherent risks associated with most such devices.

NONSIGNIFICANT RISK DEVICES

Low Power Lasers for treatment of pain
Caries Removal Solution
Daily Wear Contact Lenses and Associated Lens Care Products not intended for use directly in the eye (e.g., cleaners; disinfecting, rinsing and storage solutions)
Contact Lens Solutions intended for use directly in the eye (e.g., lubricating/rewetting solutions) using active ingredients or preservation systems with a history of prior ophthalmic/contact lens use or generally recognized as safe for ophthalmic use
Conventional Gastroenterology and Urology Endoscopes and/or
Accessories
Conventional General Hospital Catheters (long-term percutaneous, implanted, subcutaneous and intravascular)
Conventional Implantable Vascular Access Devices (Ports)

NONSIGNIFICANT RISK DEVICES

Conventional Laparoscopes, Culdoscopes, and Hysteroscopes
Dental Filling Materials, Cushions or Pads made from traditional materials and designs
Denture Repair Kits and Realigners
Digital Mammography [Note: an IDE is required when safety and effectiveness data are
    collected which will be submitted in support of a marketing application.]
Electroencephalography (e.g., new recording and analysis methods, enhanced diagnostic capabilities)
Externally Worn Monitors for Insulin Reactions
Functional Electrical Neuromuscular Stimulators
General Biliary Catheters General Urological Catheters (e.g., Foley and diagnostic catheters)
Jaundice Monitors for Infants
Magnetic Resonance Imaging (MRI) Devices within FDA specified parameters
Manual Image Guided Surgery
Menstrual Pads (Cotton or Rayon, only)
Menstrual Tampons (Cotton or Rayon, only)
Nonimplantable Electrical Incontinence Devices
Nonimplantable Male Reproductive Aids with no components that enter the vagina
Ob/Gyn Diagnostic Ultrasound within FDA approved parameters
Transcutaneous Electric Nerve Stimulation (TENS) Devices for treatment of pain
Wound Dressings, excluding absorbable hemostatic devices and dressings
    (also excluding Interactive Wound and Burn Dressings)

SIGNIFICANT RISK DEVICES

GENERAL MEDICAL USE

Catheters:
- Urology - urologic with anti-infective coatings
- General Hospital - except for conventional long-term percutaneous, implanted, subcutaneous and intravascular
- Neurological - cerebrovascular, occlusion balloon
- Cardiology - transluminal coronary angioplasty, intra-aortic balloon with control system
- Collagen Implant Material for use in ear, nose and throat, orthopedics, plastic surgery,
- urological and dental applications
- Surgical Lasers for use in various medical specialties
- Tissue Adhesives for use in neurosurgery, gastroenterology, ophthalmology, general and plastic surgery, and cardiology

ANESTHESIOLOGY

Breathing Gas Mixers
Bronchial Tubes
Electroanesthesia Apparatus
Epidural and Spinal Catheters
Epidural and Spinal Needles
Esophageal Obturators
Gas Machines for anesthesia or analgesia
High Frequency Jet Ventilators greater than 150 BPM
Rebreathing Devices
Respiratory Ventilators
Tracheal Tubes

CARDIOVASCULAR

Aortic and Mitral Valvuplasty Catheters
Arterial Embolization Devices Cardiac Assist Devices: artificial heart (permanent implant and
    short term use), cardiomyoplasty devices, intra-aortic balloon pumps, ventricular assist devices
Cardiac Bypass Devices: oxygenators, cardiopulmonary non-roller blood pumps, closed c
    hest devices
Cardiac Pacemaker/Pulse Generators: antitachycardia, esophageal, external transcutaneous,
    implantable
Cardiopulmonary Resuscitation (CPR) Devices
Cardiovascular/Intravascular Filters
Coronary Artery Retroperfusion Systems
Coronary Occluders for ductus arteriosus, atrial and septal defects
Coronary and Peripheral Arthrectomy Devices
Extracorporeal Membrane Oxygenators (ECMO)
Implantable Cardioverters/Defibrillators
Laser Coronary and Peripheral Angioplasty Devices
Myoplasty Laser Catheters
Organ Storage/Transport Units
Pacing Leads
Percutaneous Conduction Tissue Ablation Electrodes
Peripheral, Coronary, Pulmonary, Renal, Vena Caval and Peripheral Stents
Replacement Heart Valves
RF Catheter Ablation and Mapping Systems
Ultrasonic Angioplasty Catheters
Vascular and Arterial Graft Prostheses
Vascular Hemostasis Devices

DENTAL

Absorbable Materials to aid in the healing of periodontal defects and other maxillofacial
    applications
Bone Morphogenic Proteins with and without bone, e.g., Hydroxyapatite (HA)
Dental Lasers for hard tissue applications
Endosseous Implants and associated bone filling and augmentation materials used in
    conjunction with the implants
Subperiosteal Implants
Temporomandibular Joint (TMJ) Prostheses

EAR, NOSE, AND THROAT

Auditory Brainstem Implants
Cochlear Implants
Laryngeal Implants
Total Ossicular Prosthesis Replacements

continued

GASTROENTEROLOGY AND UROLOGY

Anastomosis Devices
Balloon Dilation Catheters for benign prostatic hyperplasia (BPH)
Biliary Stents
Components of Water Treatment Systems for Hemodialysis
Dialysis Delivery Systems
Electrical Stimulation Devices for sperm collection
Embolization Devices for general urological use
Extracorporeal Circulation Systems
Extracorporeal Hyperthermia Systems
Extracorporeal Photopheresis Systems
Femoral, Jugular and Subclavian Catheters
Hemodialyzers
Hemofilters
Implantable Electrical Urinary Incontinence Systems
Implantable Penile Prostheses
Injectable Bulking Agents for incontinence
Lithotripters (e.g., electrohydraulic extracorporeal shock-wave, laser, powered mechanical, ultrasonic)
Mechanical/Hydraulic Urinary Incontinence Devices
Penetrating External Penile Rigidity Devices with components that enter the vagina
Peritoneal Dialysis Devices
Peritoneal Shunt
Plasmapheresis Systems
Prostatic Hyperthermia Devices
Urethral Occlusion Devices
Urethral Sphincter Prostheses
Urological Stents (e.g., ureteral, prostatG)

GENERAL AND PLASTIC SURGERY

Absorbable Adhesion Barrier Devices
Absorbable Hemostatic Agents
Artificial Skin and Interactive Wound and Burn Dressings
Injectable Collagen
Implantable Craniofacial Prostheses
Repeat Access Devices for surgical procedures
Sutures

GENERAL HOSPITAL

Implantable Vascular Access Devices (Ports) - if new routes of administration or new design
Infusion Pumps (implantable and closed-loop - depending on the infused drug)

NEUROLOGICAL

Electroconvulsive Therapy (ECT) Devices
Hydrocephalus Shunts
Implanted Intracerebral/Subcortical Stimulators
Implanted Intracranial Pressure Monitors
Implanted Spinal Cord and Nerve Stimulators and Electrodes

FDA Information Sheets

OBSTETRICS AND GYNECOLOGY

Antepartum Home Monitors for Non-Stress Tests
Antepartum Home Uterine Activity Monitors
Catheters for Chorionic Villus Sampling (CVS)
Catheters Introduced into the Fallopian Tubes
Cervical Dilation Devices

Contraceptive Devices:

- Cervical Caps
- Condoms (for men) made from new materials (e.g., polyurethane)
- Contraceptive In Vitro Diagnostics (IVDs)
- Diaphragms
- Female Condoms
- Intrauterine Devices (IUDs)
- New Electrosurgical Instruments for Tubal Coagulation
- New Devices for Occlusion of the Vas Deferens
- Sponges
- Tubal Occlusion Devices (Bands or Clips)

Devices to Prevent Post-op Pelvic Adhesions
Embryoscopes and Devices intended for fetal surgery
Falloposcopes and Falloposcopic Delivery Systems
Intrapartum Fetal Monitors using new physiological markers
New Devices to Facilitate Assisted Vaginal Delivery
Thermal Systems for Endometrial Ablation

OPHTHALMICS

Class III Ophthalmic Lasers
Contact Lens Solutions intended for direct instillation (e.g., lubrication/rewetting solutions) in
    the eye using new active agents or preservatives with no history of priorophthalmic/contact
    lens use or not generally recognized as safe for ophthalmic use
Corneal Implants
Corneal Storage Media
Epikeratophakia Lenticules
Extended Wear Contact Lens
Eye Valve Implants (glaucoma implant)
Intraocular Lenses (IOLs) [21 CFR part 813]
Keratoprostheses Retinal Reattachment Systems: fluids, gases, perfluorocarbons,
    perfluorpropane, silicone oil, sulfur hexafluoride, tacks
Viscosurgical Fluids

ORTHOPEDICS AND RESTORATIVE

Bone Growth Stimulators
Calcium Tri-Phosphate Hydroxyapatite
Ceramics Collagen and Bone Morphogenic Protein Meniscus Replacements
Implantable Prostheses (ligament, tendon, hip, knee, finger)
Computer Guided Robotic Surgery

RADIOLOGY

Boron Neutron Capture Therapy
Hyperthermia Systems and Applicators

Your comments and suggestions for additional examples are welcome and should be sent to:

Program Operation Staff (HFZ-403)
Office of Device Evaluation
Center for Devices and Radiological Health
Food and Drug Administration
9200 Corporate Blvd.
Rockville, MD 20850
(301) 594-1190

Emergency Use of Unapproved Medical Devices

For the purpose of this information sheet, an unapproved medical device is defined as a device that is used for a purpose or condition for which the device requires, but does not have, an approved application for premarket approval under section 515 of the Federal Food, Drug, and Cosmetic Act [21 U.S.C. 360(e)]. An unapproved device may be used in human subjects only if it is approved for clinical testing under an approved application for an Investigational Device Exemption (IDE) under section 520(g) of the Act [21 U.S.C. 360(j)(g)] and 21 CFR part 812. Medical devices that have not received marketing clearance under section 510(k) of the FD&C Act are also considered unapproved devices which require an IDE.

The Food and Drug Administration (FDA) recognizes that emergencies arise where an unapproved device may offer the only possible life-saving alternative, but an IDE for the device does not exist, or the proposed use is not approved under an existing IDE, or the physician or institution is not approved under the IDE. Using its enforcement discretion, FDA has not objected if a physician chooses to use an unapproved device in such an emergency, provided that the physician later justifies to FDA that an emergency actually existed.

Requirements for Emergency Use

Each of the following conditions must exist to justify emergency use:
1. the patient is in a life-threatening condition that needs immediate treatment;
2. no generally acceptable alternative for treating the patient is available; and
3. because of the immediate need to use the device, there is no time to use existing procedures to get FDA approval for the use.

FDA expects the physician to determine whether these criteria have been met, to assess the potential for benefits from the unapproved use of the device, and to have substantial reason to believe that benefits will exist. The physician may not conclude that an "emergency" exists in advance of the time when treatment may be needed based solely on the expectation that IDE approval procedures may require more time than is available. Physicians should be aware that FDA expects them to exercise reasonable foresight with respect to potential emergencies and to make appropriate arrangements under the IDE procedures far enough in advance to avoid creating a situation in which such arrangements are impracticable.

In the event that a device is to be used in circumstances meeting the criteria listed above, the device developer should notify the Center for Devices and Radiological Health (CDRH), Program Operation Staff by telephone (301-594-1190) immediately after shipment is made. [Note: an unapproved device may not be shipped in anticipation of an emergency.] Nights and weekends, contact the Division of Emergency and Epidemiological Operations (202-857-8400).

FDA would expect the physician to follow as many subject protection procedures as possible. These include:

1. obtaining an independent assessment by an uninvolved physician;
2. obtaining informed consent from the patient or a legal representative;
3. notifying institutional officials as specified by institutional policies;
4. notifying the Institutional Review Board (IRB); and
5. obtaining authorization from the IDE holder, if an approved IDE for the device exists.

After-use Procedures

After an unapproved device is used in an emergency, the physician should:

1. report to the IRB within five days [21 CFR 56.104(c)] and otherwise comply with provisions of the IRB regulations [21 CFR part 56];
2. evaluate the likelihood of a similar need for the device occurring again, and if future use is likely, immediately initiate efforts to obtain IRB approval and an approved IDE for the device's subsequent use; and
3. if an IDE for the use does exist, notify the sponsor of the emergency use, or if an IDE does not exist, notify FDA of the emergency use (CDRH Program Operation Staff 301-594-1190) and provide FDA with a written summary of the conditions constituting the emergency, subject protection measures, and results.

Subsequent emergency use of the device may not occur unless the physician or another person obtains approval of an IDE for the device and its use. If an IDE application for subsequent use has been filed with FDA and FDA disapproves the IDE application, the device may not be used even if the circumstances constituting an emergency exist. Developers of devices that could be used in emergencies should anticipate the likelihood of emergency use and should obtain an approved IDE for such uses.

Exception From Informed Consent Requirement

Even for an emergency use, the investigator is required to obtain informed consent of the subject or the subject's legally authorized representative unless both the investigator and a physician who is not otherwise participating in the clinical investigation certify in writing all of the following [21 CFR 50.23(a)]:

(1) The subject is confronted by a life-threatening situation necessitating the use of the test article.
(2) Informed consent cannot be obtained because of an inability to communicate with, or obtain legally effective consent from, the subject.
(3) Time is not sufficient to obtain consent from the subject's legal representative.
(4) No alternative method of approved or generally recognized therapy is available that provides an equal or greater likelihood of saving the subject's life.

If, in the investigator's opinion, immediate use of the test article is required to preserve the subject's life, and if time is not sufficient to obtain an independent physician's determination that the four conditions above apply, the clinical investigator should make the determination and, within 5 working days after the use of the article, have the determination reviewed and evaluated in writing by a physician who is not participating in the clinical investigation. The investigator must notify the IRB within 5 working days after the use of the test article [21 CFR 50.23(c)].

Exception from Informed Consent for Planned Emergency Research

The conduct of planned research in life-threatening emergent situations where obtaining prospective informed consent has been waived, is provided by 21 CFR 50.24. The research plan must be approved in advance by FDA and the IRB, and publicly disclosed to the community in which the research will be conducted. Such studies are usually not eligible for the emergency approvals described above. The information sheet "Exception from Informed Consent for Studies Conducted in Emergency Settings: Regulatory Language and Excerpts from Preamble," is a compilation of the wording of 21 CFR 50.24 and pertinent portions of the preamble from the October 2, 1996 Federal Register.

Also see these FDA Information Sheets:
"Exception from Informed Consent for Studies Conducted in Emergency Settings: Regulatory Language and Excerpts from Preamble"
"Emergency Use of an Investigational Drug or Biologic"

THIS PAGE INTENTIONALLY LEFT BLANK

FDA Operations

FDA Institutional Review Board Inspections

Background

Since 1971, FDA regulations have required that studies involving investigational new drugs and biologics performed on human subjects in institutions (including hospitals, nursing homes, mental institutions, and prisons) receive review and approval by an Institutional Review Board (IRB). Medical devices have required IRB review since 1976.

FDA developed the Bioresearch Monitoring Program and began an expanded review of IRB operations in April 1977. The Bioresearch Monitoring Program, which encompasses not only IRBs, but also clinical investigators, research sponsors, monitors, and non-clinical (animal) laboratories, is primarily intended to ensure the quality and integrity of data submitted to FDA for regulatory decisions, as well as to protect human subjects of research. For this reason, the IRB regulations note that FDA may inspect IRBs and review and copy IRB records [21 CFR 56.115(b)].

IRB Review Program

Under the Bioresearch Monitoring Program, FDA conducts on-site procedural reviews of IRBs. These reviews are conducted to determine whether an IRB is operating in accordance with its own written procedures as well as in compliance with current FDA regulations affecting IRBs. These regulations include 21 CFR part 50 (Informed Consent), part 56 (Standards for IRBs), part 312 (Investigational New Drugs), and part 812 (Investigational Devices).

When an IRB is selected for a procedural review, an investigator from one of the Agency's District Offices will contact a responsible individual at the institution, usually the IRB chairperson, and arrange a mutually acceptable time for the visit. When the field investigators arrive at the institution, they will show FDA credentials (photo ID) and present a "Notice of Inspection" form to the responsible official. This is done simply to let those persons at the institution know that the investigators are duly authorized representatives of FDA conducting official business. The investigator will interview appropriate persons and obtain information about the IRB's policies and procedures. Then, using one or more studies which are subject to FDA regulations, the investigator will examine the IRB's performance by tracking these studies through the review process used by the IRB. The IRB procedures and membership rosters will be examined to see whether they conform to current Agency regulations. The FDA investigator may request copies of records of IRB membership, IRB procedures and guidelines, minutes of meetings at which the studies were reviewed and discussed, material on the studies submitted by the clinical investigator to the IRB, and any other materials pertaining to these studies. Copies of these materials become part of the field investigator's report to FDA Headquarters.

After the inspection has been completed, the investigator will conduct an "exit interview" with a responsible institutional representative and/or the IRB chairperson. At this interview, the investigator will review the findings, clarify any misunderstandings that might exist, describe any deviations from the current regulations, and may suggest corrective actions. A written Form FDA-483 (Notice of Observations) may be left with the institution.

After the investigator returns to the District Office, a written report is prepared. This report is forwarded to FDA Headquarters for evaluation. When the evaluation is completed, a letter may be sent to the IRB chairperson or other responsible institutional official. If the regulations have not been followed, the letter may suggest methods to achieve compliance and ask the IRB to correct its procedures. If serious deviations were observed, a written response assuring adequate correction is usually required. A follow-up inspection may be also conducted. FDA may take administrative actions against IRBs and/or their institutions for noncompliance with the regulations [21 CFR part 56 subpart E].

Additional Information

A copy of the FDA Compliance Program Guidance Manual for IRB Inspections (Program 7348.809) is available to the public by writing to:

Freedom of Information Staff (HFI-35)
Food and Drug Administration
5600 Fishers Lane
Rockville, Maryland 20857

Contact Person for Inspection Problems

If, during the course of an inspection, questions arise that the FDA field investigator has not answered, the Director of the District Office may be contacted. The name and telephone number of the District Director is available from the field investigator and is also on the Notice of Inspection (Form FDA-482).

FDA Inspections of Clinical Investigators

Background

The FDA Bioresearch Monitoring Program involves site visits to clinical investigators, research sponsors, contract research organizations, Institutional Review Boards (IRBs), and nonclinical(animal) laboratories. All FDA product areas, i.e., drugs, biologics, medical devices, radiological products, foods, and veterinary drugs, are involved in the Bioresearch Monitoring Program. While program procedures differ slightly depending upon product type, all inspections have as their objective ensuring the quality and integrity of data and information submitted to FDA as well as the protection of human research subjects.

Clinical Investigator Inspection Programs

FDA carries out three distinct types of clinical investigator inspections: (1) study-oriented inspections; (2) investigator-oriented inspections; and (3) bioequivalence study inspections. Bioequivalence study inspections are conducted because one study may be the sole basis for a drug's marketing approval. The bioequivalence study inspection differs from the other inspections in that it requires participation by an FDA chemist or an investigator knowledgeable about analytical evaluations. The other two types of inspections are discussed in more detail below.

Study-oriented Inspections

FDA field offices conduct study-oriented inspections on the basis of assignments developed by headquarters staff. Assignments are based almost exclusively on studies that are important to product evaluation, such as new drug applications and product license applications pending before the Agency.

When a clinical investigator, who has participated in the study being examined, is selected for an inspection, the FDA investigator from the FDA District Office will contact the clinical investigator to arrange a mutually acceptable time for the visit. Upon arrival, the FDA investigator will show FDA credentials (photo ID) and present a "Notice of Inspection" form to the clinical investigator. FDA credentials let the clinical investigator know that the FDA investigator is a duly authorized FDA representative.

If, during the course of an FDA inspection, a clinical investigator has any questions that the FDA investigator has not answered, either the Director of the District Office or the Center that initiated the inspection may be contacted. The name and telephone number of the District Director and the specific Center contact person are available from the FDA investigator.

The investigation consists of two basic parts. First, determining the facts surrounding the conduct of the study:

- who did what,
- the degree of delegation of authority,
- where specific aspects of the study were performed,
- how and where data were recorded,
- how test article accountability was maintained,
- how the monitor communicated with the clinical investigator, and
- how the monitor evaluated the study's progress.

Second, the study data is audited. The FDA investigator compares the data submitted to the Agency and/or the sponsor with all available records that might support the data. These records may come from the physician's office, hospital, nursing home, laboratories and other sources. FDA may also examine patient records that predate the study to determine whether the medical condition being studied was, in fact, properly diagnosed and whether a possibly interfering medication had been given before the study began. The FDA investigator may also review records covering a reasonable period after completion of the study to determine if there was proper follow-up, and if all signs and symptoms reasonably attributable to the product's use had been reported.

Investigator-oriented Inspections

An investigator-oriented inspection may be initiated because an investigator conducted a pivotal study that merits in-depth examination because of its singular importance in product approval or its effect on medical practice. An inspection may also be initiated because representatives of the sponsor have reported to FDA that they are having difficulty getting case reports from the investigator, or that they have some other concern with the investigator's work. In addition, the Agency may initiate an inspection, if a subject in a study complains about protocol or subject rights violations. Investigator-oriented inspections may also be initiated because clinical investigators have participated in a large number of studies or have done work outside their specialty areas. Other reasons include safety or effectiveness findings that are inconsistent with those of other investigators studying the same test article; too many subjects with a specific disease given the locale of the investigation are claimed; or laboratory results that are outside the range of expected biological variation.

Once the Agency has determined that a investigator-oriented inspection should be conducted, the procedures are essentially the same as in the study-oriented inspection except that the data audit goes into greater depth, covers more case reports, and may cover more than one study. If the investigator has repeatedly or deliberately violated FDA regulations or has submitted false information to the sponsor in a required report, FDA will initiate actions that may ultimately determine that the clinical investigator is not to receive investigational products in the future.

Inspection Findings

At the end of an inspection, the FDA investigator will conduct an "exit interview" with the clinical investigator. At this interview, the FDA investigator will discuss the findings from the inspection, clarify any misunderstandings that might exist, and may issue a written Form FDA-483 (Inspectional Observations) to the clinical investigator. Following the inspection, the FDA field investigator prepares a written report and submits it to headquarters for evaluation.

After the report has been evaluated, FDA headquarters usually issues a letter to the clinical investigator. The letter is usually one of three types:

(1) a notice that no significant deviations from the regulations were observed. This letter does not require any response from the clinical investigator.
(2) an informational letter that identifies deviations from regulations and good investigational practice. This letter may, or may not require a response from the clinical investigator. If a response is requested, the letter will describe what is necessary and give a contact person for questions.

(3) a "Warning Letter" identifying serious deviations from regulations requiring prompt correction by the clinical investigator. The letter will give a contact person for questions. In these cases, FDA may inform both the study sponsor and the reviewing IRB of the deficiencies. The Agency may also inform the sponsor if the clinical investigator's procedural deficiencies indicate ineffective monitoring by the sponsor. In addition to issuing these letters, FDA may take other courses of action, i.e., regulatory and/or administrative sanctions.

Additional Information

A copy of the FDA Compliance Program Guidance Manual for Clinical Investigator Inspections (Program 7348.811), the document used by the FDA investigator to conduct the inspection, is available by writing to:

Freedom of Information Staff (HFI-35)
Food and Drug Administration
5600 Fishers Lane
Rockville, Maryland 20857.

Also see this FDA Information Sheet
"Clinical Investigator Regulatory Sanctions"

THIS PAGE INTENTIONALLY LEFT BLANK

Clinical Investigator Regulatory Sanctions

This information sheet focuses on the applicability of regulatory sanctions to clinical investigators participating in studies involving investigational new drugs, antibiotics, biologics, medical devices, medical foods or food additives. [Note: Although this information sheet refers to human subjects in the context of an Investigational New Drug Application (IND), analogous principles apply to animal subjects in an Investigational New Animal Drug Application (INAD).]

Regulations do not state that hearings will be held at FDA headquarters. Investigators may suggest another location. Hearings can be denied if investigators fail to submit any information that raises any question of fact.

The Disqualification Process

Informal Conference or Written Explanation

FDA may disqualify clinical investigators from receiving investigational drugs, biologics and devices only when the investigator has repeatedly or deliberately violated the Agency's regulations, or has submitted false information to the sponsor in a required report. The appropriate FDA Center will send the investigator a written notice describing the noncompliance or false submission and offer the investigator an opportunity to respond to the notice at an informal conference or in writing. The Agency will specify a time period within which the investigator must respond. While the conference is informal, a transcript may be made, and the investigator may have legal representation.

If the investigator offers a timely and satisfactory explanation for the noncompliance, and the Center accepts, the process is terminated and the investigator is so notified in writing. If, however, the investigator offers an explanation that the Center rejects, or if the investigator fails to respond within the specified time period, FDA will offer the investigator an opportunity for an informal regulatory "Part 16" hearing under the Agency's regulations [21 CFR part 16] to determine whether the investigator should remain eligible to receive investigational test articles.

Notice of an Opportunity for Hearing on Proposed Disqualifications

FDA initiates a Part 16 hearing when it sends the investigator a written Notice of Opportunity for Hearing. The Notice specifies the allegations and other relevant information that is the subject of the hearing. If the investigator does not respond within the time period specified in the letter, FDA considers the offer to have been refused, and no informal hearing will be held.The Commissioner will then consider the information available to FDA to determine whether the investigator should be disqualified.

If a hearing is requested, the Commissioner will designate a presiding officer from the Office of Health Affairs (OHA), and the hearing will take place at a mutually agreeable time at FDA headquarters. If agreement cannot be reached, however, the presiding officer will designate a hearing date acceptable to FDA.

Part 16 Hearing and Final Order on Disqualification

Before the hearing, FDA gives the investigator notice of the matters to be considered at the hearing which includes a comprehensive statement of the basis for the proposal to disqualify the investigator and a general summary of the information that the Center will present. The

Center and the investigator exchange written notices of any published articles or written information to be presented or relied upon at the hearing. If it seems unreasonable to expect the other party to have, or to be able to obtain, a copy of a particular document, a copy of the document is provided. The investigator or the Center may each file a motion for summary decision.

Part 16 hearings are informal, and the rules of evidence do not apply. Any participant may comment upon or rebut all data, information, and views presented. The presiding officer conducts the hearing. The hearing begins with Center staff giving a complete statement of the action that is the subject of the hearing and describing the information and reasons supporting disqualification. They may present any oral or written information relevant to the hearing. The investigator, who may be represented by legal counsel, then may present any oral or written information relevant to the hearing.

After the hearing, the OHA presiding officer prepares a written report. This report includes a recommended decision and the reasons for the recommendation. The administrative record of the hearing includes all written material presented at the hearing and the hearing transcript. The parties are given the opportunity to review and comment on the presiding officer's report. The report and the comments of the parties are transmitted to the Commissioner who considers them along with the administrative record to determine whether the investigator should be disqualified. The Commissioner issues a written decision giving the basis for the action taken.

Actions Upon Disqualification

If the Commissioner determines that the investigator has repeatedly or deliberately failed to comply with the regulatory requirements, or has deliberately or repeatedly submitted false information to the sponsor in any required report, the Commissioner will:

(1) Notify the investigator and the sponsor(s) of any investigation(s) in which the investigator has participated that the investigator is not entitled to receive investigational drugs, biologics or devices. The notification will include a statement explaining the basis for this determination.

(2) Notify the sponsors of studies conducted under each IND, IDE or each approved application containing data reported by the investigator that the Agency will not accept the investigator's work in support of claims of safety and efficacy without validating information establishing that the study results were unaffected by the investigator's misconduct.

(3) After the investigator's data are eliminated from consideration, determine whether the data remaining can support a conclusion that studies under the IND or IDE may continue. If the Commissioner determines that the remaining data are inadequate, the sponsor will be notified and will have an opportunity for a regulatory hearing under 21 CFR part 16. If a danger to public health exists, however, the Commissioner will terminate the IND or IDE immediately and notify the sponsor of the determination. The sponsor will then have an opportunity for a Part 16 regulatory hearing to determine whether the IND or IDE should be reinstated.

(4) After the investigator's data are eliminated from consideration, determine whether the continued approval of the product is justified. If it is not, the Commissioner will move to withdraw approval in accordance with the applicable provisions of the Federal Food, Drug, and Cosmetic Act.

The action to be taken with regard to an ongoing clinical investigation conducted by a disqualified investigator is made on a case-by-case basis. FDA considers the nature of the clinical investigation, the number of subjects involved, the risks to the subjects from discontinuation of the study, and the need for involvement of an acceptable investigator. If another investigator accepts responsibility for the investigation, FDA may allow an investigation to continue. If not, further use of the test article is deferred until another investigator is identified. If this deferment could create a life-threatening situation, FDA may permit a subject to continue to receive or use a test article without a further written statement from the disqualified investigator. The investigator can bring such cases to the Agency's attention during the regulatory hearing, so that the Commissioner may consider this option.

Public Disclosure of Information Regarding Disqualification

A danger to the public health includes not only the subjects' safety in any study in question, but also the safety of subjects in other studies in which the investigator is involved.

The Notice of Opportunity for a Hearing letter is available under the Freedom of Information provisions but is not placed on public display. FDA will notify other government agencies of a proposed disqualification whenever the Agency deems such notification to be appropriate.

If the Agency notifies other parties of its preliminary findings prior to final disqualification, FDA will provide a description of these findings, state that the Agency has yet to reach a final decision on whether the investigator should be disqualified, and will not recommend that action be taken by the third party. If the disqualification proceeding does not result in a disqualification or a consent agreement, FDA will so advise those third parties that had been contacted. A copy of each notification will be sent to the investigator.

If the Agency gives notice of the disqualification of a specific investigator to a third party, FDA will provide a copy of the final disqualification order, explain its legal meaning, and state that FDA is not advising or recommending that the person notified take any action upon the matter. A copy of each notification will be sent to the investigator. The list of investigators who are ineligible to receive investigational new drugs, biologics and devices or who have agreed to some restriction of use of investigational drugs, biologics and devices (see below) is not considered to be a "notice" as discussed above.

Reinstatement of a Disqualified Investigator

Investigators who have been disqualified may be reinstated if the Commissioner determines that the investigators have presented adequate assurances that they will employ investigational drugs, biologics and devices in compliance with FDA regulations. The Agency's reinstatement guidelines, entitled "Procedures for Reinstating Eligibility of Disqualified Clinical Investigators to Receive Investigational Articles" are available by writing to the FOI Staff at the address given below.

Consent Agreements

In addition to an opportunity for an informal conference or to respond in writing to Center allegations, the Center for Drug Evaluation and Research, the Center for Biologics Evaluation and Research, and the Center for Devices and Radiological Health offer investigators the opportunity to enter into a consent agreement whereby the investigator agrees to meet certain conditions mutually acceptable to both FDA and the investigator. This agreement obviates the need to proceed further with the disqualification process. Consent agreements generally take one of two forms: (1) the individual agrees to refrain from further studies with FDA regulated test articles or (2) the individual agrees to specific restrictions in the use of investigational products, such as oversight by an individual acceptable to both the investigator and to the Agency. The consent agreement option remains available to the clinical investigator at all stages of the disqualification process. Most actions have been settled by consent agreements.

Criminal Prosecutions

After a Part 16 proceeding, a final order or entry into a consent agreement constitutes final Agency administrative action. This, however, does not preclude institution of criminal proceedings against an investigator. Those investigators referred for criminal prosecution are generally clinical investigators who have knowingly or willingly submitted false information to a research sponsor.

Additional Information

FDA maintains a list of investigators who are ineligible to receive FDA regulated test articles or who have agreed to some restriction of use of FDA regulated test articles. This list is regularly updated and is not considered to be a "notice" of disqualification (see above). The list is available to the public by writing to the following FDA office.

Freedom of Information Staff (HFI-35)
Food and Drug Administration
5600 Fishers Lane
Rockville, MD 20857.

The list is also available on the internet at:
www.fda.gov/ora/compliance_ref/bimo/dis_res_assur.htm

Appendix A

A List of Selected FDA Regulations
<u>Relating to the Protection of Human Subjects</u>

This list contains Food and Drug Administration (FDA) regulations that specifically relate to the protection of human subjects in clinical investigations. The citations selected below are only a few of the FDA regulations (contained in nine volumes) that apply to clinical investigations and govern the development and approval of drugs, biologics, and devices. The regulations are contained in Title 21 of the Code of Federal Regulations (CFR), which can be purchased from the Superintendent of Documents, Attn: New Orders, P.O. Box 371954, Pittsburgh, PA 15250-7954; (202-512-1800, fax: 202-512-2233)

I. FDA HUMAN SUBJECT PROTECTIONS

Part 50 - Protection of Human Subjects (Informed Consent)
Part 56 - Institutional Review Boards

II. SUBSTANCES AND ARTICLES REGULATED BY FDA

<u>Foods</u>
Part 71 - Color Additives
Part 171 - Food Additive Petitions
Part 180 - Food Additives (Interim)

<u>Drugs</u>
Part 312 - Investigational New Drug Application
Part 314 - New Drug Applications
Part 320 - Bioavailability and Bioequivalence Requirements
Part 330 - Over-the-Counter Human Drugs
Part 361.1 - Radioactive Drugs for Certain Research Uses

<u>Biologics</u>
Part 312 - Investigational New Drug Application
Part 601 - Licensing
Part 630 - Additional Standards for Viral Vaccines

<u>Medical Devices</u>
Part 812 - Investigational Device Exemptions
Part 814 - Premarket Approval of Medical Devices

<u>Radiological Health</u>
Part 361.1 - Radioactive Drugs for Certain Research Uses
Part 1010 - Performance Standards for Electronic Products

III. RELATED FDA PROCEDURES

Part 10 - General Agency Administrative Procedures
Part 16 - Regulatory Hearings before the FDA
Part 20 - Public Information

IV. STATUTES PROVIDING AUTHORITY FOR REGULATIONS LISTED ABOVE:

Biological Control Act of 1902/Virus, Serum and Toxin Act of 1902
Food, Drug and Cosmetic Act of 1938 (as amended)
Public Health Service Act of 1944 (as amended)
Food Additive Amendments of 1958
Color Additives Amendment of 1960
New Drug Amendments of 1962
Radiation Control for Public Health and Safety Act of 1968
National Research Act of 1974
Medical Device Amendments of 1976
Safe Medical Devices Act of 1990
Device Amendments of 1992
FDA Modernization Act of 1997

Appendix B and Appendix C have been ommitted in order to reduce duplication Please see Appendix A

Appendix D

Clinical Investigations Which May Be Reviewed Through Expedited Review Procedures Set Forth in FDA Regulations

This notice contains a list of research activities which institutional review boards may review through the expedited review procedures set forth in FDA regulations for the protection of human research subjects. This list will be amended as appropriate and current list will be published periodically in the Federal Register. Research activities with human subjects involving no more than minimal risk and involving one or more of the following categories (carried out through standard methods), may be reviewed by an IRB through the expedited review procedures authorized in 21 CFR 56.110.

(1) Collection of hair and nail clippings in a non-disfiguring manner; of deciduous teeth; and of permanent teeth if patient care indicates a need for extraction.

(2) Collection of excreta and external secretions including sweet and uncannulated saliva; of placenta at delivery; and of amniotic fluid at the time of rupture of the membrane before or during labor.

(3) Recording of data from subjects who are 18 years of age or older using non-invasive procedures routinely employed in clinical practice. This category includes the use of physical sensors that are applied either to the surface of the body or at a distance and do not involve input of matter or significant amounts of energy into the subject or an invasion of the subject's privacy. It also includes such procedures as weighing, electrocardiography, electroencephalography, thermography detection of naturally occurring radioactivity, diagnostic echography, and electroretinography. This category does not include exposure to electromagnetic radiation outside the visible range (for example, x-rays or microwaves).

(4) Collection of blood samples by venipuncture, in amounts not exceeding 450 milliliters in an eight week period and no more often than two times per week from subjects who are 18 years of age or older and who are in good health and not pregnant.

(5) Collection of both supra- and subgingival dental plague and calculus, provided the procedure is not more invasive than routine prophylactic scaling of the teeth, and the process is accomplished in accordance with accepted prophylactic techniques.

(6) Voice recordings made for research purposes such as investigations of speech defects.

(7) Moderate exercise by healthy volunteers.

(8) The study of existing data, documents, records, pathological specimens, or diagnostic specimens.

(9) Research on drugs or devices for which an investigational new drug exemption or an investigational device exemption is not required.

[Federal Register Vol 46, No. 17 Tuesday, January 27, 1981, 46 FR 8960]

Appendix E

Significant Differences in FDA and HHS Regulations for Protection of Human Subjects

The Department of Health and Human Services (HHS) regulations [45 CFR part 46] apply to research involving human subjects conducted by the HHS or funded in whole or in part by the HHS. The Food and Drug Administration (FDA) regulations [21 CFR parts 50 and 56] apply to research involving products regulated by the FDA. Federal support is not necessary for the FDA regulations to be applicable. When research involving products regulated by the FDA is funded, supported or conducted by FDA and/or HHS, both the HHS and FDA regulations apply.

IRB Regulations

Section Numbers	Description
ß 56.102 (FDA) ß 46.102 (HHS)	FDA definitions are included for terms specific to the type of research covered by the FDA regulations (test article, application for research or marketing permit, clinical investigation). A definition for emergency use is provided in the FDA regulations.
ß 56.104 (FDA) ß 46.116 (HHS)	FDA provides exemption from the prospective IRB review requirement for "emergency use" of test article in specific situations. HHS regulations state that they are not intended to limit the provision of emergency medical care.
ß 56.105 (FDA) ß 46.101 (HHS)	FDA provides for sponsors and sponsor-investigators to request a waiver of IRB review requirements (but not informed consent requirements). HHS exempts certain categories of research and provides for a Secretarial waiver.
ß 56.109 (FDA) ß 46.109 (HHS) ß 46.117(c)(HHS)	Unlike HHS, FDA does not provide that an IRB may waive the requirement for signed consent when the principal risk is a breach of confidentiality because FDA does not regulate studies which would fall into that category of research. (Both regulations allow for IRB waiver of documentation of informed consent in instances of minimal risk.)
ß 56.110 (FDA) ß 46.110 (HHS)	The FDA list of investigations eligible for expedited review (published in the Federal Register) does not include the studies described in category 9 of the HHS list because these types of studies are not regulated by FDA

ß 56.114 (FDA) ß 46.114 (HHS)	FDA does not discuss administrative matters dealing with grants and contracts because they are irrelevant to the scope of the Agency's regulation. (Both regulations make allowances for review of multi-institutional studies.)
ß 56.115 (FDA) ß 46.115 (HHS)	FDA has neither an assurance mechanism nor files of IRB membership. Therefore, FDA does not require the IRB or institution to report changes in membership whereas HHS does require such notification.
ß 56.115(c) (FDA)	FDA may refuse to consider a study in support of a research or marketing permit if the IRB or the institution refuses to allow FDA to inspect IRB records. HHS has no such provision because it does not issue research or marketing permits.
ß 56.120 —— ß 56.124 (FDA)	FDA regulations provide sanctions for non-compliance with regulations.

Informed Consent Regulations

Section Numbers	Description
ß 50.23 (FDA)	FDA, but not HHS, provides for an exception from the informed consent requirements in emergency situations. The provision is based on the Medical Device Amendments of 1976, but may be used in investigations involving drugs, devices, and other FDA regulated products in situations described in ß 50.23.
ß 46.116(c)&(d) (HHS)	HHS provides for waiving or altering elements of informed consent under certain conditions. FDA has no such provision because the types of studies which would qualify for such waivers are either not regulated by FDA or are covered by the emergency treatment provisions (ß 50.23).
ß 50.25(a)(5) (FDA) ß 46.116(a)(5) (HHS)	FDA explicitly requires that subjects be informed that FDA may inspect the records of the study because FDA may occasionally examine a subject's medical records when they pertain to the study. While HHS has the right to inspect records of studies it funds, it does not impose that same informed consent requirement.
ß 50.27(a)	FDA explicitly requires that consent forms be dated as well as signed by the subject or the subject's legally authorized representative. The HHS regulations do not explicitly require consent forms to be dated.

Appendix F

The Belmont Report

Ethical Principles and Guidelines for
the Protection of Human Subjects of Research

The National Commission for the Protection
Of Human Subjects of Biomedical and Behavioral Research

April 18, 1979

Ethical Principles and Guidelines for Research Involving Human Subjects

Scientific research has produced substantial social benefits. It has also posed some troubling ethical questions. Public attention was drawn to these questions by reported abuses of human subjects in biomedical experiments, especially during the Second World War. During the Nuremberg War Crime Trials, the Nuremberg code was drafted as a set of standards for judging physicians and scientists who had conducted biomedical experiments on concentration camp prisoners. This code became the prototype of many later codes intended to assure that research involving human subjects would be carried out in an ethical manner.

The codes consist of rules, some general, others specific, that guide the investigators or the reviewers of research in their work. Such rules often are inadequate to cover complex situations; at times they come into conflict, and they are frequently difficult to interpret or apply. Broader ethical principles will provide a basis on which specific rules may be formulated, criticized and interpreted.

Three principles, or general prescriptive judgments, that are relevant to research involving human subjects are identified in this statement. Other principles may also be relevant. These three are comprehensive, however, and are stated at a level of generalization that should assist scientists, subjects, reviewers and interested citizens to understand the ethical issues inherent in research involving human subjects. These principles cannot always be applied so as to resolve beyond dispute particular ethical problems. The objective is to provide an analytical framework that will guide the resolution of ethical problems arising from research involving human subjects.

This statement consists of a distinction between research and practice, a discussion of the three basic ethical principles, and remarks about the application of these principles.

A. Boundaries Between Practice and Research

It is important to distinguish between biomedical and behavioral research, on the one hand, and the practice of accepted therapy on the other, in order to know what activities ought to undergo review for the protection of human subjects of research. The distinction between research and practice is blurred partly because both often occur together (as in research designed to evaluate a therapy) and partly because notable departures from standard practice are often called "experimental" when the terms "experimental" and "research" are not carefully defined.

For the most part, the term "practice" refers to interventions that are designed solely to enhance the well being of an individual patient or client and that have a reasonable expectation of success. The purpose of medical or behavioral practice is to provide diagnosis, preventive treatment or therapy to particular individuals. By contrast, the term "research" designates an activity designed to test an hypothesis, permit conclusions to be drawn, and thereby to develop or contribute to generalizable knowledge (expressed, for example, in theories, principles, and statements of relationships). Research is usually described in a formal protocol that sets forth an objective and a set of procedures designed to reach that objective.

When a clinician departs in a significant way from standard or accepted practice, the innovation does not, in and of itself, constitute research. The fact that a procedure is "experimental," in the sense of new, untested or different, does not automatically place it in the category of research. Radically new procedures of this description should, however, be made the object of formal research at an early stage in order to determine whether they are safe and effective. Thus, it is the responsibility of medical practice committees, for example, to insist that a major innovation be incorporated into a formal research project.

Research and practice may be carried on together when research is designed to evaluate the safety and efficacy of a therapy. This need not cause any confusion regarding whether or not the activity requires review; the general rule is that if there is any element of research in an activity, that activity should undergo review for the protection of human subjects.

B. Basic Ethical Principles

The expression "basic ethical principles" refers to those general judgments that serve as a basic justification for the many particular ethical prescriptions and evaluations of human actions. Three basic principles, among those generally accepted in our cultural tradition, are particularly relevant to the ethic of research involving human subjects: the principles of respect for persons, beneficence and justice.

1. *Respect for Persons.* Respect for persons incorporates at least two ethical convictions; first, that individuals should be treated as autonomous agents, and second, that persons with diminished autonomy are entitled to protection. The principle of respect for persons thus divides into two separate moral requirements: the requirement to acknowledge autonomy and the requirement to protect those with diminished autonomy.

   An autonomous person is an individual capable of deliberation about personal goals and of acting under the direction of such deliberation. To respect autonomy is to give weight to autonomous persons' considered opinions and choices while refraining from obstructing their actions unless they are clearly detrimental to others. To show lack of respect for an autonomous agent is to repudiate that person's considered judgments, to deny an individual the freedom to act on those considered judgments, or to withhold information necessary to make a considered judgment, when there are no compelling reasons to do so.

However, not every human being is capable of self-determination. The capacity for self-determination matures during an individual's life, and some individuals lose this capacity wholly or in part because of illness, mental disability, or circumstances that severely restrict liberty. Respect for the immature and the incapacitated may require protecting them as they mature or while they are incapacitated.

Some persons are in need of extensive protection, even to the point of excluding them from activities which may harm them; other persons require little protection beyond making sure they undertake activities freely and with awareness of possible adverse consequences. The extent of protection afforded should depend upon the risk of harm and the likelihood of benefit. The judgment that any individual lacks autonomy should be periodically reevaluated and will vary in different situations.

In most cases of research involving human subjects, respect for persons demands that subjects enter into the research voluntarily and with adequate information. In some situations, however, application of the principle is not obvious. The involvement of prisoners as subjects of research provides an instructive example. On the one hand, it would seem that the principle of respect for persons requires that prisoners not be deprived of the opportunity to volunteer for research. On the other hand, under prison conditions they may be subtly coerced or unduly influenced to engage in research activities for which they would not otherwise volunteer. Respect for persons would then dictate that prisoners be protected. Whether to allow prisoners to "volunteer" or to "protect" them presents a dilemma. Respecting persons, in most hard cases, is often a matter of balancing competing claims urged by the principle of respect itself.

2. *Beneficence*. Persons are treated in an ethical manner not only by respecting their decisions and protecting them from harm, but also by making efforts to secure their well being. Such treatment falls under the principle of beneficence. The term "beneficence" is often understood to cover acts of kindness or charity that go beyond strict obligation. In this document, beneficence is understood in a stronger sense. as an obligation. Two general rules have been formulated as complementary expressions of beneficent actions in this sense: (1) do not harm and (2) maximize possible benefits and minimize possible harms.

The Hippocratic maxim "do no harm" has long been a fundamental principle of medical ethics. Claude Bernard extended it to the realm of research, saying that one should not injure one person regardless of the benefits that might come to others. However, even avoiding harm requires learning what is harmful; and, in the process of obtaining this information, persons may be exposed to risk of harm. Further, the Hippocratic Oath requires physicians to benefit their patients "according to their best judgment." Learning what will in fact benefit may require exposing persons to risk. The problem posed by these imperatives is to decide when it is justifiable to seek certain benefits despite the risks involved, and when the benefits should be foregone because of the risks.

The obligations of beneficence affect both individual investigators and society at large, because they extend both to particular research projects and to the entire enterprise of research. In the case of particular projects, investigators and members of their institutions are obliged to give forethought to the maximization of benefits and the reduction of risk that might occur from the research investigation. In the case of scientific research in general, members of the larger society are obliged to give forethought the longer term benefits and risks that may result from the improvement of knowledge and from the development of novel medical. psychotherapeutic. and social procedures.

The principle of beneficence often occupies a well-defined justifying role in many areas of research involving human subjects. An example is found in research involving children. Effective ways of treating childhood diseases and fostering healthy development are benefits that serve to justify research involving children - even when individual research subjects are not direct beneficiaries. Research also makes is possible to avoid the harm that may result from the application of previously accepted routine practices that on closer investigation turn out to be dangerous. But the role of the principle of beneficence is not always so unambiguous. A difficult ethical problem remains, for example, about research that presents more than minimal risk without immediate prospect of direct benefit to the children involved. Some have argued that such research is inadmissible, while others have pointed out that this limit would rule out much research promising great benefit to children in the future. Here again, as with all hard cases, the different claims covered by the principle of beneficence may come into conflict and force difficult choices.

3. *Justice*. Who ought to receive the benefits of research and bear its burdens? This is a question of justice, in the sense of "fairness in distribution" or "what is deserved." An injustice occurs when some benefit to which a person is entitled is denied without good reason or when some burden is imposed unduly. Another way of conceiving the principle of justice is that equals ought to be treated equally. However, this statement requires explication. Who is equal and who is unequal? What considerations justify departure from equal distribution? Almost all commentators allow that distinctions based on experience, age, deprivation, competence, merit and position do sometimes constitute criteria justifying differential treatment for certain purposes. It is necessary, then, to explain in what respects people should be treated equally. There are several widely accepted formulations of just ways to distribute burdens and benefits. Each formulation mentions some relevant property on the basis of which burdens and benefits should be distributed. These formulations are (1) to each person an equal share, (2) to each person according to individual need, (3) to each person according to individual effort, (4) to each person according to societal contribution, and (5) to each person according to merit.

Questions of justice have long been associated with social practices such as punishment, taxation and political representation. Until recently these questions have not generally been associated with scientific research. However, they are foreshadowed even in the earliest reflections on the ethics of research involving human subjects. For example, during the 19th and early 20th centuries the burdens of serving as research subjects fell largely upon poor ward patients, while the benefits of improved medical care flowed primarily to private patients. Subsequently, the exploitation of unwilling prisoners as research subjects in Nazi concentration camps was condemned as a particularly flagrant injustice. In this country, in the 1940's, the Tuskegee syphilis study used disadvantaged, rural black men to study the untreated course of a disease that is by no means confined to that population. These subjects were deprived of demonstrably effective treatment in order not to interrupt the project, long after such treatment became generally available.

Against this historical background, it can be seen how conceptions of justice are relevant to research involving human subjects. For example, the selection of research subjects needs to be scrutinized in order to determine whether some classes (e.g., welfare patients, particular racial and ethnic minorities, or persons confined to institutions) are being systematically selected simply because of their easy availability, their compromised position, or their manipulability, rather than for reasons directly related to the problem being studied. Finally, whenever research supported by public funds leads to the development of therapeutic devices and procedures, justice demands both that these not provide advantages only to those who can afford them and that such research should not unduly involve persons from groups unlikely to be among the beneficiaries of subsequent applications of the research.

C. Applications

Applications of the general principles to the conflict of research leads to consideration of the following requirements: informed consent, risk/benefit assessment, and the selection of subjects of research.

1. *Informed Consent.* Respect for persons requires that subjects, to the degree that they are capable, be given the opportunity to choose what shall or shall not happen to them. This opportunity is provided when adequate standards for informed consent are satisfied. While the importance of informed consent is unquestioned, controversy prevails over the nature and possibility of an informed consent. Nonetheless, there is widespread agreement that the consent process can be analyzed as containing three elements: information, comprehension and voluntariness.

   Information. Most codes of research establish specific items for disclosure intended to assure that subjects are given sufficient information. These items generally include: the research procedure, their purposes, risks and anticipated benefits, alternative procedures (where therapy is involved), and a statement offering the subject the opportunity to ask questions and to withdraw at any time from the research. Additional items have been proposed, including how subjects are selected, the person responsible for the research, etc.

   However, a simple listing of items does not answer the question of what the standard should be for judging how much and what sort of information should be provided. One standard frequently invoked in medical practice, namely the information commonly provided by practitioners in the field or in the locale, is inadequate since research takes place precisely when a common understanding does not exist. Another standard, currently popular in malpractice law, requires the practitioner to reveal the information that reasonable persons would wish to know in order to make a decision regarding their care. This, too, seems insufficient since the research subject, being in essence a volunteer, may wish to know considerably more about risks gratuitously undertaken than do patients who deliver themselves into the hand of a clinician for needed care. It may be that a standard of "the reasonable volunteer" should be proposed: the extent and nature of information should be such that persons, knowing that the procedure is neither necessary for their care nor perhaps fully understood, can decide whether they wish to participate in the furthering of knowledge. Even when some direct benefit to them is anticipated, the subjects should understand clearly the range of risk and the voluntary nature of participation.

A special problem of consent arises where informing subjects of some pertinent aspect of the research is likely to impair the validity of the research. In many cases, it is sufficient to indicate to subjects that they are being invited to participate in research of which some features will not be revealed until the research is concluded. In all cases of research involving incomplete disclosure, such research is justified only if it is clear that (1) incomplete disclosure is truly necessary to accomplish the goals of the research, (2) there are no undisclosed risks to subjects that are more than minimal. and (3) there is an adequate plan for debriefing subjects, when appropriate, and for dissemination of research results to them. Information about risks should never be withheld for the purpose of eliciting the cooperation of subjects, and truthful answers should always be given to direct questions about the research. Care should be taken to distinguish cases in which disclosure would destroy or invalidate the research from cases in which disclosure would simply inconvenience the investigator.

Comprehension. The manner and context in which information is conveyed is as important as the information itself. For example, presenting information in a disorganized and rapid fashion, allowing too little time for consideration or curtailing opportunities for questioning, all may adversely affect a subject's ability to make an informed choice.

Because the subject's ability to understand is a function of intelligence, rationality, maturity and language, it is necessary to adapt the preservation of the information to the subject's capabilities. Investigators are responsible for ascertaining that the subject has comprehended the information. While there is always an obligation to ascertain that the information about risk to subjects is complete and adequately comprehended, when the risks are more serious, that obligation increases. On occasion, it may be suitable to give some oral or written tests of comprehension.

Special provision may need to be made when comprehension is severely limited — for example, by conditions of immaturity or mental disability. each class of subjects that one might consider as incompetent (e.g., infants and young children, mentally disabled patients, the terminally ill and the comatose) should be considered on its own terms. Even for these persons, however, respect requires giving them the opportunity to choose to the extent they are able, whether or not to participate in research. The objections of these subjects to involvement should be honored, unless the research entails pro-providing them a therapy unavailable elsewhere. Respect for persons also requires seeking the permission of other parties in order to protect the subjects from harm. Such persons are thus respected both by acknowledging their own wishes and by the use of third parties to protect them from harm.

The third parties chosen should be those who are most likely to understand the incompetent subject's situation and to act in that person's best interest. The person authorized to act on behalf of the subject should be given an opportunity to observe the research as it proceeds in order to be able to withdraw the subject from the research, if such action appears in the subject's best interest.

Voluntariness. An agreement to participate in research constitutes a valid consent only if voluntarily given. This element of informed consent requires conditions free of coercion and undue influence. Coercion occurs when an overt threat of harm is intentionally presented by one person to another in order to obtain compliance. Undue influence, by contrast, occurs through an offer of an excessive, unwarranted, inappropriate or improper reward or other overture in order to obtain compliance. Also, inducements that would ordinarily be acceptable may become undue influences if the subject is especially vulnerable.

Unjustifiable pressures usually occur when persons in positions of authority or commanding influence — especially where possible sanctions are involved - urge a course of action for a subject. A continuum of such influencing factors exists, however, and it is impossible to state precisely where justifiable persuasion ends and undue influence begins. But undue influence would include actions such as manipulating a person's choice through the controlling influence of a close relative and threatening to withdraw health services to which an individual would otherwise be entitled.

2. Assessment of Risks and Benefits. The assessment of risks and benefits requires a careful arrayal of relevant data, including, in some cases, alternative ways of obtaining the benefits sought in the research. Thus, the assessment presents both an opportunity and a responsibility to gather systematic and comprehensive information about proposed research. For the investigator, it is a means to examine whether the proposed research is properly designed. For a review committee, it is a method for determining whether the risks that will be presented to subjects are justified. For prospective subjects, the assessment will assist the determination whether or not to participate.

The Nature and Scope of Risks and Benefits. The requirement that research be justified on the basis of a favorable risk / benefit assessment bears a close relation to the principle of beneficence, just as the moral requirement that informed consent be obtained is derived primarily from the principle of respect for persons.

The term "risk" refers to a possibility that harm may occur. However, when expressions such as "small risk" or "high risk" are used, they usually refer (often ambiguously) both to the chance (probability) of experiencing a harm and the severity (magnitude) of the envisioned harm.

The term "benefit" is used in the research context to refer to something of positive value related to health or welfare. Unlike "risk," "benefit" is not a term that expresses probabilities. Risk is properly contrasted to probability of benefits, and benefits are properly contrasted with harms rather than risks of harm. Accordingly, so-called risk/ benefit assessments are concerned with the probabilities and magnitudes of possible harms and anticipated benefits. Many kinds of possible harms and benefits need to be taken into account. There are, for example, risks of psychological harm, physical harm, legal harm, social harm and economic harm and the corresponding benefits. While the most likely types of harms to research subjects are those of psychological or physical pain or injury, other possible kinds should not be overlooked.

Risks and benefits of research may affect the individual subjects, the families of the individual subjects, and society at large (or special groups of subjects in society). Previous codes and Federal regulations have required that risks to subjects be outweighed by the sum of both the anticipated benefit to the subject, if any, and the anticipated benefit to society in the form of knowledge to be gained from the research. In balancing these different elements, the risks and benefits affecting the immediate research subject will normally carry special weight. On the other hand, interests other than those of the subject may on some occasions be sufficient by themselves to justify the risks involved in the research, so long as the subjects' rights have been protected. Beneficence thus requires that we protect against risk of harm to subjects and also that we be concerned about the loss of the substantial benefits that might be gained from research.

The Systematic Assessment of Risks and Benefits. It is commonly said that benefits and risks must be "balanced" and shown to be "in a favorable ratio." The metaphorical character of these terms draws attention to the difficulty of making precise judgments. Only on rare occasions will quantitative techniques be available for the scrutiny of research protocols. However, the idea of systematic, nonarbitrary analysis of risks and benefits should be emulated insofar as possible. This ideal requires those making decisions about the justifiability of research to be thorough in the accumulation and assessment of information about all aspects of the research, and to consider alternatives systematically. This procedure renders the assessment of research more rigorous and precise, while making communication between review board members and investigators less subject to misinterpretation, misinformation and conflicting judgments. Thus, there should first be a determination of the validity of the presuppositions of the research; then the nature, probability and magnitude of risk should be distinguished with as much clarity as possible. The method of ascertaining risks should be explicit, especially where there is no alternative to the use of such vague categories as small or slight risk. It should also be determined whether an investigator's estimates of the probability of harm or benefits are reasonable, as judged by known facts or other available studies.

Finally, assessment of the justifiability of research should reflect at least the following considerations: (i) Brutal or inhumane treatment of human subjects is never morally justified. (ii) Risks should be reduced to those necessary to achieve the research objective. It should be determined whether it is in fact necessary to use human subjects at all. Risk can perhaps never be entirely eliminated, but it can often be reduced by careful attention to alternative procedures. (iii) When research involves significant risk of serious impairment, review committees should be extraordinarily insistent on the justification of the risk (looking usually to the likelihood of benefit to the subject - or, in some rare cases, to the manifest voluntariness of the participation). (iv) When vulnerable populations are involved in research, the appropriateness of involving them should itself be demonstrated. A number of variables go into such judgments, including the nature and degree of risk, the condition of the particular population involved, and the nature and level of the anticipated benefits. (v) Relevant risks and benefits must be thoroughly arrayed in documents and procedures used in the informed consent process.

3. Selection of Subjects. — Just as the principle of respect for persons finds expression in the requirements for consent, and the principle of beneficence in risk/ benefit assessment, the principle of justice gives rise to moral requirements that there be fair procedures and outcomes in the selection of research subjects.

Justice is relevant to the selection of subjects of research at two levels: the social and the individual. Individual justice in the selection of subjects would require that researchers exhibit fairness: thus, they should not offer potentially beneficial research only to some patients who are in their favor or select only "undesirable" persons for risky research. Social justice requires that distinction be drawn between classes of subjects that ought, and ought not, to participate in any particular kind of research, based on the ability of members of that class to bear burdens and on the appropriateness of placing further burdens on already burdened persons. Thus, it can be considered a matter of social justice that there is an order of preference in the selection of classes of subjects (e.g., adults before children) and that some classes of potential subjects (e.g., the institutionalized mentally infirm or prisoners) may be involved as research subjects, if at all, only on certain conditions.

Injustice may appear in the selection of subjects. even if individual subjects are selected fairly by investigators and treated fairly in the course of research. Thus injustice arises from social. racial, sexual and cultural biases institutionalized in society. Thus, even if individual researchers are treating their research subjects fairly, and even if IRBs are taking care to assure that subjects are selected fairly within a particular institution. unjust social patterns may nevertheless appear in the overall distribution of the burdens and benefits of research. Although individual institutions or investigators may not be able to resolve a problem that is pervasive in their social setting. They can consider distributive justice in selecting research subjects.

Some populations, especially institutionalized ones, are already burdened in many ways by their infirmities and environments. When research is proposed that involves risks and does not include a therapeutic component, other less burdened classes of persons should be called upon first to accept these risks of research, except where the research is directly related to the specific conditions of the class involved. Also, even though public funds for research may often flow in the same directions as public funds for health care, it seems unfair that populations dependent on public health care constitute a pool of preferred research subjects if more advantaged populations are likely to be the recipients of the benefits.

One special instance of injustice results from the involvement of vulnerable subjects. Certain groups, such as racial minorities, the economically disadvantaged, the very sick, and the institutionalized may continually be sought as research subjects, owing to their ready availability in settings where research is conducted. Given their dependent status and their frequently compromised capacity for free consent, they should be protected against the danger of being involved in research solely for administrative convenience, or because they are easy to manipulate as a result of their illness or socioeconomic condition.

**Appendix G
has been ommitted
in order to reduce duplication
Please see Appendix C**

Appendix H

A Self-Evaluation Checklist for IRBs

The Food and Drug Administration (FDA) has regulations that govern human subject protection aspects of research on products regulated by the Agency. In addition, other federal agencies and departments and some States have regulations that govern human subject protection. Each IRB/institution should be familiar with the laws and regulations that apply to research reviewed by the IRB. This checklist was developed to help IRBs/institutions evaluate procedures for the protection of human subjects of research.

Successful IRBs make use of written procedures that, in one way or another, cover a common core of topics. This checklist is an effort to present these topics in a systematic way. Written procedures for some of the items are not specifically required by FDA regulations (e.g., policy regarding place and time of meeting) but are appropriate to consider when comprehensive procedures are being developed.

Once an IRB/institution establishes its structure and procedures, those procedures should be followed. FDA inspections assess compliance with both the regulatory requirements and the IRB/institution's own written procedures. Since the written procedures should reflect the current processes, the written procedures should be reviewed on a regular basis and updated as necessary to remain current. FDA believes that when good procedures are developed, written, and followed, the rights and welfare of the subjects of research are more likely to be adequately protected.

Tips on checklist use:

Three "response" columns are provided — "Yes," "No," and N/A." A "Yes" means that the institution has a written policy/procedure and that it is current. A "No" may mean that a policy/procedure is lacking or needs to be updated. The "N/A" column indicates that a topic is not applicable or a procedure is not needed by the IRB.

The columns may be completed by checking the appropriate box. Instead of a check-mark, some IRBs record the date of issuance or revision date. Others have found it useful to record the policy/procedure number on the form. Any "No" responses indicate a need to write/revise policies and/or procedures.

FOOTNOTED items are referenced in the FDA regulations. ASTERISKED items are those for which WRITTEN PROCEDURES are specifically required by the FDA regulations.

A Self-Evaluation Checklist for IRBs

YES	NO	N/A		DOES THE INSTITUTION HAVE WRITTEN POLICIES OR PROCEDURES THAT DESCRIBE:
____	____	____	I.	THE INSTITUTIONAL AUTHORITY UNDER WHICH THE IRB IS ESTABLISHED AND EMPOWERED.æ
____	____	____	II.	THE DEFINITION OF THE PURPOSE OF THE IRB, i.e., THE PROTECTION OF HUMAN SUBJECTS OF RESEARCH.,,
____	____	____	III.	THE PRINCIPLES WHICH GOVERN THE IRB IN ASSURING THAT THE RIGHTS AND WELFARE OF SUBJECTS ARE PROTECTED.
			IV.	THE AUTHORITY OF THE IRB.
____	____	____		A. The scope of authority is defined, i.e., what types of studies must be reviewed.
____	____	____		B. Authority to disapprove, modify or approve studies based upon consideration of human subject protection aspects.[4]
____	____	____		* C. Authority to require progress reports from the investigators and oversee the conduct of the study.[5]
____	____	____		* D. Authority to suspend or terminate approval of a study.[6]
____	____	____		* E. Authority to place restrictions on a study.[7]
			V.	THE IRB'S RELATIONSHIP TO
____	____	____		A. The top administration of the institution.
____	____	____		B. The other committees and department chairpersons within the institution.
____	____	____		C. The research investigators.
____	____	____		D. Other institutions.
____	____	____		E. Regulatory agencies.
			VI.	THE MEMBERSHIP OF THE IRB.
____	____	____		A. Number of members.[8]
____	____	____		B. Qualification of members.[9]

YES	NO	N/A	DOES THE INSTITUTION HAVE WRITTEN POLICIES OR PROCEDURES THAT DESCRIBE:
			C. Diversity of members10(for example, representation from the community, and minority groups), including representation by:
____	____	____	— both men and women[11]
____	____	____	— multiple professions[12]
____	____	____	— scientific and non-scientific member(s)[13]
____	____	____	— not otherwise affiliated member(s)[14]
____	____	____	D. Alternate members (if used).

II. MANAGEMENT OF THE IRB.

A. The Chairperson

YES	NO	N/A	
____	____	____	— selection and appointment
____	____	____	— length of term/service
____	____	____	— duties
____	____	____	— removal

B. The IRB Members.

____	____	____	— selection and appointment
____	____	____	— length of term/service and description of staggered rotation or overlapping of terms, if used
____	____	____	— duties
____	____	____	— attendance requirements
____	____	____	—removal

C. Training of IRB Chair and members

____	____	____	— orientation
____	____	____	— continuing education
____	____	____	— reference materials (IRB library)
____	____	____	D. Compensation of IRB members.
____	____	____	E. Liability coverage for IRB members.
____	____	____	F. Use of consultants.[15]
____	____	____	G. Secretarial/administrative support staff (duties).
____	____	____	H. Resources (for example, meeting area, filing space, reproduction equipment, computers).

FDA Information Sheets

YES	NO	N/A	DOES THE INSTITUTION HAVE WRITTEN POLICIES OR PROCEDURES THAT DESCRIBE:

I. Conflict of interest policy

YES	NO	N/A	
____	____	____	— no selection of IRB members by investigators
____	____	____	— prohibition of participation in IRB deliberations and voting by investigators.[16]

VIII. FUNCTIONS OF THE IRB.

____	____	____	* A. Conducting initial and continuing review.[17]
____	____	____	* B. Reporting, in writing, findings and actions of the IRB to the investigator and the institution.[18]
____	____	____	* C. Determining which studies require review more often than annually.[19]
____	____	____	* D. Determining which studies need verification from sources other than the investigators that no material changes have occurred since previous IRB review.[20]
____	____	____	E. Ensuring prompt reporting to the IRB of changes in research activities. [21]
____	____	____	* F. Ensuring that changes in approved research are not initiated without IRB review and approval except where necessary to eliminate apparent immediate hazards. [22]

G. Ensuring prompt reporting to the IRB, appropriate institutional officials, and the FDA of:

____	____	____	* — unanticipated problems involving risks to subjects or others[23]
____	____	____	* — serious or continuing noncompliance with 21 CFR parts 50 and 56 or the requirements of the IRB[24]
____	____	____	* — suspension or termination of IRB approval.[25]
____	____	____	H. Determining which device studies pose significant or non-significant risk.

IX. OPERATIONS OF THE IRB.

| ____ | ____ | ____ | * A. Scheduling of meetings. [26] |
| ____ | ____ | ____ | B. Pre-meeting distribution to members, of, for example, place and time of meeting, agenda, and study material to be reviewed. |

YES	NO	N/A	DOES THE INSTITUTION HAVE WRITTEN POLICIES OR PROCEDURES THAT DESCRIBE:
			C. The review process
			* — description of the process ensuring that[27]
___	___	___	1) all members receive complete study documentation for review (see XI.B);
			or
___	___	___	2) one or more "primary reviewers"/"secondary reviewers" receives the complete study documentation for review, reports to IRB and leads discussion; if other members review summary information only, these members must have access to complete study documentation
___	___	___	— role of any subcommittees of the IRB
___	___	___	* — emergency use notification and reporting procedures[28]
			* — expedited review procedure[29]
___	___	___	— for approval of studies that are both minimal risk and on the FDA approved list (see Appendix A)
___	___	___	— for approval of modifications to ongoing studies involving no more than minimal risk
___	___	___	**D. Criteria for IRB approval contain all requirements of 21 CFR 56.111.**
			E. Voting requirements[30]
___	___	___	- quorum required to transact business
___	___	___	- diversity requirements of quorum (for example requiring at least one physician member when reviewing studies of FDA regulated articles)
___	___	___	- percent needed to approve or disapprove a study
___	___	___	- full voting rights of all reviewing members
___	___	___	- no proxy votes (written or telephone)
___	___	___	- prohibition against conflict-of-interest voting
___	___	___	**F. Further review/approval of IRB actions by others within the institution. (Override of disapprovals is prohibited.)[31]**

YES	NO	N/A	DOES THE INSTITUTION HAVE WRITTEN POLICIES OR PROCEDURES THAT DESCRIBE:
			G. Communication from the IRB.
____	____	____	* - to the investigator for additional information[32]
____	____	____	* - to the investigator conveying IRB decision[33]
____	____	____	* - to institution administration conveying IRB decision[34]
____	____	____	- to sponsor of research conveying IRB decision
			H. Appeal of IRB decisions.
____	____	____	- criteria for appeal
____	____	____	- to whom appeal is addressed
____	____	____	- how appeal is resolved (Override of IRB disapprovals by external body/official is prohibited.)[35]

X.   IRB RECORD REQUIREMENTS.

YES	NO	N/A	
____	____	____	A. IRB membership roster showing qualifications[36]
____	____	____	* B. Written procedures and guidelines. [37]
			C. Minutes of meetings. [38]
____	____	____	- members present (any consultants/ guests/others shown separately)
____	____	____	- summary of discussion on debated issues - record of IRB decisions
____	____	____	- record of voting (showing votes for, against and abstentions)
____	____	____	D. Retention of protocols reviewed and approved consent documents[39]
____	____	____	E. Communications to and from the IRB.[40]
____	____	____	* F. 1) Adverse reactions reports, and[41]
____	____	____	2) documentation that the IRB reviews such reports.
____	____	____	H. Records of continuing review.[42]
			I. Record retention requirements. (at least 3 years after completion for FDA studies)[43]
____	____	____	J. Budget and accounting records.
____	____	____	K. Emergency use reports.[44]
____	____	____	L. Statements of significant new findings provided to subjects.[45]

YES	NO	N/A	DOES THE INSTITUTION HAVE WRITTEN POLICIES OR PROCEDURES THAT DESCRIBE:
			XI. INFORMATION THE INVESTIGATOR PROVIDES TO THE IRB.
____	____	____	A. Professional qualifications to do the research (including a description of necessary support services and facilities).
			B. Study protocol which includes/addresses[46]
____	____	____	- title of the study.
____	____	____	- purpose of the study (including the expected benefits obtained by doing the study).
____	____	____	- sponsor of the study.
____	____	____	- results of previous related research.
____	____	____	- subject inclusion/exclusion criteria.
____	____	____	- justification for use of any special/vulnerable subject populations (for example, the decisionally impaired, children)
____	____	____	- study design (including as needed, a discussion of the appropriateness of research methods).
____	____	____	- description of procedures to be performed.
____	____	____	- provisions for managing adverse reactions.
____	____	____	- the circumstances surrounding consent procedure, including setting, subject autonomy concerns, language difficulties, vulnerable populations.
____	____	____	- the procedures for documentation of informed consent, including any procedures for obtaining assent from minors, using witnesses, translators and document storage.
____	____	____	- compensation to subjects for their participation.
____	____	____	- any compensation for injured research subjects.
____	____	____	- provisions for protection of subject's privacy.
____	____	____	- extra costs to subjects for their participation in the study.
____	____	____	- extra costs to third party payers because of subject's participation.
____	____	____	C. Investigator's Brochure (when one exists)[47]
____	____	____	D. The case report form (when one exists)

YES	NO	N/A	DOES THE INSTITUTION HAVE WRITTEN POLICIES OR PROCEDURES THAT DESCRIBE:
			E. The proposed informed consent document.[48]
____	____	____	- containing all requirements of 21 CFR 50.25(a)
____	____	____	- containing requirements of 21 CFR 50.25(b) that are appropriate to the study.
____	____	____	- meeting all requirements of 21 CFR 50.20
____	____	____	- translated consent documents, as necessary, considering likely subject population(s)
____	____	____	* F. Requests for changes in study after initiation.[49]
____	____	____	* G. Reports of unexpected adverse events.[50]
____	____	____	* H. Progress reports.[51]
____	____	____	I. Final report.
____	____	____	J. Institutional forms/reports

XII. EXEMPTION FROM PROSPECTIVE IRB REVIEW[52]

YES	NO	N/A	
____	____	____	* A. Notify IRB within 5 working days[53]
____	____	____	B. Emergency use[54]
____	____	____	C. Review protocol and consent when subsequent use is anticipated.[55]

XIII. EMERGENCY RESEARCH CONSENT EXCEPTION[56]

YES	NO	N/A	
____	____	____	A. The IRB may find that the 50.24 requirements are met[57]
____	____	____	B. The IRB shall promptly notify in writing the investigator and the sponsor when it determines it cannot approve a 50.24 study[58]
____	____	____	C. The IRB shall provide in writing to the sponsor a copy of the information that has been publically disclosed under 50.24(a)(7)(ii) and (a)(7)(iii)[59]
____	____	____	D. In order to approve an emergency research consent waiver study, the IRB must find and document:
____	____	____	(1) subjects are in a life-threatening situation, available treatments unproven or unsatisfactory and collection of scientific evidence is necessary [60]
			(2) Obtaining informed consent is not feasible because:[61]
____	____	____	- medical condition precludes consent[62]

YES	NO	N/A	DOES THE INSTITUTION HAVE WRITTEN POLICIES OR PROCEDURES THAT DESCRIBE:
____	____	____	- no time to get consent from legally authorized representative[63]
____	____	____	- prospective identity of likely subjects not reasonable[64]
			(3) Prospect of direct benefits to study subjects because:[65]
____	____	____	- life-threatining situation that necessisates treatment
____	____	____	- data support potential for direct benefit to individual subjects
____	____	____	- risk/benefit of both standard and proposed treatments reasonable
____	____	____	(4) waiver needed to carry out study
____	____	____	(5) plan defines therapeutic window, during which investigator will seek consent rather than starting without consent. Summary of efforts will be given to IRB at time of continuing review.
____	____	____	(6) IRB reviews and approves consent procedures and document. IRB reviews and approves family member objection procedures
			(7) Additional protections, including at least:
____	____	____	- consultation with community representatives
____	____	____	- public disclosure of plans, risks and expected benefits
____	____	____	- public disclosure of study results
____	____	____	- assure an independent Data Monitoring Committee established
____	____	____	- objection of family member summarized for continuing review
____	____	____	(8) Ensure procedures in place to inform at earliest feasible opportunity of subject's inclusion in the study, participation may be discontinued. Procedures to inform family the subject was in the study if subject dies.
____	____	____	(9) Separate IND or IDE required, even for marketed products.
____	____	____	(10) IRB disapproval must be documented in writing and sent to the clinical investigator and the sponsor of the clinical investigation. Sponsor must promptly disclose to FDA, other investigators and other IRBs.

FDA Information Sheets

Checklist References

æ. 21 CFR 56.109(a)
„. 21 CFR 56.101(a)
4. 21 CFR 56.109(a)
5. 21 CFR 56.108(a)(1) and 56.109(f)
6. 21 CFR 56.108(b)(3) and 56.113
7. 21 CFR 56.108(a)(1), 56.109(a) and 56.113
8. 21 CFR 56.107(a)
9. 21 CFR 56.107(a)
10. 21 CFR 56.107(a)
∞. 21 CFR 56.107(b) Only requires every nondiscriminatory effort
°æ. 21 CFR 56.107(a)
°„. 21 CFR 56.107(c)
14. 21 CFR 56.107(d)
15. 21 CFR 56.107(f) Consultant use not required by FDA regulation.
16. 21 CFR 56.107(e)
17. 21 CFR 56.108(a)(1) and 56.109(a - f)
18. 21 CFR 56.108(a)(1) and 56.109(e)
19. 21 CFR 56.108(a)(2) and 56.109(f)
20. 21 CFR 56.108(a)(2)
21. 21 CFR 56.108(a)(3)
22. 21 CFR 56.108(a)(4) and 56.115(a)(1)
23. 21 CFR 56.108(b)(1) and 56.115(a)(1)
24. 21 CFR 56.108(b)(2)
25. 21 CFR 56.108(b)(3) and 56.113
26. 21 CFR 56.108(a)(1)
27. 21 CFR 56.108(a)(1)
28. 21 CFR 56.104(c), 56.108(a)(1) and 108(b)(1)
29. 21 CFR 56.108(a)(1) and 56.110(a - c) not required if IRB does not use expedited procedures
30. 21 CFR 56.108(c) and 56.107(e - f)
31. 21 CFR 56.112
32. 21 CFR 56.108(a)(1), 56.109(a) and 56.115(a)(4)
33. 21 CFR 56.108(a)(1) and 56.109(e)
34. 21 CFR 56.108(a)(1) and 56.109(e)
35. 21 CFR 56.112
36. 21 CFR 56.115(a)(5)
37. 21 CFR 56.108(a - b) and 56.115(a)(6)
38. 21 CFR 56.115(a)(2)
39. 21 CFR 56.115(a)(1)
40. 21 CFR 56.115(a)(4)
41. 21 CFR 56.108(a) and 56.115(a)(1 and 4)
42. 21 CFR 56.115(a)(3)
43. 21 CFR 56.115(b)
44. 21 CFR 56.115(a)(4) and 56.104(c)
45. 21 CFR 56.115(a)(7)
46. 21 CFR 56.103(a) and 56.115(a)(1)

Checklist References

47. 21 CFR 56.111 (a)(2), 56.115(a)(1) and 21 CFR 312.55
48. 21 CFR 56.111(a)(4 - 5) and 56.111(a)(1)
49. 21 CFR 56.108(a)(4) and 56.115(a)(3 - 4)
50. 21 CFR 56.108(b)(1), 56.115(a)(3 - 4), 56.115(b)(1) and 56.113
51. 21 CFR 56.108(a)(1) and 56.115(a)(1, 3 and 4)
52. Not required when the scope of studies reviewed by the IRB does not include serious and life-threatening diseases or conditions.
53. 21 CFR 56.104(c) and 56.108(a)(3)
54. 21 CFR 56.102(d) and 56.108(a)(3)
55. 21 CFR 56.104(c) and 56.108(a)(3) The IRB may determine that a rapid means of approval is preferable to a preapproved protocol and consent. Also see information sheet: "Emergency Use of a Drug or Biologic."
56. 21 CFR 50.24 The IRB/instituion may determine that research in emergent settings will not be conducted or supported. When that is the case, written procedures for this section need not be prepared.
57. 21 CFR 56.109(c)(2)
58. 21 CFR 56.109(e) The written statement shall include a statement of the reasons for the IRB's determination.
59. 21 CFR 56.109(g)
60. 21 CFR 50.24(a)(1)
61. 21 CFR 50.24(a)(2)
62. 21 CFR 50.24(a)(2)(i)
63. 21 CFR 50.24(a)(2)(ii)
64. 21 CFR 50.24(a)(2)(iii)
65. 21 CFR 50.24(a)(3)
66. 21 CFR 50.24(a)(3)(i)
67. 21 CFR 50.24(a)(3)(ii)
68. 21 CFR 50.24(a)(3)(iii)
69. 21 CFR 50.24(a)(4)
70. 21 CFR 50.24(a)(5)
71. 21 CFR 50.24(a)(6) Family member objection procedures at 50.24 (a)(7)(v)
72. 21 CFR 50.24(a)(7)
73. 21 CFR 50.24(a)(7)(i)
74. 21 CFR 50.24(a)(7)(ii)
75. 21 CFR 50.24(a)(7)(iii)
76. 21 CFR 50.24(a)(7)(iv)
77. 21 CFR 50.24(a)(7)(v)
78. 21 CFR 50.24(b)
79. 21 CFR 50.24(d) The study may not begin until FDA approves the separate IND/IDE.
80. 21 CFR 50.24(e)

THIS PAGE INTENTIONALLY LEFT BLANK

Appendix I

FDA District Offices

Food and Drug Administration (FDA) District Offices are located throughout the country. IRB and other inspections are conducted by FDA District Office personnel. Problems or questions related to FDA regulated products or IRB inspections may be directed to the Director of the Investigations Branch (unless otherwise indicted), or the Bioresearch Monitoring Program Coordinator, in the appropriate District Office.

District	States Served
ATLANTA District 60 Eighth Street, N.E. Atlanta, Georgia 30309 (404) 347- 3218	Georgia, North Carolina, South Carolina
BALTIMORE District 900 Madison Avenue Baltimore, Maryland 21201-2199 (410) 962- 3590	District of Columbia, Maryland, Virginia, West Virginia
NEW ENGLAND District One Montvale Avenue Stoneham, Massachusetts 02180 (617) 279-1675, EXT 128	Connecticut, Maine, Massachusetts, New Hampshire, Rhode Island, Vermont
BUFFALO District 599 Delaware Avenue Buffalo, New York 14202 (716) 551-4461	New York (except New York City, Long Island)
CHICAGO District 300 S. Riverside Plaza 5th Floor, Suite 550 South Chicago, Illinois 60606 (312) 353-5863 EXT 132	Illinois
CINCINNATI District 1141 Central Parkway Cincinnati, Ohio 45202-1097 (513) 684-3501 EXT 130	Ohio, Kentucky

District	States Served
DALLAS District 3310 Live Oak St. Dallas, Texas 75204 (214) 655-5310 EXT 504	Arkansas, Oklahoma, Texas
DENVER District P.O. Box 25087 6th and Kipling Sts. Denver Federal Center Denver, Colorado 80225-0087 (303) 236-3051	Colorado, New Mexico, Utah
DETROIT District 1560 East Jefferson Detroit Michigan 48207-3179 (313) 226- 6260	Indiana, Michigan
FLORIDA District 555 Winderley Place Maitland, Florida 32751 (407) 475-4700	Florida
KANSAS CITY District 11630 West 80th St. Lenexa, Kansas 66285-5905 (913) 752-2423	Iowa, Kansas, Missouri, Nebraska
LOS ANGELES District 19900 MacArthur Blvd., Suite 300 Irvine, California 92612-2445 (714) 798-7769	Arizona, California (southern)
MINNEAPOLIS District 240 Hennepin Avenue Minneapolis, Minnesota 55401-1912 (612) 334-4100 EXT 162	Minnesota, Wisconsin, North Dakota, South Dakota,
NASHVILLE District 297 Plus Park Boulevard Nashville, Tennessee 37217 (615) 781- 5378	Alabama, Tennessee

District	States Served
NEW JERSEY District Waterview Corp. Center 10 Waterway Bend, 3rd Floor Parsippany, New Jersey 07054 (201) 526-6000	New Jersey
NEW ORLEANS District 4298 Elysian Fields Avenue New Orleans, Louisiana 70122 (504) 589-6344	Louisiana, Mississippi
NEW YORK District 850 Third Avenue Brooklyn, New York 11232-1593 (718)340-7000	New York City, Long Island
PHILADELPHIA District 2nd and Chestnut Streets Room 900 Philadelphia, Pennsylvania 19106 (215) 597-4390	Delaware, Pennsylvania
SEATTLE District 22201 23rd Drive S.E. P.O. Box 3012 Bothell, Washington 98041-3012 (206) 483-4941	Alaska, Idaho, Oregon, Montana, Washington
SAN FRANCISCO District 1431 Harbor Bay Parkway Alameda, California 94502-7070 (510) 337-6733	California (northern), Hawaii, Nevada, American Samoa, Guam, Pacific Trust Territory
SAN JUAN District #466 Fernandez Juncos Avenue Stop 8 1/2 San Juan, Puerto Rico 00901-3223 (809) 729- 6608	Puerto Rico

Appendix J

Important FDA Contacts for Clinical Trials

(Revised August 7, 2001)

GENERAL

Call 301-827-4000 or 301-827-3340 (GCP Staff, Office of Science Coordination and Communication, Office of the Commissioner) for:

- general questions about FDA good clinical practice regulations and policy
- general questions about the FDA clinical Bioresearch Monitoring Program, and specifically clinical investigator, Institutional Review Board (IRB), sponsor, monitor, and contract research organization programs
- questions about or suggestions relating to FDA's Information Sheets for IRB's and Clinical Investigators

Call 301-827-0425 (fax: 301-827-0482) (Division of Compliance Policy, Office of Enforcement, Office of Regulatory Affairs) for:

- questions about the overall FDA Bioresearch Monitoring Program, and specifically the Good Laboratory Practice (GLP; nonclinical laboratories) Program
- general Bioresearch Monitoring Program enforcement issues
- reports made pursuant to 21 CFR 56.108(b) and 56.113 involving an FDA-regulated product if you do not know which FDA Center has jurisdiction (e.g., drug, medical device, biological product) including:
  - unanticipated problems involving risks to subjects [21 CFR 56.108(b)(1)]
  - serious or continuing noncompliance by an investigator with FDA regulations or with the IRB's determinations [21 CFR 56.108(b)(2)]
  - suspension or termination of IRB approval of a protocol [21 CFR 56.108(b)(3)]

PRODUCT SPECIFIC

DRUG PRODUCTS—Center for Drug Evaluation and Research (CDER)

Call 301-827-4573 (Drug Information Branch, CDER) for questions about:

- the legal status of a test article (e.g., whether an article is a "drug," or whether a drug is approved for marketing)
- whether research with a marketed drug in a particular study "significantly increases the risks" (or decreases the acceptability of the risks) and therefore requires an IND
- whether an investigational new drug application (IND) is required for a drug study

Call 301-594-0020; fax: 301-594-1204 (Division of Scientific Investigations, Office of Medical Policy, CDER) for questions about:

- human subject protection regulations pertaining to drugs [21 CFR Parts 50, 56, 312 and 361]
- CDER-assigned IRB Inspections (e.g., "FDA-483s" and "Warning Letters")
- eports made pursuant to 21 CFR 56.108(b) and 56.113 involving a drug product including:
  - unanticipated problems involving risks to subjects [21 CFR 56.108(b)(1)]
  - serious or continuing noncompliance by an investigator with FDA regulations or with the IRB's determinations [21 CFR 56.108(b)(2)]
  - suspension or termination of IRB approval of a protocol [21 CFR 56.108(b)(3)]
- reporting complaints related to human subject protection/Good Clinical Practice in FDA-regulated drug trials

Call 301-594-1032; fax: 301-827-5290 (Good Clinical Practice Branch, Office of Medical Policy, CDER) for questions about:

- regulations pertaining to clinical investigators [21 CFR Part 312] CDER-assigned Clinical Investigator Inspections (e.g., "FDA-483s" and "Warning Letters")

CDER's new document e-mail list—Subscribe to receive daily or weekly notification about new material on CDER's Website.

BIOLOGICAL PRODUCTS—Center for Biologics Evaluation and Research (CBER)

Call 301-827-6221; fax: 301-827-6748 (Bioresearch Monitoring Branch, Office of Compliance and Biologics Quality, CBER) for questions about:

- the legal status of a test article (that is, whether an article is a biological drug or device, or whether an IND or IDE [investigational device exemption] is required for an investigational study)
- human subject protection regulations pertaining to biologics, including certain devices or in vitro diagnostics regulated by CBER [21 CFR Parts 50, 56, 312, 812]
- CBER-assigned IRB Inspections (e.g., "FDA-483s" and "Warning Letters")
- CBER-assigned Clinical Investigator Inspections (e.g., "FDA-483s" and "Warning Letters")
- reports made pursuant to 21 CFR 56.108(b) and 56.113 involving a biologic product including:
  - unanticipated problems involving risks to subjects [21 CFR 56.108(b)(1)]
  - serious or continuing noncompliance by an investigator with FDA regulations or with the IRB's determinations [21 CFR 56.108(b)(2)]
  - suspension or termination of IRB approval of a protocol [21 CFR 56.108(b)(3)]
- Questions about specific products or classes of products should be directed to one of the following offices:
  - Office of Therapeutics Research and Review (301-827-5101)
  - Office of Vaccines Research and Review (301-827-3070)
  - Office of Blood Research and Review (301-827-3518)

- **Call 301-827-2000**; fax: 301-827-3843 (Division of Communication and Consumer Affairs, CBER) for reporting complaints related to human subject protection/Good Clinical Practice in FDA-regulated biologics trials
- CBER list to receive all fractionated product recall and blood information documents, as well as all other new documents, including guidelines, points to consider, and other CBER information. To subscribe, send an e-mail message to: cberinfo@listmanager.fda.gov with the word "subscribe" as the first word of the subject or the first word of the first line of the message.

MEDICAL DEVICES—Center for Devices and Radiological Health (CDRH)

Call 800-638-2041 (Division of Small Manufacturers Assistance, CDRH) for copies of publications pertaining to device studies.

Call 301-594-1190; fax: 301-594-2977 (Program Operation Staff, CDRH) for questions about:

- whether an investigational device exemption (IDE) is required for a device study
- whether a device is deemed "significant risk" or "non-significant risk"
- unanticipated problems involving risks to subjects [21 CFR 56.108(b)(1)]
- suspension or termination of IRB approval of a protocol [21 CFR 56.108(b)(3)]

Call 888-463-6332 or 800-638-2041; fax: 301-443-9535 (Consumer Staff, Division of Small Manufacturers Assistance, CDRH) for questions about:

- whether a device is approved for marketing

Call 301-594-4718; fax: 301-827-6748 (Division of Bioresearch Monitoring, Office of Compliance, CDRH) for questions about:

- human subject protection regulations pertaining to devices [21 CFR Parts 50, 56, and 812]
- informed consent, standard operating procedures, records, and reports.
- serious or continuing noncompliance by an investigator with FDA regulations or with the IRB's determinations involving a medical device [21 CFR 56.108(b)(2)]
- reporting complaints related to human subject protection/Good Clinical Practice in FDA-regulated medical device trials

"Device Advice" is CDRH's self-service site for medical device and radiation emitting product information.

"Bioresearch Monitoring Advice" provides information about FDA's bioresearch monitoring program.

OTHER

Call 301-402-5552; fax 301-402-2071 [Office for Human Research Protections (OHRP) Education Program] for general information about the Department of Health and Human Services human subject regulations at 45 CFR 46.

Call 301-496-7005; fax 301-402-0527 [Office for Human Research Protections (OHRP) Assurance Program] for information about the OHRP Assurances Program.

Call 301-402-5567; fax 301-402-2071 [Office for Human Research Protections (OHRP) Compliance Oversight Program] for information about noncompliance with HHS regulations.

Appendix K

Web Sites of Interest
For Good Clinical Practice and
Clinical Trial Information

(Revised May 10, 2001)

Website	URL
FDA Websites	
Archiving Submissions in Electronic Format	www.fda.gov/cder/guidance/arcguide.pdf
Bioresearch Monitoring Information System File: Clinical Investigators, CROs and IRBs from FDA 1571 & 1572s	www.fda.gov/cder/foi/special/bmis/index.htm
CDER Guidance Documents	www.fda.gov/cder/guidance/index.htm
CDER Organizational Chart	www.fda.gov/cder/cderorg.htm
CDRH Bioresearch Monitoring	www.fda.gov/cdrh/comp/bimo.html
CDRH Device Advice	www.fda.gov/cdrh/devadvice/
CDRH Organization Structure	www.fda.gov/cdrh/organiz.html
Clinical Investigator Disqualifications Proceedings	www.fda.gov/foi/clinicaldis/
Computerized Systems Used in Clinical Trials	www.fda.gov/ora/compliance_ref/bimo/ffinalcct.doc
Drug Approvals List	www.fda.gov/cder/da/da.htm
Electronic Regulatory Submissions and Review	www.fda.gov/cder/regulatory/ersr/
Expedited Safety Reporting Requirements Oct 7, 1997 Federal Register Final rule	www.fda.gov/cder/regulatory/ click on: (federal register - GPO) (Federal register)(1997)(final rules &regulations) enter date: (on 10/07/1997) search terms: (expedited)
FDA Debarred Persons List	www.fda.gov/ora/compliance_ref/debar/
FDA Disqualified/Restricted/Assurances List for Clinical Investigators	www.fda.gov/ora/compliance_ref/bimo/dis_res_assur.htm
FDA Dockets	www.fda.gov/ohrms/dockets/
FDA Letters Providing Clinical Investigators with Notice of Initiation of Disqualification Proceedings and Opportunity to Explain	www.fda.gov/foi/nidpoe/default.html
FDA Modernization Act of 1997	www.fda.gov/cder/guidance/105-115.htm
CDRH Guidance	www.fda.gov/cdrh/modact/modguid.html
CDER-Related Documents	www.fda.gov/cder/fdama/

FDA Information Sheets

Website	URL
FDA Websites	
FD&C Act	www.fda.gov/opacom/laws/fdcact/fdctoc.htm
Freedom of Information Reading Room	www.fda.gov/foi/
Information for Health Professionals	www.fda.gov/oc/oha/
International Conference on Harmonisation	www.fda.gov/cder/guidance/ guidance.htm#International
Conference on Harmonisation Investigational Human Drugs Clinical Investigator Inspection List	www.fda.gov/cder/regulatory/investigators/default.htm
Investigational Device Exemptions (IDE) Policies and Procedures	www.fda.gov/cdrh/ode/idepolcy.pdf
Laws Enforced by FDA	www.fda.gov/opacom/laws/lawtoc.htm
MedWatch	www.fda.gov/medwatch/
National Drug Code (NDC) directory	www.fda.gov/cder/ndc/
New Drug Approval Packages	www.fda.gov/cder/foi/nda/
Orange Book (Approved Drugs)	www.fda.gov/cder/orange/adp.htm
Pediatric Medicine Page	www.fda.gov/cder/pediatric/
Pharmacy Compounding	www.fda.gov/cder/pharmcomp/
Warning letters	www.fda.gov/foi/warning.htm
Non-FDA Websites	
Clinical Trials Registry	www.clinicaltrials.gov
Government Printing Office (Federal Register, Code of Federal Regulations, Congressional Record)	www.access.gpo.gov/su_docs/
HHS Employee and Organizational Directory	directory.psc.gov
Institute of Medicine of the National Academy of Sciences	www.iom.edu/IOM/IOMHome.nsf/ Pages/human+research+protections
National Bioethics Advisory Comm.	bioethics.gov/cgi-bin/bioeth_counter.pl
National Human Research Protections Advisory Committee	ohrp.osophs.dhhs.gov/nhrpac/nhrpac.htm
Office for Human Research Protections	ohrp.osophs.dhhs.gov
OHRP IRB Guidebook	ohrp.osophs.dhhs.gov/irb/irb_guidebook.htm
PHS List of Investigators Subject to Administrative Action:	silk.nih.gov/public/cbz1bje.@www.orilist.html
Veterans' Administration Office of Research Compliance and Assurance	(ORCA) www.va.gov/orca/

Appendix C

PROTECTION OF HUMAN SUBJECTS
DECLARATION OF HELSINKI

The World Medical Association
Declaration of Helsinki

World Medical Association Declaration of Helsinki: Recommendations Guiding Medical Doctors in Biomedical Research Involving Human Subjects

Adopted by the 18th World Medical Assembly, Helsinki, Finland, 1964 and as revised by the World Medical Assembly in Tokyo, Japan in 1975, in Venice, Italy in 1983, and in Hong Kong in 1989.

Introduction

It is the mission of the physician to safeguard the health of the people. His or her knowledge and conscience are dedicated to the fulfillment of this mission.

The Declaration of Geneva of the World Medical Association binds the physician with the words, "The health of my patient will be my first consideration," and the International Code of Medical Ethics declares that, "A physician shall act only in the patient's interest when providing medical care which might have the effect of weakening the physical and mental condition of the patient."

The Purpose of biomedical research involving human subjects must be to improve diagnostic, therapeutic and prophylactic procedures and the understanding of the aetiology and pathogenesis of disease.

In current medical practice most diagnostic, therapeutic or prophylactic procedures involve hazards. This applies especially to biomedical research.

Medical progress is based on research which ultimately must rest in part on experimentation involving human subjects.

In the field of biomedical research a fundamental distinction must be recognized between medical research in which the aim is essentially diagnostic or therapeutic for a patient, and medical research, the essential object of which is purely scientific and without implying direct diagnostic or therapeutic value to the person subjected to the research.

Special caution must be exercised in the conduct of research which may affect the environment, and the welfare of animals used for research must be respected.

Because it is essential that the results of laboratory experiments be applied to human beings to further scientific knowledge and to help suffering humanity, the World Medical Association has prepared the following recommendations as a guide to every physician in biomedical

435

research involving human subjects. They should be kept under review in the future. It must be stressed that the standards as drafted are only a guide to physicians all over the world. Physicians are not relieved from criminal, civil and ethical responsibilities under the laws of their own countries.

I. Basic Principles

1. Biomedical research involving human subjects must conform to generally accepted scientific principles and should be based on adequately performed laboratory and animal experimentation and on a thorough knowledge of the scientific literature.

2. The design and performance of each experimental procedure involving human subjects should be clearly formulated in an experimental protocol which should be transmitted for consideration, comment and guidance to a specially appointed committee independent of the investigator and the sponsor provided that this independent committee is in conformity with the laws and regulations of the country in which the research experiment is performed.

3. Biomedical research involving human subjects should be conducted only by scientifically qualified persons and under the supervision of a clinically competent medical person. The responsibility for the human subject must always rest with a medically qualified person and never rest on the subject of the research, even though the subject has given his or her consent.

4. Biomedical research involving human subjects cannot legitimately be carried out unless the importance of the objective is in proportion to the inherent risk to the subject.

5. Every biomedical research project involving human subjects should be preceded by careful assessment of predictable risks in comparison with foreseeable benefits to the subject or to others. Concern for the interests of the subject must always prevail over the interests of science and society.

6. The right of the research subject to safeguard his or her integrity must always be respected. Every precaution should be taken to respect the privacy of the subject and to minimize the impact of the study on the subject's physical and mental integrity and on the personality of the subject.

7. Physicians should abstain from engaging in research projects involving human subjects unless they are satisfied that the hazards involved are believed to be predictable. Physicians should cease any investigation if the hazards are found to outweigh the potential benefits.

8. In publication of the results of his or her research, the physician is obliged to preserve the accuracy of the results. Reports of experimentation not in accordance with the principles laid down in this Declaration should not be accepted for publication.

9. In any research on human beings, each potential subject must be adequately informed of the aims, methods, anticipated benefits and potential hazards of the study and the discomfort it may entail. He or she should be informed that he or she is at liberty to abstain from participation in the study and that he or she is free to withdraw his or her consent to participation at any time. The physician should then obtain the subject's freely-given informed consent, preferably in writing.

10. When obtaining informed consent for the research project the physician should be particularly cautious if the subject is in a dependent relationship to him or her or may consent under duress. In that case the informed consent should be obtained by a physician who is not engaged in the investigation and who is completely independent of this official relationship.

11. In case of legal incompetence, informed consent should be obtained from the legal guardian in accordance with national legislation. Where physical or mental incapacity makes it impossible to obtain informed consent, or when the subject is a minor, permission from the responsible relative replaces that of the subject in accordance with national legislation. Whenever the minor child is in fact able to give a consent, the minor's consent must be obtained in addition to the consent of the minor's legal guardian.

12. The research protocol should always contain a statement of the ethical considerations involved and should indicate that the principles enunciated in the present Declaration are complied with.

II. Medical Research Combined with Professional Care (Clinical Research)

1. In the treatment of the sick person, the physician must be free to use a new diagnostic and therapeutic measure, if in his or her judgment it offers hope of saving life, reestablishing health or alleviating suffering.

2. The potential benefits, hazards and discomfort of a new method should be weighed against the advantages of the best current diagnostic and therapeutic methods.

3. In any medical study, every patient—including those of a control group, if any—should be assured of the best proven diagnostic and therapeutic method.

4. The refusal of the patient to participate in a study must never interfere with the physician-patient relationship.

5. If the physician considers it essential not to obtain informed consent, the specific reasons for this proposal should be stated in the experimental protocol for transmission to the independent committee (I,2).

6. The physician can combine medical research with professional care, the objective being the acquisition of new medical knowledge, only to the extent that medical research is justified by its potential diagnostic or therapeutic value for the patient.

III. Non-Therapeutic Biomedical Research Involving Human Subjects (Non-Clinical Biomedical Research)

1. In the purely scientific application of medical research carried out on a human being, it is the duty of the physician to remain the protector of the life and health of that person on whom biomedical research is being carried out.

2. The subjects should be volunteers—either healthy persons or patients for whom the experimental design is not related to the patient's illness.

3. The investigator or the investigating team should discontinue the research if in his/her or their judgment it may, if continued, be harmful to the individual.

4. In research on man, the interest of science and society should never take precedence over considerations related to the well-being of the subject.

GLOSSARY OF TERMS

ABBREVIATED NEW DRUG APPLICATION (ANDA)

Shortened version of a New Drug Application referencing data from other NDAs.

ACTION LETTER

A letter from the Food and Drug Administration to a sponsor indicating a decision on an application submittal. An *approvable letter* indicates the product can be approved after minor issues are resolved. A *nonapprovable letter* describes significant deficiencies in the application that require correction before the application can be considered.

ACTIVE TREATMENT

A treatment in a clinical trial where an active medication, known to be effective, is used, usually as a positive control compared to the investigational agent.

ADJUVANT

Treatment used in addition to the primary therapy.

ADME

Refers to the absorption, distribution, metabolism, and excretion of a drug compound.

ADMINISTRATIVE LOOK

a. Review of data from an ongoing nonconfirmatory study or

b. Review of data from an ongoing confirmatory trial that is used to make administrative decisions about the design of future trials, allocated manufacturing resources, and so on but NOT to modify the ongoing trial.

ADVERSE DRUG REACTION (ADR), ADVERSE REACTION

See Adverse Experience (AE).

ADVERSE EXPERIENCE (AE)

Any undesirable symptom or occurrence that a trial subject experiences during the clinical trial; it may or may not be considered related to the study agent. *Also referred to as* adverse reaction, adverse event, adverse drug reaction (ADR), side effect.

ADVISORY COMMITTEE

A committee of outside experts assembled by the Food and Drug Administration (FDA) to review data from a New Drug Application (NDA) submitted to the FDA. The committee consists of experts in the field and meets as needed. Many advisory committees to the FDA exist and differ for different therapeutic areas. The committee does not approve an NDA but only advises the FDA on the merit of the application.

AMENDMENT

to an Investigational New Drug (IND) Application: A change or addition to an IND filed with the Food and Drug Administration; generally, these include a new protocol, a change to an existing protocol, or a new investigator.

to a New Drug Application (NDA): A supplement to a pending NDA, such as a safety update or data obtained from a supplementary study.

to a protocol: A change in a study protocol requiring an amendment to the protocol.

ASCENDING DOSE

Subjects are dosed with increasingly higher doses of a drug until a maximum tolerated dose (MTD) is reached. These studies are generally Phase I studies.

AUDIT

A careful review of study data, protocol procedures, study conduct, and interim or final study reports to determine whether the conclusions are valid and whether the study has been carried out appropriately.

AUDIT TRAIL

Written record of documents, correspondences, and reports that document study conduct, such as study files, changes to Case Report Forms, and drug accountability records.

BASELINE

Measurements usually taken at the beginning of a study to serve as a reference for subsequent measurements or observations.

BIAS

Influencing study by factors other than the treatment being tested.

BIOAVAILABILITY

Determination of amount of drug detectable in blood (or other body tissues) at various times after administration.

BIOEQUIVALENCE

Term used to describe comparable activity of one drug compound to another (usually a generic product to a product that has received approval). If bioequivalence can be demonstrated, the product does not have to undergo extensive clinical trials to demonstrate safety and efficacy.

BIOPHARMACEUTICAL

Refers to pharmaceutical products developed using biotechnology.

BIRA

British Institute of Regulatory Affairs.

BLINDING

Characteristic of a controlled study design to deter bias in interpretation of reported results. In a *double-blind* study, neither the patient nor the investigator knows which treatment the patient receives. In a *single-blind* study, the patient or observer does not know which treatment is being received. In a *triple-blind* study, the investigator, patient, and sponsor all are blinded to the study medication. The term *open study* refers to a trial where all parties may know the treatment the patient receives.

CANDA

Computer-Assisted New Drug Application; a method of filing a New Drug Application with the Food and Drug Administration where much of the information is transmitted electronically.

CASE HISTORY RECORD

The hospital chart, medical office file, or patient record containing medical and demographic information on the study subject. The "source document" used to verify the authenticity of the information recorded in the Case Report Form.

CASE REPORT FORM (CRF)

Form designed specifically for each protocol to collect data on each subject enrolled in a clinical trial. Information collected on the CRF is determined by the study protocol. *Also referred to as* Case Record Form.

CAUSALITY

Relationship between the adverse experience and the test agent in terms defined in the protocol (e.g., not reasonably attributable, possibly attributable, or reasonably attributable).

CBER

Center for Biologics Evaluation and Research, branch of the Food and Drug Administration.

CDER

Center for Drug Evaluation and Research, branch of the Food and Drug Administration.

CDRH

Center for Devices and Radiological Health, branch of the Food and Drug Administration.

CFR

Code of Federal Regulations.

CLINICAL INVESTIGATION

According to Title 21, Part 312.3, of the U.S. Code of Federal Regulations, "means any experiment in which a drug is administered or dispensed to, or used involving, one or more human subjects." *See also* Clinical Trial.

CLINICAL INVESTIGATOR

See Investigator.

CLINICAL RESEARCH ASSOCIATE (CRA)

A qualified individual working with the sponsor to oversee the progress of a clinical trial, the liaison between the sponsor and the investigator/site. *Also referred to as* Clinical Research Scientist (CRS), Medical Research Associate (MRA), and Monitor.

CLINICAL RESEARCH COORDINATOR (CRC)

Usually a nurse or other health professional, this individual is the study site's organizer of day-to-day conduct of study activities, including completing Case Report Forms, maintaining study files, and assisting the investigator. *Also referred to as* study coordinator, study nurse, research coordinator, or clinical coordinator.

CLINICAL STUDY AGREEMENT (CSA)

See Contract.

CLINICAL TRIAL

The systematic investigation of the effects of materials (e.g., investigational drugs, devices) or methods (e.g., surgery, radiation) on a disease state conducted according to a formal study plan (protocol). Generally, a clinical trial refers to the evaluation of treatment methods (drugs, surgery, etc.), although methods of prevention, detection, or diagnosis may also be the object of a clinical trial.

In the pharmaceutical industry, clinical trials are typically a systematic study of a medicinal product or device in human subjects (patients or nonpatient volunteers) in order to discover or verify the effects of and identify adverse reactions to investigational products in order to ascertain the efficacy and safety of the investigational agents. Also, clinical trials may study the absorption, distribution, metabolism, and excretion as well as the pharmacodynamic interaction of investigational agents.

CLINICAL TRIAL EXEMPTION [CT(X)]

A means of obtaining rapid approval of clinical trials by submitting only summary data (chemistry, pharmacology, toxicology, and volunteer studies).

CODE BREAKER

A sealed envelope or label that contains the identity of the test agent for each study subject; should be opened only under emergency or unusual circumstances, as specified by the protocol and/or the study sponsor. *Also called* unblinding envelope.

COINVESTIGATOR/SUBINVESTIGATOR

A physician or qualified individual who assists the Principal Investigator in the conduct of the clinical trial; listed under item 6 of the FDA Statement of Investigator (Form FDA 1572). *Subinvestigator* is the preferred term of the FDA.

COMBINATION THERAPY

The use of two or more modes of treatment—surgery, radiotherapy, drug therapy—in combination, alternately or together, to treat a disease.

COMPARATIVE STUDY

A study design where the investigative agent is compared to another treatment (e.g., placebo, active treatment).

COMPASSIONATE USE

Circumstances under which certain Food and Drug Administration regulations may be exempt to allow the use of an investigational agent for a single patient.

COMPLIANCE

Patient: A term referring to the degree to which the patient has followed the instructions and dosing requirements of the protocol.

Protocol: Refers to adherence to the procedures defined in the study protocol.

CONFIDENTIALITY AGREEMENT

An agreement between two parties (an investigator and sponsor or sponsor and CRO, etc.) where it is agreed that information regarding the study and investigational agent will be kept strictly confidential.

CONFIRMATORY STUDY

Any clinical study designed to provide the substantial evidence of efficacy required for regulatory approval. These studies are typically double blind with a randomized control group.

CONTRACT (Clinical Study Agreement, [CSA], Clinical Trial Agreement [CTA])

A document signed and dated by the investigator (or institution representative) and sponsor representative that delineates agreements on financial matters and delegation/distribution of responsibilities.

CONTRACT RESEARCH ORGANIZATION (CRO)

An independent organization that contracts with the sponsor to assume some of the sponsor's responsibilities for conducting clinical trials. According to Title 21, Part 312.2, of the U.S. Code of Federal Regulations, "means a person that assumes, as an independent contractor with the sponsor, one or more obligations of a sponsor, e.g., design of a protocol, selection or monitoring of investigations, evaluation of reports, and preparation of materials to be submitted to the Food and Drug Administration."

CONTRAINDICATION

An indication or condition in which it is recommended that a drug NOT be administered.

CONTROL GROUP

The group of patients receiving the standard treatment or placebo used for comparison to results obtained in the "treatment group," the group of patients undergoing the experimental treatment regimen.

CONTROLLED CLINICAL TRIAL

A study design that compares the investigational drug with either placebo or with another treatment known to be effective against the disease in which subjects are randomly allocated to treatment groups.

COORDINATING CENTER

A clinical research site that will coordinate activities and data management in multi-center trials.

COORDINATING INVESTIGATOR

Investigator assigned to coordinate other investigators at different study sites in a multi-center trial.

CPMP

Committee for Proprietary Medicinal Products.

CROSSOVER DESIGN

A study design that has each patient participate in two or more treatments in a specified order.

CURRICULUM VITAE (CV)

Prepared by an investigator to summarize his/her training and expertise; similar to a resume.

DATA

Information obtained from a clinical trial, usually on a Case Report Form (CRF) or clinical lab electronic file.

DATA AND SAFETY MONITORING BOARD (DSMB)

An independent group that will review data from ongoing blinded clinical trials to evaluate excessive risk or profound efficacy. The DSMB can stop the clinical trial for excessive toxicity or if evidence is adequate to show treatment is beneficial. *Also called* IDMC, Independent Data Monitoring Committee.

DATA AUDIT

Comparison of source documentation of original data to the data transcribed on a Case Report Form as a check for discrepancies.

DATA EDIT

Comparison of data, manually or automated, to detect incorrect information for the purpose of clarification and quality control of the database.

DATA MANAGEMENT

All data-processing activities, automated and manual, beginning with data collection and transcription through the generation of tables and charts.

DECLARATION OF HELSINKI

See Helsinki, Declaration of.

DHHS

Department of Health and Human Services.

DOCUMENTATION

All records in any form (documents, electronic files and optical records) describing methods and conduct of a clinical trial, as well as factors affecting the trial and action taken.

DOSE–RANGING STUDIES

A study design to evaluate the effect and/or safety of different doses of an investigational agent.

DOUBLE BLIND

See Blinding.

DROPOUT

A subject who does not complete all of the protocol-required parameters for a clinical trial.

DSMB

Data and Safety Monitoring Board.

EC

European Community.

EFFICACY

A measure of a drug's ability to ameliorate the signs and/or symptoms of a disease.

EFPIA

European Federation of Pharmaceutical Industries and Associations.

EMEA

European Agency for the Evaluation of Medicinal Products.

ENDPOINT

A predetermined event (per protocol) that indicates a patient's completion of the trial, either by disease state (cure, progression) or by completion of all study visits.

ETHICS COMMITTEE (EC)

An independent group of medical and nonmedical professionals whose purpose is to verify that the clinical trial is performed safely, with integrity, and with respect to the rights of the human subjects. Most countries require that an EC provide a statement of its opinion on any research involving human subjects. The Ethics Committee is the European Union equivalent of the U.S. Institutional Review Board.

EU

European Union.

EVALUABLE PATIENT

Patient in a clinical trial who has satisfied all protocol requirements and may be evaluated for safety and efficacy in the analysis.

FINAL REPORT (by Investigator)

Each investigator is required to summarize the clinical trial (per specifications of their Institutional Review Board [IRB]) and submit it to the IRB and sponsor in a final report.

FINAL STUDY REPORT (*or* Final Medical Report)

A complete and comprehensive description of the completed study, including descriptions of experimental materials and methods, presentation and evaluation of the results, statistical analyses, and a critical discussion of the results.

FOOD AND DRUG ADMINISTRATION (FDA)

The federal agency responsible for regulating the sale of food, drugs, and cosmetics in the United States.

GOOD CLINICAL PRACTICE (GCP)

Ethical and scientific standards by which clinical trials are designed, implemented, and reported to insure that the data are scientifically sound and that the rights of the subjects are protected. Refer to Title 21, Parts 50, 56, 312, 314, 812, and 813 of the U.S. Code of Federal Regulations.

GOOD LABORATORY PRACTICE (GLP)

Regulations pertaining to research laboratories. Refer to Title 21, Part 58, of the U.S. Code of Federal Regulations.

GOOD MANUFACTURING PRACTICE (GMP)

That part of pharmaceutical quality assurance that ensures that products are consistently produced and controlled in conformity with quality standards appropriate for their intended use and as required by the product specification. Refer to Title 21, Part 211, of the U.S. Code of Federal Regulations.

HELSINKI, DECLARATION OF

International document concerning the ethical conduct of clinical trials.

ICH

International Conference on Harmonisation.

IDE

Investigational Device Exemption.

IDMC

See Data and Safety Monitoring Board.

INDEMNIFICATION

A legal document indicating protection or exemption from liability for compensation or damages from a third party; usually protects an investigator and/or hospital or institution from claims made by the study subject (or relatives) that harm was caused to the subject as a result of participation in the clinical trial.

INDICATION

The disease state or medical problem being evaluated with the study agent.

INFORMATION AMENDMENT

Refers to an amendment to the Investigational New Drug Application (not necessarily to a specific protocol) that provides additional information, such as the addition of a new investigator.

INFORMED CONSENT FORM

Form used to confirm a trial subject's willingness to participate voluntarily in a study. The subject or legal representative signs the form after all appropriate information about the trial, including objectives, potential benefits and risks, and subject rights and responsibilities, has been explained and all subject questions have been answered. The Informed Consent Form and information must be reviewed and approved by the Institutional Review Board.

INSPECTION

An official audit conducted by regulatory authorities of the Food and Drug Administration, sponsor, or cooperative group at the site of investigation and/or the sponsor.

The purpose of the inspection is to verify adherence to applicable regulations and guidelines, including those of Good Clinical Practice.

INSTITUTIONAL REVIEW BOARD (IRB)

An independent body of medical and nonmedical members established according to requirements outlined in Title 21, Part 56, of the U.S. Code of Federal Regulations. The IRB, usually institution specific, is responsible for the initial and continuing approval of research involving human subjects, as well as for verifying the protection of safety and rights of those human subjects.

INVESTIGATIONAL AGENT OR PRODUCT

A pharmaceutical product, placebo, or device being used in an investigational clinical trial.

INVESTIGATIONAL NEW DRUG

According to Title 21, Part 312.2, of the U.S. Code of Federal Regulations, "means a new drug, antibiotic drug, or biological drug that is used in a clinical investigation. The term also includes a biological product that is used in vitro for diagnostic purposes."

INVESTIGATIONAL NEW DRUG (IND) APPLICATION

In the United States, process by which investigational new drugs are registered with the Food and Drug Administration for administration to human subjects in clinical trials; includes information on pharmacology, chemistry, toxicology, previous clinical studies results, and future study proposals.

INVESTIGATOR (Principal Investigator)

As the leader of the investigational team, this individual (usually a physician or dentist) is responsible for conducting the clinical trial and ensuring the safety and welfare of the study subjects. The investigator signs the Statement of Investigator form (Form FDA 1572). According to Title 21, Part 312.3, of the U.S. Code of Federal Regulations, "means an individual who actually conducts a clinical investigation (i.e., under whose immediate direction the drug is administered or dispensed to a subject). In the event an investigation is conducted by a team of individuals, the investigator is the responsible leader of the team. "Subinvestigator" includes any other member of that team."

INVESTIGATOR'S BROCHURE

Collection of all relevant information on the investigational product known prior to the start of a particular clinical trial, including preclinical data such as chemical, pharmaceutical, toxicological; pharmacokinetic and pharmacodynamic data in animals and in man; and the results of earlier clinical trials. The data should support the justification for the proposed trial and evaluate safety or precautions. The brochure should be updated on a continual basis as new information is gathered.

LABORATORY CERTIFICATION

A certificate given to a laboratory indicating that the laboratory is capable of performing all tests as required by use of a proficiency testing program. The certification is usually renewed on a biannual or annual basis after appropriate inspection and testing.

MARKETING AUTHORIZATION APPLICATION (MAA)

A complete dossier of information, including chemical, pharmaceutical, biological, and clinical data, which is sent to a regulatory authority to support a request for marketing authorization in the European Union.

MAXIMUM TOLERATED DOSE (MTD)

The dose determined to be the highest dose to give subjects without unacceptable side effects.

MONITOR

See Clinical Research Associate (CRA).

MONITORING

A contact by the sponsor with an investigator or member of the investigative staff that serves to further the progress of a clinical trial.

MOU

Memo of Understanding. A document between the Food and Drug Administration and other regulatory agencies that allows mutual inspection.

MULTICENTER TRIAL

A clinical trial conducted according to a single protocol at various investigational sites by various investigators.

NEW DRUG APPLICATION (NDA)

The complete dossier of information submitted to the Food and Drug Administration to request marketing authorization for an investigational agent. The contents include chemical, pharmaceutical, biological, and clinical data.

NEW MOLECULAR ENTITY (NME)

(*also referred to as* New Chemical Entity [NCE])

An active ingredient of a drug preparation that has not been previously marketed in the United States.

NIH

National Institutes of Health. A federal agency under the Department of Health and Human Services that is composed of several institutes and centers dedicated to specific areas of medical and health research.

NONCLINICAL STUDIES

See Preclinical Studies.

OPEN LABEL STUDY

A study in which the treatment schedules, drug treatment, and doses are known to both the investigator and the subject.

OHRP

Office of Human Research Protection (DHHS).

OHRT

Office of Human Research Trials (FDA).

OHSP

Office of Human Subject Protection (NIH).

OUTCOME

A result, condition, or event associated with individual study subjects used to assess efficacy.

PACKAGE INSERT

Refers to the prescribing information supplied with a marketed pharmaceutical product and summarizes known information about dosing, safety, and indications.

PARALLEL STUDY DESIGN

A study design where subjects are randomized to one treatment plan for the duration of the trial (as opposed to "crossover" design).

PATIENT INFORMATION SHEET

European Community equivalent of Informed Consent Form; may also refer to the information provided to subjects prior to signing an Informed Consent Form. *See also* Informed Consent Form. May also refer to instructions to patients for administration of investigational agents.

PHARMACEUTICAL PRODUCT

Any substance or combination of substances that has a therapeutic, prophylactic, or diagnostic purpose intended to modify physiological functions and presented in a dosage form suitable for administration to humans.

PHARMACODYNAMICS (PD)

The science involving the pharmacology of the interaction of drugs in a physiological environment.

PHARMACOKINETICS (PK)

The science involving the absorption, distribution, metabolism, and elimination (ADME) of drugs.

PHARMACOECONOMIC STUDY

The study of a specific treatment in relation to the economic benefits of the treatment.

PHARMACOEPIDEMIOLOGY

The study of the use of drugs in the general population and large numbers of specific group types.

PHARMACOLOGY

The science involving drugs—their sources, appearance, chemistry, actions, and uses.

PHASE I

The first clinical trials conducted after filing an Investigational New Drug Application. Generally aimed at establishing safety, pharmacokinetics, and doses with a small number of normal volunteers. *See also* Title 21, Part 312.21, of the U.S. Code of Federal Regulations.

PHASE II

After Phase I studies, Phase II studies are the first look at efficacy in a given indication. They are usually randomized, tightly controlled studies using a relatively small number of carefully selected patients. *See also* Title 21, Part 312.21, of the U.S. Code of Federal Regulations.

PHASE III

Clinical trials where the number of subjects is expanded and the inclusion criteria are less stringent to gain experience with the investigational agent in a large number of patients. Also, specific patient populations, such as geriatrics and pediatrics, may be investigated. *See also* Title 21, Part 312.21, of the U.S. Code of Federal Regulations.

PHASE IV

Phase IV trials are often referred to as "postmarketing studies" and are done for a variety of reasons: to place the drug in the market, to make marketing claims, or to conduct pharmacoeconomic studies and quality-of-life studies. New formulations of the drug or new indications must be investigated as Phase I/II clinical trials.

PLACEBO

An inactive substance made to appear identical to the test agent in appearance and taste used as a control in clinical studies.

PMA

Premarket Approval Application. Refers to devices and application for marketing.

POSTMARKETING SURVEILLANCE

Monitoring by the sponsor of the use of a drug in the general population after approved for marketing to evaluate adverse events.

PRECLINICAL STUDIES

Studies done prior to human clinical trials and aimed at establishing information about a new drug, such as absorption, distribution, metabolism, elimination, toxicity, and carcinogenicity. Preclinical studies may continue after studies in humans are underway. *Also referred to as* Nonclinical Studies.

PRINCIPAL INVESTIGATOR

See Investigator.

PROTOCOL

A detailed plan for the investigation of an experimental agent, treatment, or procedure. This document explains the background, rationale, and objectives of the trial and specifically outlines the design, methodology, organization, and condition of conducting the study.

PROTOCOL–SPECIFIED (PLANNED) ADMINISTRATIVE LOOK

An administrative review of the data in a study that is planned as an integral part of the protocol.

RANDOMIZATION

A method by which study subjects are assigned to a treatment group to obtain equal, comparable treatment groups.

RANDOMIZATION CODE

Investigational agent randomization to a treatment group by subject number. In addition to indicating the randomization on prepackaged drug for individual subjects, subjects may be randomized to a treatment arm by prerandomized numbers sealed in envelopes or randomization cards.

REGIONAL CLINICAL RESEARCH ASSOCIATE (RCRA)

Monitors located in geographic regions. *See also* Clinical Research Associate (CRA).

REGULATORY FILE

See Study files.

RISK–BENEFIT RATIO

The relationship between the risks and benefits of a given treatment or procedure. Institutional Review Boards determine that the risks in a study are reasonable with respect to the potential benefits.

SAFETY

Refers to the evaluation of the safeness of a drug when used in humans. May refer to establishing safety of a new drug in an investigational trial or surveillance of the safety of a marketed drug.

SERIOUS ADVERSE EXPERIENCE (SAE)

Any experience that suggests a significant hazard, contraindication, side effect, or precaution. This includes, but is not limited to, any experience that is fatal, life-threatening, permanently or significantly disabling, or requires inpatient hospitalization or prolongation of hospitalization. In addition, congenital anomaly, occurrence of malignancy, and overdose are always regarded as serious.

SIDE EFFECT

See Adverse Experience (AE).

SINGLE BLIND

See Blinding.

SOURCE DATA

See Source Document.

SOURCE DOCUMENT

Trial subject's medical chart, laboratory report, nurses' notes, or any official record documenting original observations or activities. Source documents are used for verification of the data entered on the Case Report Form.

SOURCE DOCUMENT VERIFICATION

Process of assuring the validity and completeness of the data recorded in the Case Report Form (CRF) by comparing the information in the source document to that recorded on the CRF.

SPONSOR

Individual or organization that takes responsibility for initiation, organization, and management of a clinical trial. According to Title 21, Part 312.2, of the U.S. Code of Federal Regulations, "means a person who takes responsibility for and initiates a clinical investigation. The sponsor may be an individual or pharmaceutical company, government agency, academic institution, private organization, or other organization. The sponsor does not actually conduct the investigation unless the sponsor is a sponsor-investigator. A person other than an individual that uses one or more of its own employees to conduct an investigation that it has initiated is a sponsor, not a sponsor-investigator, and the employees are investigators."

SPONSOR–INVESTIGATOR

According to Title 21, Part 312.3, of the U.S. Code of Federal Regulations, "means an individual who both initiates and conducts an investigation and under whose immediate direction the investigational drug is administered or dispensed. The term does not include any person other than an individual. The requirements applicable to a sponsor-investigator under this part include both those applicable to an investigator and a sponsor."

STAGING

A process categorizing the extent of disease for enrollment into a clinical trial.

STANDARD OPERATING PROCEDURE

Detailed, written instructions for the management of clinical trials.

STANDARD TREATMENT

The treatment currently being used for an indication and considered to be of proven effectiveness.

STATEMENT OF INVESTIGATOR (SOI) FORM (*also called* Form FDA 1572)

FDA–required document for all clinical trials conducted as part of a U.S. Investigational New Drug (IND) Application to register the investigator to do research for the IND; signed by investigator to indicate his/her acceptance of key responsibilities of the clinical trial; contains information about the trial, investigator(s), and key responsibilities.

STRATA/STRATIFICATION

Subgroup of subjects selected by certain variables usually at baseline.

STUDY ARM

One part, segment, or specific treatment group of a study.

STUDY COORDINATOR

See Clinical Research Coordinator (CRC).

STUDY DRUG

The investigational agent(s) being studied in a particular clinical trial; may be in solid, liquid, or gas (such as anesthetic) form.

STUDY FILES

The files located at the study site that pertain to an investigator's documentation of a clinical trial. *Also referred to as* Regulatory File.

STUDY PROCEDURES MANUAL

Document that gives practical guidelines for conducting a specific protocol and clinical trial. It may include guidelines for completing Case Report Forms, handling adverse experiences, collecting lab specimens, and so on.

SUBINVESTIGATOR

A qualified individual (usually a physician or dentist) who assists the Primary Investigator in the conduct of a clinical trial. *See also* Coinvestigator/Subinvestigator and Investigator (Principal Investigator).

SUBJECT

A human being (patient or nonpatient volunteer) participating in a clinical trial. They may be a

a. healthy person volunteering in a trial,

b. person with a condition *unrelated* to the use of the investigational product,

c. person whose condition *is related* to the use of the investigational product, or

d. recipient of the study drug being tested or a control.

According to Title 21, Part 312.2, of the U.S. Code of Federal Regulations, "means a human who participates in an investigation, either as a recipient of the investigational new drug or as a control. A subject may be a healthy human or a patient with a disease."

SURROGATE OUTCOME

The use of a test or measurement instead of a clinical event as an outcome of a clinical trial.

TEAR–OFF LABELS

Refers to a specific type of label for investigational agent where a portion of the package label can be torn off and may contain specific information about the investigational agent, e.g., the identity of the randomized code.

THERAPEUTIC

Pertaining to treatment.

TOXICOLOGY

The study of the toxic pharmacology of a compound.

TREATMENT GROUP

Group of patients receiving the experimental treatment regimen.

TREATMENT INVESTIGATIONAL NEW DRUG (Treatment IND)

A mechanism by which a drug is approved for treatment use and made available to patients before it has been approved by the Food and Drug Administration for sale.

TRIPLE BLIND

See Blinding.

UNCONTROLLED

A study where the investigational treatment is not being compared (by study design) to a concurrent treatment as a control.

UNEXPECTED ADVERSE EXPERIENCE

Defined by Food and Drug Administration regulations as any adverse experience that is *not* identified in nature, severity, or frequency in the current Investigator's Brochure or in the risk information described in the investigational plan or in the current Investigational New Drug Application.

UNPLANNED ADMINISTRATIVE LOOK

An administrative review of data that was not anticipated in the protocol but arose after the trial started. Generally, this is done to analyze the data if the drug seems to be extremely effective or toxic. The need for unplanned administrative looks may arise occasionally, but these should be done sparingly since they can interfere with the statistical effectiveness of the results of the trial.

WHO

World Health Organization.

ABBREVIATIONS

Following is a brief list of common abbreviations used in the conduct of clinical trials.

ACRP	Association of Clinical Research Professionals
ADE	Adverse Drug Experience
ADME	absorption, distribution, metabolism, excretion
ADR	Adverse Drug Reaction
AE	Adverse Event, Adverse Experience
ANDA	Abbreviated New Drug Application
BSA	body surface area
CANDA	Computer-Assisted New Drug Application
CBER	Center for Biologics Evaluation and Research
CCRA	Certified Clinical Research Associate
CCRC	Certified Clinical Research Coordinator
CDC	Centers for Disease Control
CDER	Center for Drug Evaluation and Research
CFR	Code of Federal Regulations
CLIA	Clinical Laboratory Improvements Amendments
COSTART	Coding Symbols for a Thesaurus of Adverse Reaction Terms
CRA	Clinical Research Associate
CRC	Clinical Research Coordinator
CRF	Case Report Form, Case Record Form
CRO	Contract Research Organization
CTA	Clinical Trial Agreement
CTX	Clinical Trial Exemption
CV	curriculum vitae
DHHS	Department of Health and Human Services

DIA	Drug Information Association
DOD	Department of Defense
DRG	diagnosis-related groups
	Division of Research Grants (NIH)
DSMB	Data and Safety Monitoring Board
EC	European Commission
EIR	Establishment Inspection Report
FDA	Food and Drug Administration
FDLI	Food and Drug Law Institute
GCP	Good Clinical Practice
GLP	Good Laboratory Practice
GMP	Good Manufacturing Practice
HHS	(U.S. Department of) Health and Human Services
ICH	International Conference on Harmonization
IDE	Investigational Device Exemption
IDMC	Independent Data Monitoring Committee
IND	Investigational New Drug (Application)
IRB	Institutional Review Board
JCAHO	Joint Commission for the Accreditation of Hospitals
MTD	maximum tolerated dose
NAF	notice of adverse findings
NAI	no action indicated
NCE	New Chemical Entity
NDA	New Drug Application
NIH	National Institutes of Health
NME	New Molecular Entity
OAI	official action indicated
OHRP	Office of Human Research Protection (formerly OPRR)
OHRT	Office of Human Research Trials
OHSP	Office of Human Subject Protection
OIG	Office of the Inspector General
OPRR	Office of Protection from Research Risks
OSHA	Occupational Safety and Health Administration

OTC	over-the-counter
PD	pharmacodynamics
PERI	Pharmaceutical Education and Research Institute
PhRMA	Pharmaceutical Research and Manufacturers of America
PHS	Public Health Service
PI	Principal Investigator
	package insert
PK	pharmacokinetics
PLA	Product License Application
PMA	Premarket Approval (application)
PPI	patient package insert
RDE	remote data entry
SMO	site management organization
SOP	Standard Operating Procedure
SUD	sudden unexpected death
TIND	Treatment Investigational New Drug
TMO	trial management organization
VAI	voluntary action indicated
WHO	World Health Organization
WHOART	World Health Organization Adverse Reaction Terminology

Resources for additional abbreviations:

Glossary: Acronyms, Abbreviations, and Initials. *Applied Clinical Trials,* Vol. 9 (12), pp. 24–28, December 2000.

The PF (Pitfalls) of B (Brevity). A. Papke, *Journal of Clinical Research and Drug Development,* Vol. 7 (2), pp. 77–86, 1993.

Medical Abbreviations: 15,000 Conveniences at the Expense of Communications and Safety, 10th ed. N. M. Davis, Neil M. Davis Associates, Huntingdon Valley, Penn, 2001. http://www.neilmdavis.com.

Index

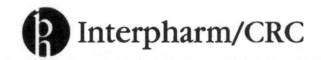 Interpharm/CRC

Order Form

TITLE	Author/Editor	Pub Date	Cat. no.	ISBN #	QTY	PRICE	TOTAL
BIOMEDICAL SCIENCE MICROBIOLOGY							
Microbiological Assay: A Rational Approach	Hewitt, William	22-Aug-03	PH1824	0849318246		$229.95 / £154.00	
PHARMACEUTICAL SCIENCE & REGULATION BIOTECHNOLOGY/BIOPHARMACEUTIC							
International Biotechnology, Bulk Chemical, and Pharmaceutical GMPs, Fifth Edition	Anisfeld, Michael H.	31-May-99	PH1838	0849318386		$329.95 / £220.00	
Pharmaceutical Biotechnology, Second Edition	Groves, Michael J.	18-Oct-03	PH1873	0849318734		$139.95 / £93.00	
PHARMACEUTICAL SCIENCE & REGULATION CLEANING & STERILIZATION							
Aseptic Pharmaceutical Manufacturing II: Applications for the 1990s	Groves, Michael J.	31-May-95	PH4775	0935184775		$179.00 / £120.00	
Clean Room Design: Minimizing Contamination Through Proper Design	Ljungqvist, Bengt	30-Nov-96	PH0329	1574910329		$89.95 / £59.99	
Cleaning and Cleaning Validation: A Biotechnology Perspective	Voss, Jon	30-Jun-96	PH9507	0939459507		$249.95 / £167.00	
Cleaning Validation: A Practical Approach	Bismuth, Gil	31-Jan-00	PH1082	1574911082		$199.95 / £133.00	
Cleanroom Microbiology for the Non-Microbiologist	Carlberg, David M.	30-Apr-95	PH4732	0935184732		$119.00 / £79.99	
Fluid Sterilization by Filtration: The Filter Integrity Test and Other Filtration Topics, Second Edition	Johnston, Peter R.	31-Jul-97	PH0396	1574910396		$149.95 / £99.00	
Isolation Technology: A Practical Guide	Coles, Tim	31-Dec-97	PH0590	1574910590		$199.95 / £133.00	
Sterile Facility Product Design and Project Management, Second Edition	Odum, Jeffrey N.	25-Nov-03	PH1874	0849318742		$189.95 / £127.00	
Sterilization of Drugs and Devices: Technologies for the 21st Century	Nordhauser, Fred M.	30-Apr-98	PH0604	1574910604		$239.95 / £160.00	
Sterilization of Medical Devices	Booth, Anne F.	30-Nov-98	PH0876	1574910876		$179.95 / £120.00	
Validated Cleaning Technologies for Pharmaceutical Manufacturing	LeBlanc, Destin A.	28-Feb-00	PH1663	1574911163		$199.95 / £133.00	
PHARMACEUTICAL SCIENCE & REGULATION CLINICAL TRIALS							
Clinical Development: Strategic, Pre-Clinical, and Regulatory Issues	Steiner, Janice	30-Nov-96	PH0280	1574910280		$149.95 / £99.00	
Clinical Research Coordinator Handbook: GCP Tools and Techniques, Second Edition	Dresser, Michelle	01-Oct-01	PH1236	1574911236		$189.95 / £127.00	
Clinical Research Monitor Handbook: GCP Tools and Techniques, Second Edition	Rosenbaum, Deborah	30-Jun-98	PH1252	1574911252		$189.95 / £127.00	
Drug Regimen Compliance	Metry, Jean-Michel	30-Jan-98	PH1227	0471971227		$169.95 / £113.00	
GCP Harmonization Handbook, The	Maynard, Donald E.	31-May-96	PH0132	1574910132		$199.95 / £133.00	
Handbook of SOPs for Good Clinical Practice, A	Maynard, Donald E.	28-Feb-96	PH0094	1574910094		$229.95 / £153.00	
International Clinical Trials: A Guidebook and Compendium of National Drug Laws, Two-Volume Set	Brunier, Dominique	30-Jun-99	PH0949	1574910949		$379.95 / £253.00	
International Medical Device Clinical Investigations: A Practical Approach	Pieterse, Herman	30-Dec-98	PH054X	157491054X		$0.00 / £0.00	
International Medical Device Clinical Investigations: A Practical Approach, Second Edition	Pieterse, Herman	30-Apr-99	PH085X	157491085X		$239.95 / £160.00	
Outsourcing in Clinical Drug Development	Drucker, Roy	15-Aug-02	PH1120	1574911120		$249.95 / £167.00	
Physician Investigator Handbook: GCP Tools and Techniques, Second Edition	Rosenbaum, Deborah	01-Jan-02	PH1244	1574911244		$189.95 / £127.00	
Practical Clinical Trials Resource Guide	Rosenbaum, Deborah	25-Jul-03	PH1870	084931870X		$69.95 / £46.99	

TITLE	Author/Editor	Pub Date	Cat. no.	ISBN #	QTY	PRICE	TOTAL
Practical Guide to Clinical Data Management	Prokscha, Susanne	31-Jan-99	PH0434	1574910434		$209.95 / £140.00	
Veterinary Clinical Trials From Concept to Completion	Dent, Nigel	31-Dec-01	PH121X	157491121X		$249.95 / £167.00	

PHARMACEUTICAL SCIENCE & REGULATION COMPUTER SOFTWARE

TITLE	Author/Editor	Pub Date	Cat. no.	ISBN #	QTY	PRICE	TOTAL
Computer Validation Compliance: A Quality Assurance Perspective	Double, Mary Ellen	31-Jan-94	PH4481	0935184481		$199.95 / £133.00	
Electronic Communication Technologies: A Practical Guide for Healthcare Manufacturers	Mitchard, Mervyn	30-May-98	PH0698	1574910698		$239.95 / £160.00	
Good Computer Validation Practices: Common Sense Implementation	Stokes, Teri	31-May-94	PH4554	0935184554		$249.95 / £167.00	
Practical Computer Validation Handbook	McDowall, R.D.	25-Nov-03	PH1880	0849318807		$269.95 / £180.00	
Software Development and Quality Assurance for the Healthcare Manufacturing Industries, Third edition	Mallory, Steven R.	31-Jul-02	PH1368	1574911368		$239.95 / £160.00	
Software Quality Assurance SOPs for Healthcare Manufacturers, Second Edition	Mallory, Steven R.	30-Jun-02	PH135X	157491135X		$219.95 / £147.00	
Software Quality Assurance: A Guide for Developers and Auditors	Smith, Howard T. Gar	30-Jun-97	PH0493	1574910493		$229.95 / £153.00	
Survive and Thrive Guide to Computer Validation, The	Stokes, Teri	31-May-98	PH0671	1574910671		$239.95 / £160.00	
Validating Corporate Computer Systems: Good IT Practice for Pharmaceutical Manufacturers	Wingate, Guy	31-May-00	PH1171	1574911171		$239.95 / £160.00	
Validation of Computerized Analytical and Networked Systems	Huber, Ludwig	01-Oct-01	PH1333	1574911333		$269.95 / £180.00	
Validation of Computerized Analytical Systems	Huber, Ludwig	31-May-95	PH4759	0935184759		$199.95 / £133.00	

PHARMACEUTICAL SCIENCE & REGULATION DRUG DEVELOPMENT

TITLE	Author/Editor	Pub Date	Cat. no.	ISBN #	QTY	PRICE	TOTAL
Advances in Drug Discovery Techniques	Harvey, Alan L.	15-Aug-98	PH5095	0471975095		$169.95 / £113.00	
Drug Development Programme Management	Lead, Barbara Ann	31-Oct-00	PH1112	1574911112		$209.95 / £140.00	
Good Pharmaceutical Freeze-Drying Practice	Cameron, Peter	30-Jun-97	PH0310	1574910310		$199.95 / £133.00	
Guide to Pharmaceutical Particulate Science, A	Hickey, Anthony	13-Mar-03	PH1844	1574911422		$239.95 / £160.00	
Injectable Drug Development: Techniques to Reduce Pain and Irritation	Gupta, Pramod K.	31-May-99	PH0957	1574910957		$249.95 / £167.00	
Lyophilization: Introduction and Basic Principles	Jennings, Thomas A.	31-Aug-99	PH0817	1574910817		$219.95 / £147.00	
Pharmaceutical Preformulation and Formulation: A Practical Guide from Candidate Drug Selection to Commercial Dosage Form	Gibson, Mark	01-Aug-01	PH1201	1574911201		$269.95 / £180.00	
Separations Technology: Pharmaceutical and Biotechnology Applications	Olson, Wayne P.	01-Jun-95	PH4724	0935184724		$179.95 / £120.00	
Sustained-Release Injectable Products	Senior, Judy	31-Mar-00	PH1015	1574911015		$259.95 / £173.00	
Transdermal and Topical Drug Delivery Systems	Ghosh, Tapash K.	30-Jun-97	PH0418	1574910418		$239.95 / £160.00	
Twenty-First Century Pharmaceutical Development	Blaisdell, Peter	31-Oct-00	PH1023	1574911023		$239.95 / £160.00	
Water-Insoluble Drug Formulation	Liu, Rong	30-Sep-00	PH1058	1574911058		$269.95 / £180.00	

PHARMACEUTICAL SCIENCE & REGULATION LABORATORY

TITLE	Author/Editor	Pub Date	Cat. no.	ISBN #	QTY	PRICE	TOTAL
Automated Microbial Identification and Quantitation: Technologies for the 2000s	Olson, Wayne P.	31-Jan-96	PH4821	0935184821		$197.00 / £133.00	
Calibration in the Pharmaceutical Laboratory	Kowalski, Tony	31-Dec-01	PH0922	1574910922		$229.95 / £153.00	
Data Acquisition and Measurement Techniques	Munoz-Ruiz, Angel	30-Jun-98	PH068X	157491068X		$229.95 / £153.00	
GLP Essentials: A Concise Guide to Good Laboratory Practice, Second Edition (5-pack)	Anderson, Milton A.	30-Jun-02	PH1384	1574911384		$129.95 / £87.00	
GLP Quality Audit Manual, Third edition	Anderson, Milton A.	22-Jun-00	PH1066	1574911066		$189.95 / £127.00	
International Stability Testing	Mazzo, David J.	31-Aug-98	PH0787	1574910787		$219.95 / £147.00	
Managing the Analytical Laboratory: Plain and Simple	Nilsen, Clifford	31-May-96	PH0159	1574910159		$179.95 / £120.00	
Microbial Limit and Bioburden Tests: Validation Approaches and Global Requirements	Clontz, Lucia	31-Oct-97	PH0620	1574910620		$189.95 / £127.00	
Rapid Microbiological Methods in the Pharmaceutical Industry	Easter, Martin	13-Mar-03	PH1414	1574911414		$239.95 / £160.00	
The QC Laboratory Chemist: Plain and Simple	Nilsen, Clifford	30-Apr-97	PH0531	1574910531		$159.95 / £107.00	
Validation and Qualification in Analytical Laboratories	Huber, Ludwig	31-Oct-98	PH0809	1574910809		$219.95 / £147.00	

PHARMACEUTICAL SCIENCE & REGULATION MANAGEMENT

TITLE	Author/Editor	Pub Date	Cat. no.	ISBN #	QTY	PRICE	TOTAL
Continuous Improvement in the Healthcare Manufacturing Industry: A Practical Guide	Bland, Valerie	30-Sep-99	PH099X	157491099X		$169.95 / £113.00	
Good Technical Management Practices: A Complete Menu	Tingstad, James Edwa	31-Aug-98	PH0868	1574910868		$149.95 / £99.00	

TITLE	Author/Editor	Pub Date	Cat. no.	ISBN #	QTY	PRICE	TOTAL
Pharmaceutical Marketing: A Practical Guide	Dogramatzis, Dimitri	01-Oct-01	PH118X	157491118X		$249.95 / £167.00	
Total R & D Management: Strategies and Tactics for 21st Century Healthcare Manufacturers	Dabbah, Roger	15-Jul-98	PH071X	157491071X		$169.95 / £113.00	
PHARMACEUTICAL SCIENCE & REGULATION MANUFACTURING & ENGINEERING							
Biotechnology and Biopharmaceutical Manufacturing, Processing, and Preservation	Avis, Kenneth E.	31-Mar-96	PH0167	1574910167		$179.95 / £120.00	
Control of Particulate Matter Contamination in Healthcare Manufacturing	Barber, Thomas A.	31-Oct-99	PH0728	1574910728		$219.95 / £147.00	
Cryopreservation: Applications in Pharmaceuticals and Biotechnology	Avis, Kenneth E.	30-Sep-99	PH0906	1574910906		$199.95 / £133.00	
How to Develop and Manage Qualification Protocols for FDA Compliance	Cloud, Phil	31-Aug-99	PH0981	1574910981		$249.95 / £167.00	
Pharmaceutical Manufacturing Change Control	Turner, Simon G.	31-Jan-99	PH0965	1574910965		$199.95 / £133.00	
Pharmaceutical Unit Operations: Coating	Avis, Kenneth E.	31-Aug-98	PH0825	1574910825		$189.95 / £127.00	
Pharmaceutical Water: System Design, Operation, and Validation	Collentro, William V	30-Sep-98	PH0272	1574910272		$229.95 / £153.00	
Sterile Pharmaceutical Products: Process Engineering Applications	Avis, Kenneth E.	31-Oct-95	PH4813	0935184813		$189.00 / £127.00	
Sterile Product Facility Design and Project Management	Odum, Jeffrey N.	01-Nov-96	PH0205	1574910205		$159.95 / £107.00	
PHARMACEUTICAL SCIENCE & REGULATION MEDICAL DEVICES							
Electromagnetic Compatibility in Medical Equipment: A Guide for Designers and Installers	Kimmel, William D.	01-Oct-95	PH4805	0935184805		$169.95 / £113.00	
Medical Device and Equipment Design: Usability Engineering and Ergonomics	Wiklund, Michael E.	15-Feb-95	PH4694	0935184694		$139.95 / £93.00	
Metered Dose Inhaler Technology	Purewal, Tol S.	31-Dec-97	PH0655	1574910655		$199.95 / £133.00	
Pharmacetical Applications in the European Union: A Guide Through the Registration Maze	Lowe, Cheng Yee	28-Feb-98	PH0647	1574910647		$179.95 / £120.00	
Practical Design Control Implementation for Medical Devices	Justiniano, Jose	13-Mar-03	PH1279	1574911279		$249.95 / £167.00	
Validation for Medical Device and Diagnostic Manufacturers, Second Edition	Desain, Carol V.	30-Sep-97	PH0639	1574910639		$199.95 / £133.00	
PHARMACEUTICAL SCIENCE & REGULATION PHARMACEUTICAL SCIENCE							
Pharmaceutical Microbiology	Cundell, Anthony	26-Sep-03	PH1872	0849318726		$119.95 / £79.99	
PHARMACEUTICAL SCIENCE & REGULATION QUALITY ASSURANCE							
Audit by Mail: Time and Cost Effective GMP Audit Tool	Lyall, John	31-Oct-95	PH466X	093518466X		$249.95 / £167.00	
Biotechnology: Quality Assurance and Validation	Avis, Kenneth E.	31-Oct-98	PH0892	1574910892		$179.95 / £120.00	
Compliance Auditing for Pharmaceutical Manufacturers: A Practical Guide to In-Depth Systems Auditing	Ginsbury, Karen	01-Aug-94	PH4600	0935184600		$249.95 / £167.00	
GMP Compliance, Productivity, and Quality: Achieving Synergy in Healthcare Manufacturing	Vinay, Bhatt	30-Jun-98	PH0779	1574910779		$229.95 / £153.00	
GMP/ISO Quality Audit Manual for Healthcare Manufacturers and their Suppliers, Sixth Edition (Volume 3 - Software Package)	Steinborn, Leonard	11-Jul-03	PH1848	0849318483		$199.95 / £133.00	
GMP/ISO Quality Audit Manual for Healthcare Manufacturers and their Suppliers, Sixth Edition, (Volume 1 - Checklists)	Steinborn, Leonard	20-Jun-03	PH1846	0849318467		$279.95 / £187.00	
GMP/ISO Quality Audit Manual for Healthcare Manufacturers and their Suppliers, Sixth Edition, (Volume 2 - Regulations)	Steinborn, Leonard	11-Jul-03	PH1847	0849318475		$299.95 / £200.00	
Pharmaceutical Quality Systems	Schmidt, Oliver	30-Apr-00	PH1090	1574911090		$189.95 / £127.00	
Pre-Production Quality Assurance for Healthcare Manufacturers	Hough, G. William	30-Jun-97	PH0450	1574910450		$169.95 / £113.00	
Quality and GMP Auditing: Clear and Simple	Vesper, James L.	31-Jul-97	PH0558	1574910558		$119.95 / £79.99	
Quality Assurance Compliance: Procedures for Pharmaceutical and Biotechnology Manufacturers	Peine, Ira C.	01-Feb-94	PH4511	0935184511		$199.95 / £133.00	
Quality Systems and GMP Regulations for Device Manufacturers: A Practical Guide to US, European, and ISO Requirements	Kuwahara, Steven	31-Mar-98	PH426X	087389426X		$139.95 / £93.00	
PHARMACEUTICAL SCIENCE & REGULATION REGULATIONS & STANDARDS							
Compact Regs CFR 21: Part 820 April 2002 Revision, Quality System Regulation (10-pack)	Drug Administration,	30-Sep-02	PH1834	0849318343		$79.95 / £52.99	
Compact Regs CFR 21: Parts 807, 812, & 814, Medical Device Approval (10-pack)	Drug Administration,	15-Aug-02	PH1837	0849318378		$99.95 / £66.99	
Compact Regs CFR 21: Part 11, Electronic Records; Electronic Signatures (10-pack)	Drug Administration,	30-Sep-02	PH1826	0849318262		$79.95 / £52.99	

TITLE	Author/Editor	Pub Date	Cat. no.	ISBN #	QTY	PRICE	TOTAL
Compact Regs CFR 21: Part 26, Mutual Recognition: US and the European Community (10-pack)	Drug Administration,	31-Dec-01	PH1827	0849318270		$79.95 / £52.99	
Compact Regs CFR 21: Part 58, Good Laboratory Practice for Nonclinical Laboratory Studies (10-pack)	Drug Administration,	30-Sep-02	PH1828	0849318289		$79.95 / £52.99	
Compact Regs CFR 21: Part 606, Current Good Manufacturing Practice for Blood and Blood Components (10-pack)	Drug Administration,	30-Sep-02	PH1832	0849318327		$99.95 / £66.99	
Compact Regs CFR 21: Part 820, Quality System Regulation (10-pack)	Drug Administration,	31-Dec-01	PH1833	0849318335		$79.95 / £52.99	
Compact Regs CFR 21: Parts 110 and 111, cGMP in Manufacturing, Packaging, or Holding Human Food, cGMP for Dietary Supple	Drug Administration,	31-Dec-01	PH1829	0849318297		$79.95 / £52.99	
Compact Regs CFR 21: Parts 201, 202, and 203, Prescription Drug Labeling, Advertising and Marketing (10-pack)	Drug Administration,	31-Dec-01	PH1830	0849318300		$99.95 / £66.99	
Compact Regs CFR 21: Parts 210 and 211, Pharmaceutical and Bulk Chemical GMPs (10-pack)	Drug Administration,	30-Sep-02	PH1831	0849318319		$99.95 / £66.99	
Compact Regs CFR 21: Parts 50, 56, and 312, Good Clinical Practices (GCP) (10-pack)	Drug Administration,	30-Sep-02	PH1836	084931836X		$129.95 / £87.00	
FDA-Speak: A Glossary and Agency Guide	Snyder, Dean E.	01-Oct-01	PH1295	1574911295		$199.95 / £133.00	
Good Drug Regulatory Practices: A Regulatory Affairs Quality Manual	Dumitriu, Helene I.	30-Sep-97	PH0515	1574910515		$169.95 / £113.00	
International Labeling Requirements	Sidebottom, Charles	26-Jun-03	PH1850	0849318505		$269.95 / £180.00	
International Pharmaceutical Registration	Chalmers, Alan A.	01-Jun-00	PH1031	1574911031		$289.95 / £193.00	
Interpharm Master Keyword Guide to 21 CFR Regulations of the U.S. Food and Drug Administration: 2001-2002 Edition	Drug Administration,	01-Mar-01	PH1406	1574911406		$119.95 / £79.99	
Interpharm Master Keyword Guide: 21 CFR Regulations of the Food and Drug Administration, 2002-2003 Edition		15-May-03	PH1851	0849318513		$199.95 / £133.00	
Understanding Biopharmaceuticals: Manufacturing and Regulatory Issues	Grindley, June	31-Dec-99	PH0833	1574910833		$229.95 / £153.00	
Write It Down: Guidance for Preparing Documentation that Meets Regulatory Requirements	Gough, Janet	01-Oct-99	PH0884	1574910884		$199.95 / £133.00	
PHARMACEUTICAL SCIENCE & REGULATION TRAINING							
Documentation Systems: Clear and Simple	Vesper, James L.	30-Sep-97	PH0507	1574910507		$119.95 / £79.99	
Quality Rules in Active Pharmaceutical Ingredients Manufacture: American Edition (5-pack)	Sharp, John	30-Jun-02	PH1392	1574911392		$99.95 / £66.99	
Quality Rules in Medical Device Manufacture: Revised American Edition (5-pack)	Sharp, John	30-Jun-02	PH1376	1574911376		$99.95 / £66.99	
Quality Rules in Packaging: Revised American Edition, 5-pack	Sharp, John	30-Jun-02	PH1325	1574911325		$99.95 / £66.99	
Quality Rules in Sterile Products: Revised American Edition (5-pack)	Sharp, John	30-Jun-02	PH1341	1574911341		$99.95 / £66.99	
Quality Rules: A Short Guide to Drug Products GMP, Revised American Edition (5-pack)	Sharp, John	30-Jun-01	PH1317	1574911317		$99.95 / £66.99	
Training for the Healthcare Manufacturing Industries: Tools and Techniques to Improve Performance	Vesper, James L.	30-Aug-93	PH4430	0935184430		$129.95 / £87.00	
PHARMACEUTICAL SCIENCE & REGULATION VALIDATION							
Computer Systems Validation: Concepts and Case Studies	Wingate, Guy	12-Sep-03	PH1871	0849318718		$269.95 / £180.00	
How to Sell Validatable Equipment to Pharmaceutical Manufacturers	Kopp, Erik	30-Sep-99	PH0973	1574910973		$239.95 / £160.00	
Pharmaceutical Equipment Validation: The Ultimate Qualification Guidebook	Cloud, Phil	31-Aug-98	PH0795	1574910795		$229.95 / £153.00	
Validating Automated Manufacturing and Laboratory Applications: Putting Principles into Practice	Wingate, Guy	30-Jun-97	PH037X	157491037X		$249.95 / £167.00	
Validation Fundamentals: How To, What To, When To Validate	Gibson, William	30-Apr-98	PH0701	1574910701		$139.95 / £93.00	
Validation of Active Pharmaceutical Ingredients, Second edition	Berry, Ira R.	31-Dec-01	PH1198	1574911198		$249.95 / £167.00	

When you order online! www.crcpress.com

BILL TO

Name_____

Title_____

Organization_____

Address_____

City_____ State/Province_____

Country_____ Zip/Postal Code_____

Daytime Phone ( )_____ Fax ( )_____
(Required to Process Order)

e-Mail Address_____

SHIP TO *(Fill in only if different than billing information)*

Name_____

Title_____

Organization_____

Address_____

City_____ State/Province_____

Country_____ Zip/Postal Code_____

Daytime Phone ( )_____ Fax ( )_____

e-Mail Address_____

Please send me the following:

CAT. NO.	QUANTITY	TITLE	PRICE EACH	TOTAL PR

Subtotal _____

Tax _____

Shipping Charges _____

Order Total _____

_____

❑ I've enclosed check #_____ payable to CRC Press LLC

❑ Bill My Company (Purchase Order Attached)

❑ Please Charge to: ❑ MasterCard ❑ VISA ❑ American Express

CARD NO.

Exp. Date

Signature _____

Phone Number _____
(Signature and Phone Number Required to Process Order)

SHIPPING AND HANDLING

Region	Delivery Time	First Title	Additional Title
USA/Canada	3-5 Days	$5.99	$1.99
America/Asia/Australia	7-14 Days	$9.99	$3.99
Europe	3-5 Days	£2.99	£0.99
Middle East/Africa	7-21 Days	£4.99	£2.99

For priorit mail servic please cont your neare CRC PRES office.

ORDERING LOCATIONS

In North & South America, Asia, and Australasia:
CRC PRESS
2000 N.W. Corporate Blvd.
Boca Raton, FL 33431-9868, USA
Tel: 1-800-272-7737 • Fax: 1-800-374-3401
From Outside the Continental U.S.
Tel: 1-561-994-0555 • Fax: 1-561-361-6018
e-mail: orders@crcpress.com

In Europe, Middle East, and Africa:
CRC PRESS / ITPS
Cheriton House, North Way
Andover, Hants, SP10 5BE, UK
Tel: 44 (0) 1264 342932
Fax: 44 (0) 1264 342788
e-mail: crcpress@itps.co.uk

Corporate Offices

CRC PRESS
2000 N.W. Corporate Blvd.
Boca Raton, FL 33431-9868, USA
Tel: 1-800-272-7737 • Fax: 1-800-374-3401
From Outside the Continental U.S.
Tel: 1-561-994-0555 • Fax: 1-561-361-6018
e-mail: orders@crcpress.com

CRC PRESS UK
23-25 Blades Court, Deodar Road
London SW15 2NU, UK
Tel: 44 (0) 20 8875 4370
Fax: 44 (0) 20 8871 3443
e-mail: enquiries@crcpress.com

www.crcpress.com